Emulsions –
A Fundamental and Practical Approach

NATO ASI Series

Advanced Science Institutes Series

A Series presenting the results of activities sponsored by the NATO Science Committee, which aims at the dissemination of advanced scientific and technological knowledge, with a view to strengthening links between scientific communities.

The Series is published by an international board of publishers in conjunction with the NATO Scientific Affairs Division

A Life Sciences	Plenum Publishing Corporation
B Physics	London and New York
C Mathematical and Physical Sciences	Kluwer Academic Publishers Dordrecht, Boston and London
D Behavioural and Social Sciences	
E Applied Sciences	
F Computer and Systems Sciences	Springer-Verlag
G Ecological Sciences	Berlin, Heidelberg, New York, London,
H Cell Biology	Paris and Tokyo
I Global Environmental Change	

NATO-PCO-DATA BASE

The electronic index to the NATO ASI Series provides full bibliographical references (with keywords and/or abstracts) to more than 30000 contributions from international scientists published in all sections of the NATO ASI Series.
Access to the NATO-PCO-DATA BASE is possible in two ways:

– via online FILE 128 (NATO-PCO-DATA BASE) hosted by ESRIN,
Via Galileo Galilei, I-00044 Frascati, Italy.

– via CD-ROM "NATO-PCO-DATA BASE" with user-friendly retrieval software in English, French and German (© WTV GmbH and DATAWARE Technologies Inc. 1989).

The CD-ROM can be ordered through any member of the Board of Publishers or through NATO-PCO, Overijse, Belgium.

Emulsions – A Fundamental and Practical Approach

edited by

Johan Sjöblom
Department of Chemistry,
University of Bergen,
Bergen, Norway

Springer-Science+Business Media, B.V.

Proceedings of the NATO Advanced Research Workshop on
Emulsions – A Fundamental and Practical Approach
Bergen, Norway
June 24–25, 1991

Library of Congress Cataloging-in-Publication Data

```
NATO Advanced Research Workshop on "Emulsions--a Fundamental and
  Practical Approach" (1991 : Bergen, Norway)
   Emulsions--a fundamental and practical approach : proceedings of
the NATO Advanced Research Workshop on "Emulsions--a Fundamental and
Practical Approach", held in Bergen, Norway, June 24-25, 1991 /
edited by Johan Sjöblom.
      p.   cm. -- (NATO ASI series. Series C, Mathematical and
physical sciences ; vol. 363)
   "Published in cooperation with NATO Scientific Affairs Division."
   Includes bibliographical references and index.
   ISBN 978-94-010-5085-2      ISBN 978-94-011-2460-7 (eBook)
   DOI 10.1007/978-94-011-2460-7
   1. Emulsions--Congresses.   I. Sjöblom, Johan, 1953-   .
II. North Atlantic Treaty Organization.  Scientific Affairs
Division.  III. Title.  IV. Series: NATO ASI series.  Series C,
Mathematical and physical sciences ; no. 363.
TP156.E6N37  1991
660'.294514--dc20                                    91-47207
```

ISBN 978-94-010-5085-2

CONTENTS

vi

PREFACE

To control stability of an emulsified system is to control fundamental processes like sedimentation (or creaming), flocculation, coalescence and Ostwald ripening. In these processes, a knowledge of fundamental physico-chemical properties of stabilizers, (surfactants or polymers) either as monomers or in an aggregated form is required. During the NATO ARW on "Emulsions - A Fundamental and Practical Approach" organized on June, 24. and 25. 1991 in Bergen, Norway, attention was focussed on emulsions from both theoretical and practical aspects. The workshop gathered 95 participants from 14 different countries.

The lectures at the workshop covered from a fundamental point of view general aspects on stability, interfacial adsorption mechanisms, interfacial rheology, direct measurements of surface forces and bulk rheological properties of emulsions, and self-diffusion properties as measured by means of NMR. With regard to applications the fields of food, crude oil and pharmaceutical emulsions were covered.

For the food emulsions a central topic is the role of the proteins at the W/O interface, their conformations and mechanisms by which they can be replaced at the interface (competitive adsorption).

For water-in-crude oil emulsions the mechanisms behind the resolution of water are of large technical importance. Characterizations of the stabilizing asphaltene fraction, physico-chemical properties of destabilizing surfactants and the interplay between asphaltenes and waxes at the W/O interface were discussed.

Stuctures of pharmaceutical emulsions and creams were characterized as well as nonionic vesicle drug administration systems. In addition fluorocarbon emulsions acting as blood substitutes were also presented.

In the poster session a large variety of topics was covered. Stability mechanisms were viewed from Langmuir monolyer studies of xanthan polymers as well as interfacially active crude oil components, from interfacial pressure studies, from equilibrium phase diagrams, from dielectric spectroscopy measurements of the stabilizing interfacial membrane, from depletion interaction and from electrolyte effect on highly concentrated water-in-oil emulsions. Technical systems presented included fluorocarbon emulsions as blood substitutes, commercial intravenous emulsions, margarines and tablespreads, crude oil emulsions, emulsion polymerization and alkyd emulsions.

The workshop was financed by NATO Scientific Affairs Division (Brussels), by the companies Berol Nobel Ab (Stenungsund, Sweden), Houm A/S (Oslo, Noway), KSV Instruments Ltd (Helsinki, Finland), Norsk Hydro A/S (Bergen, Norway) and Statoil A/S (Stavanger, Norway). Following associations also contributed: The Norwegian Research Council for Science and The Humanities (NAVF, Oslo, Norway), VISTA (Oslo, Norway) and International

Association for Colloid and Interface Scientists (IACIS).
Locally support was obtained from University of Bergen.

To all the lecturers, participants and sponsors - THANK YOU
for your contributions in making the NATO ARW a successful
workshop.

Johan Sjöblom
Department of Chemistry
University of Bergen
N-5007 Bergen
Norway

CONTRIBUTORS

M.P.Aronson
Unilever Research, United States,
Edgewater, New Jersey, 07020, U.S.A.

R.Aveyard
School of Chemistry, University of Hull,
HU6 7RX, England

B.Balinov
Physical Chemistry 1, Chemical Center,
Lund University, Sweden

B.Bergenståhl
Institute for Surface Chemistry,
Box 5607, S-11486 Stockholm, Sweden

B.P.Binks
School of Chemistry, University of Hull,
HU6 7RX, England

J.A.Bouwstra
Center for Bio-Pharmaceutical Science,
University of Leiden, The Netherlands

H.K.Christenson
Department of Applied Mathematics,
Australian National University,
Canberra, A.C.T. 2601, Australia

A.A.Christy
Department of Chemistry, University of Bergen,
N-5007 Bergen, Norway

P.Claesson
Institute for Surface Chemistry,
S-11486 Stockholm, Sweden

E.Dickinson
Procter Department of Food Science,
University of Leeds, Leeds LS2 9JT, UK

J.C.Earnshaw
Department of Pure and Applied Mathematics,
The Queen's University of Belfast,
BT7 1NN, Northern Ireland

P.D.Fletcher
School of Chemistry, University of Hull
HU6 7RX, England

S.E.Friberg
Center for Advanced Materials Processing
and Department of Chemistry,
Clarkson University, Potsdam, NY 13699-5810 USA

P.Fäldt
Institute for Surface Chemistry,
Box 5607, S-11486 Stockholm, Sweden

G.S.Gooris
Center for Bio-Pharmaceutical Science,
University of Leiden, The Netherlands

H.E.J.Hofland
Center for Bio-Pharmaceutical Science,
University of Leiden, The Netherlands

M.Jansson,
Institute for Surface Chemistry,
S-11486 Stockholm, Sweden

H.E.Junginger
Center for Bio-Pharmaceutical Science,
Leiden University, The Netherlands

R.J.Kaufman
Hemagen/PFC, St. Louis, MO 63146, U.S.A.

N.Krogh
Grindsted Products, Edwin Rahrs Vej 38,
DK-8220 Brabrand, Denmark

K.Larsson
Department of Food Technology, University of Lund,
Box 124, S-22100 Lund, Sweden

B.Lassen
Institute for Surface Chemistry,
Box 5607, S-11486 Stockholm, Sweden

J.R.Lu
Physical Chemistry Laboratory,
University of Oxford, OX1 3QZ, England

I.Lönnqvist
Physical Chemistry 1, Chemical Center,
Lund University, Sweden

A.C.McLauglin
Department of Pure and Applied Physics,
The Queen's University of Belfast,
BT7 1NN, Northern Ireland

A.J.McMahon
BP Research, Sunbury Research Centre,
Middlesex, TW16 7LN, England

L.Mingyuan
Department of Chemistry, University of Bergen,
N-5007 Bergen, Norway

J.Sjöblom
Department of Chemistry, University of Bergen,
N-5007 Bergen, Norway

P.Stenius
Department of Forest Products Technology,
Helsinki University of Technology, Finland

O.Söderman
Physical Chemistry 1, Chemical Center,
Lund University, Sweden

Th.F.Tadros
I.C.I. Jealott's Research Station, Berkshire,
RG12 6EY, England

D.T.Wasan
Illinois Institute of Technology, Chicago,
Illinois 60616, U.S.A

X.Ye
School of Chemistry, University of Hull,
HU6 7RX, England

EMULSION STABILITY

Stig E. Friberg
Center for Advanced Materials Processing and
Department of Chemistry
Clarkson University
Potsdam, N.Y. 13699-5810

Abstract. Emulsion stability is reviewed with emphasis on more complex systems. The analysis of the stability of the traditional two–liquid emulsions is focussed on two factors; the rheology of the continuous phase and a barrier between the dispersed droplets. It is demonstrated that an increase of viscosity of the continuous phase of the emulsion is not a viable alternative to increase the half-life of the emulsion to an acceptable level for practical applications. The continuous phase must show an – albeit small – yield value. The barrier function is evaluated using a square function to demonstrate the relative importance of barrier width, location, and height. The main part of the review is centered on the properties of three–phase emulsions of which three examples are chosen to illustrate influence on emulsification and stability by the third phase.

1. Introduction

The traditional view of emulsion stability (1,2) was concerned with systems of two liquids and emulsion stability could be related to the interfacial properties and to the colloidal forces across a thin liquid film after the hydrodynamic conditions of the latter had been taken into consideration.

However, a large number of emulsions, in fact, contain more than two phases; especially so in the areas of pharmacy, food and cosmetics. The importance of the third phase was recognized early (3) and the IUPAC definition of an emulsion includes a third phase (4).

With this relation in mind, this short review of emulsion stability will treat as an introduction two–phase emulsions for which emulsion stability will be defined using the consecutive stages of destabilization. For these simple systems, a distinction will be made between the relative influence on the stability by the rheology of the continuous phase and by an energy barrier between the droplets. This introduction will be followed by three special cases of three–phase emulsions. In each case the decisive importance of the third phase for specific properties of the emulsion will be demonstrated.

1

J. Sjöblom (ed.), Emulsions – A Fundamental and Practical Approach, 1–24.
© 1992 Kluwer Academic Publishers.

2

2. TWO-PHASE EMULSIONS

The stability of a two-phase emulsion is defined differently depending on the application and, hence, depends on the stage of destabilization. One finds four interrelated steps. The flocculation, Fig. 1, leads to aggregation of droplets, Fig. 1A, while the subsequent coalescence leads to an emulsion with a large variation in droplet size, Fig. 1B.

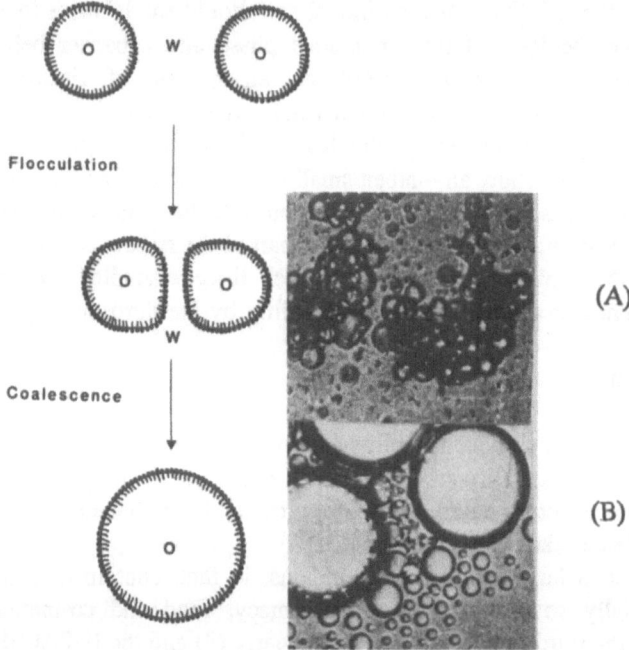

Figure 1. The primary destabilization of an emulsion is the flocculation that leaves aggregates of droplets (A). Subsequent coalescence leaves an emulsion with a wide distribution of droplet size (B).

The droplet or aggregate size is decisive for creaming or sedimentation rate leading to a highly concentrated emulsion, Fig. 2A, and finally, to phase separation, Fig. 2B.

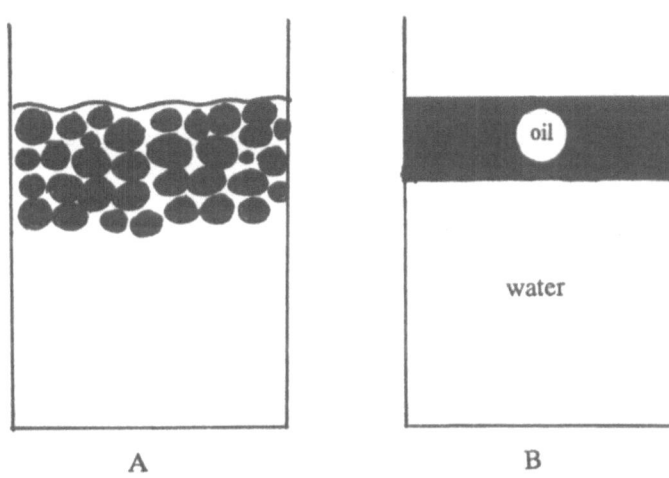

Figure 2. Creaming of emulsion droplets leaves a layer of concentrated emulsion on top of the original emulsion (A). Coalescence in this layer leads to phase separation (B).

It should be observed that the creaming leads to enhanced flocculation due to the difference in creaming velocity of droplets with varied size.

$$v = 2\Delta\rho r^2 g / 9\eta \qquad [1]$$

The square dependence on the radius means that larger droplets move much faster and an increase of flocculation rate takes place similar to that caused by shear.

The definition of emulsion stability may use any of these four phenomena; the choice depends on the application. So for example, would the stability of an emulsion as a blood substitute be carefully screened for the first destabilization step, because droplet size is of decisive importance for its performance. Beverage emulsions, on the other hand, are judged mostly from the point of view of creaming, because a concentrated emulsion layer on top of the beverage makes the product less appealing to the customer. Thirdly, a concentrated emulsion, such as a cream or lotion, is evaluated mainly for phase separation.

The present preliminary discussion is limited to the first step, the flocculation. The second step, the coalescence of the thin liquid film, is treated by Professor Wasan (5). The first step, on the other hand, is governed by flocculation kinetics and the following section illustrates the factors governing that process.

2.1 Flocculation kinetics

An unprotected emulsion is extremely unstable. The flocculation (and coalescence) rate is governed by the Schmoluchovski rate equation.

$$\frac{du}{dt} - 16\pi Drn^2 - \frac{8kT}{3\eta}n^2 \qquad [2]$$

which the half life

$$t_{1/2} - \frac{3\eta}{8kTn} \qquad [3]$$

An unprotected emulsion of "average" conditions; oil-in-water, $r=1\mu m$, water to oil volume ratio = 1/1 and $1cm^3$, has a half-life of 0.8 seconds! This value must be increased by a factor of 10^8 in order to obtain an emulsion with a half-life of 3 years, which is an acceptable level of stability for most practical applications.

Transforming this value to an enhanced viscosity would require a value of 10^6P. A viscosity at this level is found in waxes or molten sulfur and the futility of using viscosity as such for stabilization of an emulsion is obvious. Instead, stabilization by changed rheological properties requires rigidity, a yield value, in the continuous phase. It should be realized that the yield value may be extremely small; it does, as a matter of fact, not need to be noticeable during normal handling (pouring) of the emulsion.

The second possibility to increase the half-life of the emulsion is to introduce an energy barrier between the droplets. The expression now reads

$$t_{1/2} = 3B\eta/8kTn \qquad [4]$$

in which

$$B - 2r \int_{2r}^{\infty} e^{v(kT)}d\ell/\ell^2 \qquad [5]$$

The influence of the properties of the barrier is best illustrated by the use of a square function

$V(\ell) = 0 \quad \ell < xr \text{ and } \ell > yr$
$V(\ell) = V \text{ (constant)} \quad xr < \ell < yr$

The expression B in [5] now becomes

$$B = 1 + 2(e^v-1)(y-x)/xy \qquad [6]$$

No barrier (V=0) leads to equation [4] being equal to equation [3], which is correct.

The effect of the width of the barrier is found by a variation of y with x remaining constant. The expression goes from zero (y=x) to 1/x for y = ∞. The influence of barrier width is limited. The effect of the location of the barrier is obtained by a constant value for y−x while varying the value of both. Increasing the distance (ℓ) between the droplet surface and the barrier causes a strong reduction of its influence ∝ ℓ^2. The height, finally, is best estimated by a numerical example.

y = 3r and x = 2r give

$$B = 1 + (e^v - 1)/3 \qquad [7]$$

The half−life for the earlier emulsion example, but with a barrier, is given in Table I.

TABLE I

Half−life for an emulsion (O/W = 1/1, r = 1μm, volume = 1cm^3) with a constant barrier V(kT) reaching 1 radius out from the droplet surface.

V(kT)	$t_{1/2}$
0	0.8 sec
10	2.0 hrs
15	1.3 days
17.5	154 days
20	5.1 yrs
50	$5.5 \cdot 10^{13}$ yrs

The barrier height is obviously the decisive factor and a height of 20kT is sufficient value. With regard to the value for 50kT it should be observed that the half−life value is of the order of 100,000 times the age of the universe.

With these fundamental conditions clarified it is useful to examine the nature and effect of two kinds of barriers; the electric double layer and adsorbed polymers.

6

2.2 Electric Double Layer

The surface charge of emulsion droplets together with the counter ions form an electric double layer. When two droplets approach each other the overlap of the counter ions give rise to a repulsion potential which for two flat surfaces and a high electric surface potential may be approximated by

$$V_{Eldl} = \frac{64nkT}{K}e^{-2Kd} \qquad [8]$$

This repulsion potential is counter acted by the Van der Waals attraction potential volume corresponding approximation is

$$V_{vdw} = -A/48\pi d^2 \qquad [9]$$

The maximum of the total potential, Fig. 3, decides the stability; according to Table I a maximum of 20kT is appropriate.

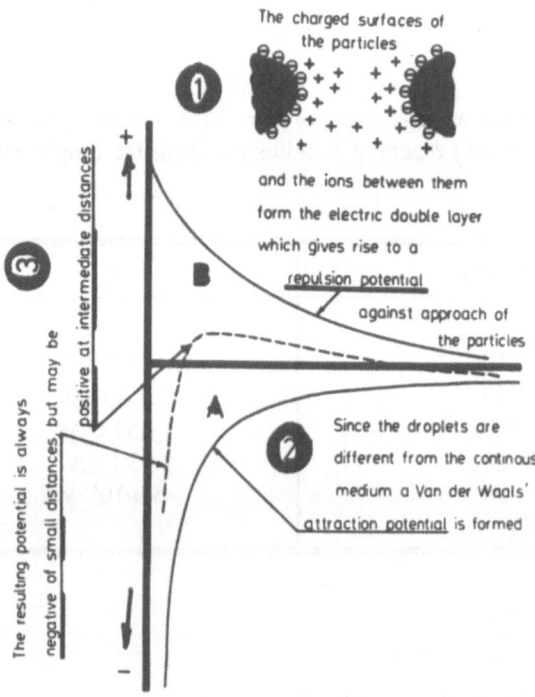

Figure 3. The DLVO theory refers colloidal stability to the distance dependence of two independent potentials.

Addition of salts, increase of K in equation [8], reduces the value of the repulsion potential, Fig. 4, and instability is encountered. It should be observed that even small variations of salt concentrations will change the repulsion potential in the most pronounced manner for emulsion droplets.

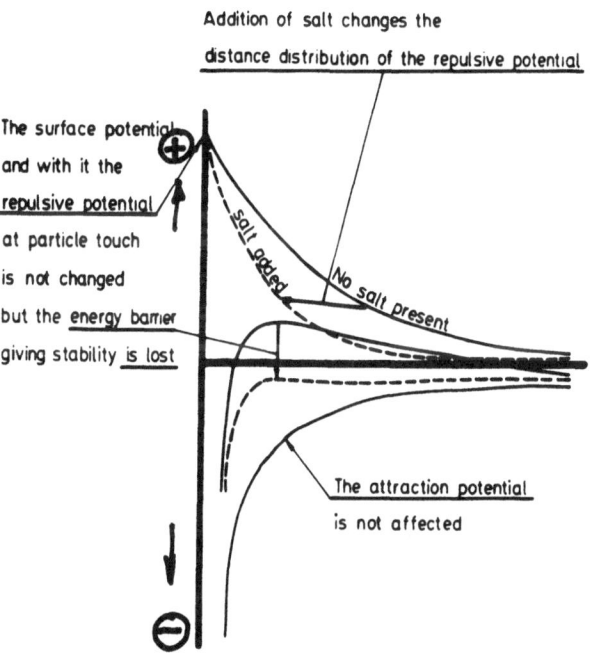

Figure 4. The presence of an electrolyte may affect the maximum of the total potential at unchanged surface potential.

The maximum for the combined potential from eq:s [8] and [9] is located approximately at

$$Kd=1 \qquad [10]$$

with

$$K = 3.1 \cdot 10^7 z c^{0.5} (cm^{-1}) \qquad [11]$$

a concentration of monovalent counter ions of 0.01M results in

$$d = 3.2 \cdot 10^{-7} cm$$

Putting $A = 10^{-12}$ erg in eq. 9 the absolute value of the potentials [8] and [9] become $V = 1.58 \cdot 10^4 \kappa T/\mu m^2$.

A change in c from 0.01M to 0.0099M causes an increase of the potential from the electric double layer potential by $79\kappa T/\mu m^2$.
Considering that an energy barrier of 20kT is sufficient for stabilization it is easily realized that a small change in salt concentration will have a drastic effect on the stability and that a critical coagulation concentration of the salt is an appropriate dimension.

With these factors in mind the influence of the surface potential from ionic surfactants may be summarized.

The electric double layer serves to stabilize water continuous emulsions provided the concentration of counter ions is below certain critical values. As a rough approximation it may be stated that the counter ion concentrations should be below $10^{-2}M$. It is essential to realize that stable O/W emulsions with counter ion concentrations in excess of these values are stabilized by some other factor and that adding salt to destabilize such systems is not optimal.

2.3 Adsorbed Polymers

Polymers have so far found limited use to stabilize emulsions. This is not due to lack of performance; publications in the area point to excellent stabilization of emulsions by polymers (6–8). Instead the use has been limited, because the adsorption of polymers to emulsion droplets has not been well understood.

A polymer is adsorbed in the form of loops, tails, and trains, Fig.5. The stabilizing action depends on the presence of sufficiently long tails or loops, as explained in Fig.6. The stabilizing action of the protruding part of the polymer is extremely efficient (8); a single loop or tail gives a barrier of approximately 20kT.

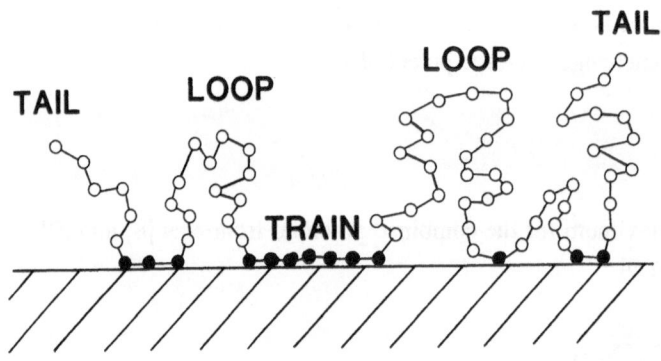

Figure 5. A polymer can be adsorbed in the form of trains, loops, and tails.

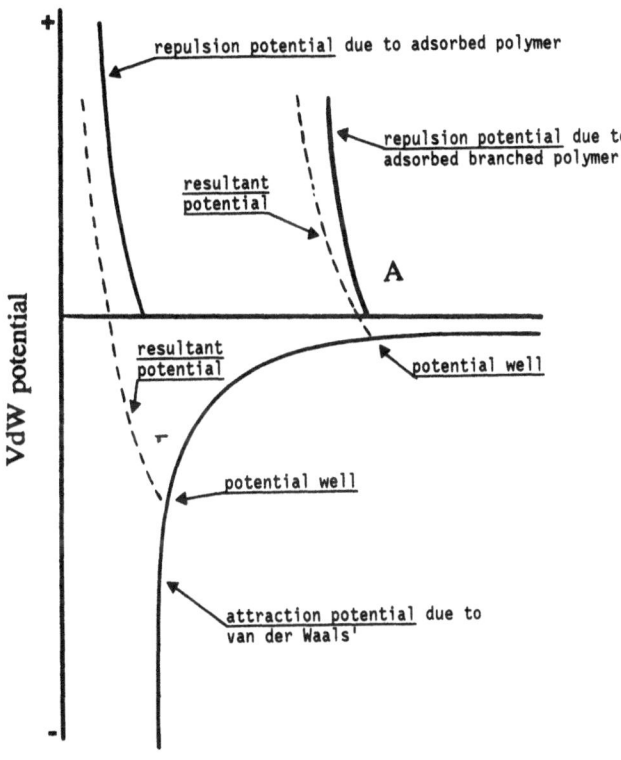

Figure 6. The chain length of an adsorbed polymer must be sufficient (A) to prevent the van der Waals potential from creating too deep a minimum (B).

However, one problem is the sensitivity of the adsorption to the properties of the polymer versus the environment. This has been described in an excellent manner by Clayfield and Lumb in early calculations of polymer conformation at an interface (9). Clayfield computed the fraction of adsorbed polymer as a function of its total adsorption energy. Adsorption energies lower than the optimal range, Fig. 7, gave no adsorption of the polymer. It is immediately evident that a polymer that does not adsorb can have no stabilizing action at the interface. At an adsorption energy higher than the optimal range, Fig.7, the polymer adheres with all its groups at the interface. Such an adsorption is also without stabilization effect. The protective action reaches only a short distance into the continuous medium and the van der Waals potential has already reached such large negative values that the flocculated state is permanent. The potential well is so deep that the droplets remain attached to each other. For the same reason, a minimum molecular weight is necessary to obtain stability because the protective action of the polymer must reach far out to prevent the van der Waals potential from becoming dominant.

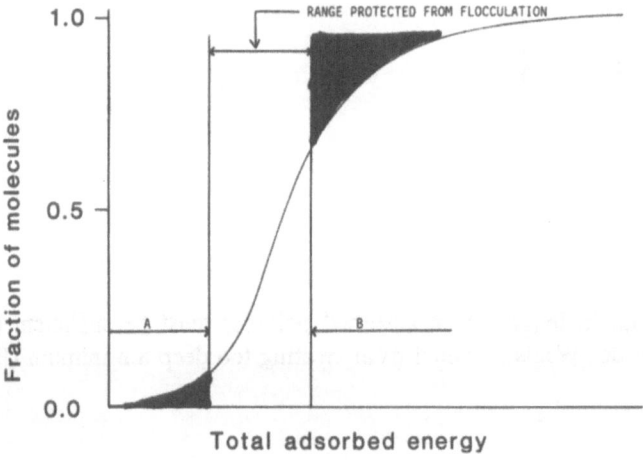

Figure 7. With too small an adsorption energy, the polymer will not adsorb at the interface (A). Too high an adsorption (B) leads to a train adsorption alone. Stabilization is achieved only in the limited range between the two.

The solution to the problem lies with a different kind of polymer, the so-called block copolymers consisting of one block of one polymer to which two or more blocks of a different polymer are attached. The two polymer blocks are chosen to be selectively soluble in the aqueous and oil phase respectively. With sufficient molecular weight in the polymer part that is soluble in the continuous phase, excellent stability is achieved.

3. THREE–PHASE EMULSIONS

Three–phase emulsions are more common than is generally believed; a great number of pharmaceutical and cosmetic lotions and creams contain three phases, often nolens volens.

A knowledge about their properties is necessary to get a better grasp of the sometimes puzzling behavior of emulsions. They also provide opportunities to formulate emulsions with properties outside the scope of normal two–phase emulsions.

For this discussion, three main categories have been chosen in which the third phase is respectively a solid, a liquid, or a liquid crystal. These three cases impart different properties to the emulsion and together provide a chart of the manner in which the three–phase emulsions may behave.

Figure 8. Solid particles stabilize an emulsion when they are adsorbed at the interface.

3.1 Third Phase a Solid

Solid particles are often part of an emulsion formulation, and the solid particles may be used to stabilize the emulsion.

The key factor for the use of particles as a stabilizing agent is their wetting by the two phases. They will stabilize the emulsion if they are located at the interface between the two liquids, Fig.8, serving as a mechanical barrier to prevent the coalescence of the droplets. The protection is based on the energy to expel the particles from the interface into the dispersed droplets. This energy depends on the contact angle and is easily calculated for the case of spherical particles located at the oil/water interface. The expression is:

$$\Delta E = \pi r_p^2 \gamma_{O/W} (1-\cos\Theta)^2 \tag{12}$$

The value for ΔE grows monotonously with Θ but experimental investigations (10) have shown 90° to be practical maximum. Higher values lead to the expulsion of the particle into the continuous phase.

The repulsive force of the particles when forced into the dispersed phase significantly exceeds that of the van der Waals attraction force. Assuming the contact angle to be constant during expulsion from the interface, the combined van der Waals and wetting force becomes:

$$\frac{dE}{dh} = 2\pi\gamma_{O/W}(h - r_p \cos\theta) \tag{13}$$

Fig.9 is drawn for particles of 50 radius acting on a flat interface of size 10^4 ^{2}. The total force is repulsive even for contact angles of 60°; for 90° angles the repulsion force is extremely strong. Hence, the stabilization is excellent, bearing in mind that thousands of such particles are adsorbed on to one emulsion droplet. These values make it evident that the contact angles must be observed rather carefully to obtain a value close to 90°.

A contact angle smaller than 90° between the water and the solid particles, Fig.10, means that the interfacial free energy is too great between the the solid material and the oil. To increase the contact angle between the aqueous phase and the powder, an oil–soluble surface active agent may be added. This surfactant should not be water soluble at all and it should bind strongly to the solid surface. Such a surfactant will bind strongly to the solid–liquid interface at a low concentration in the liquid phases. Use of as low a concentration as possible is essential because higher concentrations also lower the interfacial tension at the water–oil interface. A reduction of the O/W interfacial tension has a disadvantage because it makes the contact angle Θ more sensitive to small differences between $\gamma_{W/S}$ and $\gamma_{O/S}$.

If the water droplet does not spread and its contact angle is in excess of 90°, the surfactant is added to the aqueous phase. The adjustments are identical to the ones described for the oil phase.

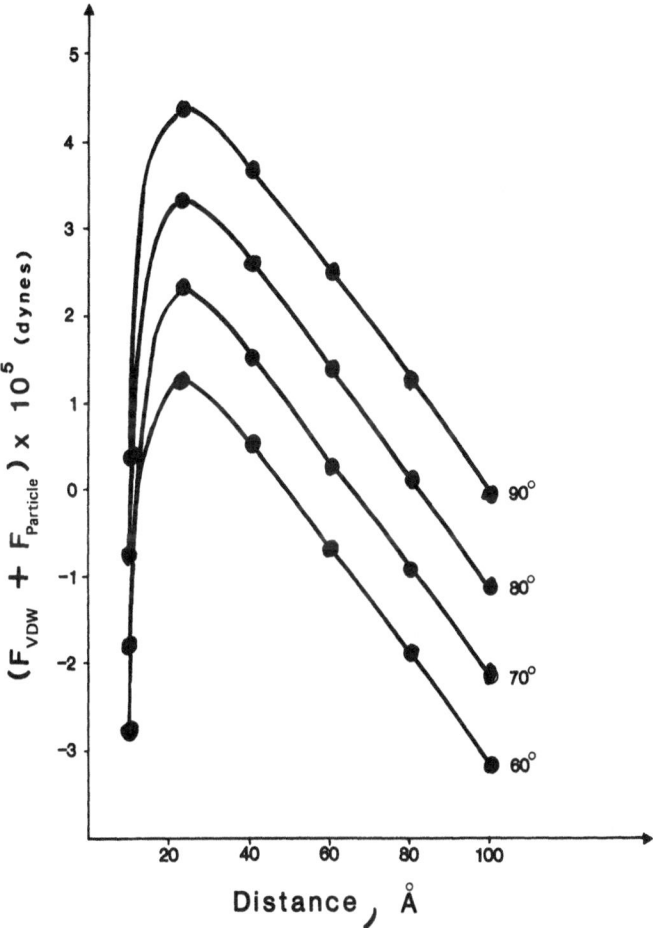

Figure 9. The combination of wetting forces, $F_{particle}$, and van der Waals, F_{VDW}, becomes strongly repulsive for high contact angles.

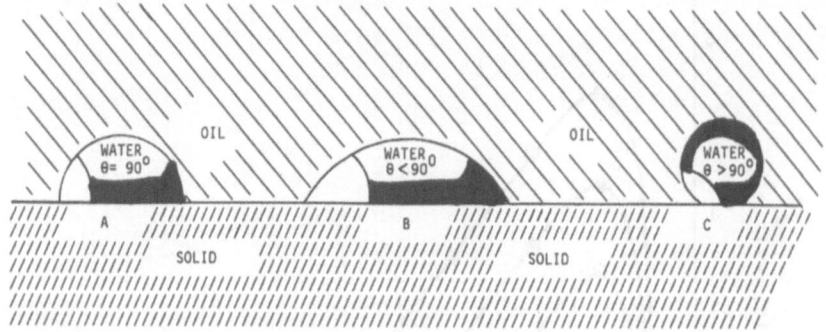

Figure 10. An oil/water/solid contact angle of Θ=90° means that the particles as such are optimal stabilizers (A). If the contact angle toward water is less than 90° (B), an oil–soluble surfactant is added. If Θ exceeds (C), a water soluble surfactant is added to reduce the interfacial free energy between water and the solid, reducing Θ.

For emulsions produced in huge quantities or batched repeatedly, it is preferred to change the surface of the solid particles through a chemical reaction. As earlier, a small contact angle in the aqueous phase means too great an interfacial energy between the oil phase and the solid particles. The surface of the solid particle must be made less polar, which may be achieved by esterification of polar groups on the particle surface, by silanization, or by similar processes. In other case, a large contact angle toward the aqueous phase, is corrected by making the particle surface more polar through oxidation, hydrolysis, or other reactions.

3.2 Third Phase a Liquid

This case has been well know for 20 years (11) for nonionic surfactants of the polyethylene glycol alkylaryl ether type. It is a useful case because it allows low energy emulsification (12) by using the strong temperature dependence of the colloidal association structures in the water–surfactant–hydrocarbon systems.

The basic background is as follows: At low temperatures, below the cloud point, (the temperature above which a nonionic surfactant of this kind ceases to be water soluble at low concentration), these surfactants are preferentially water soluble and, hence, are heavily partitioned toward the aqueous phase in an emulsion. At high temperatures, well above the cloud point, the surfactant's solubility in water is extremely small, and it is then partitioned almost entirely into the oil. At some intermediate temperature, the HLB temperature (13) or the phase inversion temperature (PIT) (14), a third phase, the surfactant phase, an isotropic liquid phase, (15) appears between the oil and the water, Fig. 11.

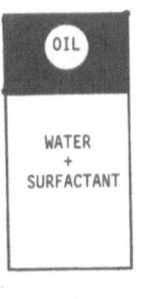  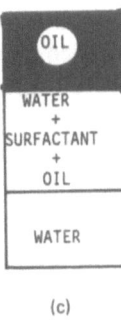

Figure 11. Water, hydrocarbon, and a nonionic surfactant form two phases at low (a) and high (b) temperatures. In a small temperature range between (a) and (b), three liquid phases are found (c) in the HLB temperature range.

The key to low energy emulsification is to emulsify at this HLB temperature (60°C) and to follow the emulsification by rapid cooling to about 25 to 30°C if an O/W emulsion is desired. A W/O emulsion requires the opposite condition: the HLB temperature should be close to 0°C (14).

The following process is used for the selection of the emulsifier. Equal amounts of the oil and the aqueous phases with all the components of the formulation pre–added are mixed with 4% of the emulsifiers to be tested in a series of samples. The samples are left thermostated at the HLB temperature to separate. After complete separation, the emulsifiers that give systems separating into three transparent layers are selected for emulsification and determination of emulsion stability; the rest, giving two phases, are discarded.

At this stage one fact cannot be overemphasized. An emulsion separation into one oil phase on top and one aqueous phase on the bottom part separated by an emulsion layer is not a three–phase system. To find whether there are two layers or three, the central emulsion layer must be separated.

The emulsifiers giving separation into three layers are used for emulsification. The emulsions are extremely unstable at the HLB temperature (14), and cooling to room temperature must be rapid. This can be achieved by spreading the emulsion in a thin layer on a cold plate or by letting it pass between cold rollers. For an O/W emulsion with sufficiently low O/W volume ratio, the following method may be useful. Before emulsification an amount corresponding to half the total emulsion is removed from the aqueous phase and cooled close to 0°C. The remaining part of the emulsion is emulsified at 55°C and stirred directly into the cooled aqueous part. The average droplet size of these emulsions is less than 1μm with very little stirring. With these small droplets the stability is good and can be improved further by addition of small amounts of an ionic surfactant (~0.05–0.1%).

These emulsions have been known for a long time (14), and low energy emulsification has been promoted by Lin and others (16). One of the more interesting future developments is in the area of separation using microemulsions. A microemulsion stabilized by a nonionic surfactant shows a strong temperature dependence of its solubilization pattern changing the hydrocarbon solubilization from the level of 50% to 5% with a temperature reduction of 25°C. This means that a reversible extraction/separation process uses a temperature change of only 25°C. There is no distillation involved and a comparison with the normal solvent extraction – with its expensive distillation processes to remove solvent – demonstrates a pronounced economic advantate for the microemulsion process (17). The extraction process of hydrocarbons has been shown to contain several phases. The latest development is to apply these surfactants in the separation of products from biotechnological processes.

3.3 Third Phase a Liquid Crystal

In addition to micelles and microemulsions droplets, surfactants may form liquid crystals, of which the one with a layered structure, Fig.12, is the most important one for emulsion science.

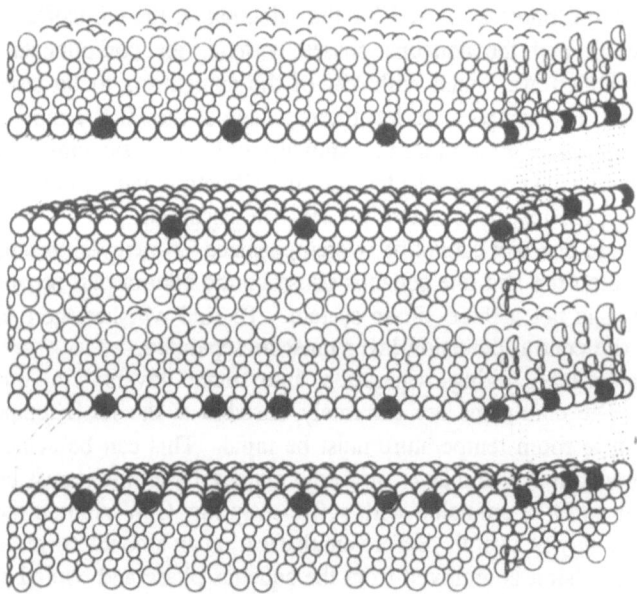

Figure 12. In a lamellar liquid crystal the surfactants form double layers separated by thin water layers.

The introduction of liquid crystals as a stabilizing element for emulsions occurred in 1969, when it was found that the sudden stabilization at emulsifier concentration in excess 2.5% of a water/p–xylene emulsion by a commercial octaethylene glycol nonylphenyl ether was due to the formation of a liquid

crystalline phase in the emulsion (18). Later investigations confirmed the strong stabilizing action of these structures (19).

The detection of the liquid crystal is based primarily on its anisotropic optical properties. An isotropic solution is black when placed between crossed polarizers and viewed against a light source, while the liquid crystal is radiant.

The structure of the phase may be identified by optical microscopy with the sample between crossed polarizers when the lamellar liquid crystal has a pattern of "oily streaks" and Maltese crosses, Fig. 13A, while the hexagonal array of cylinders give a different optical pattern, Fig.13B.

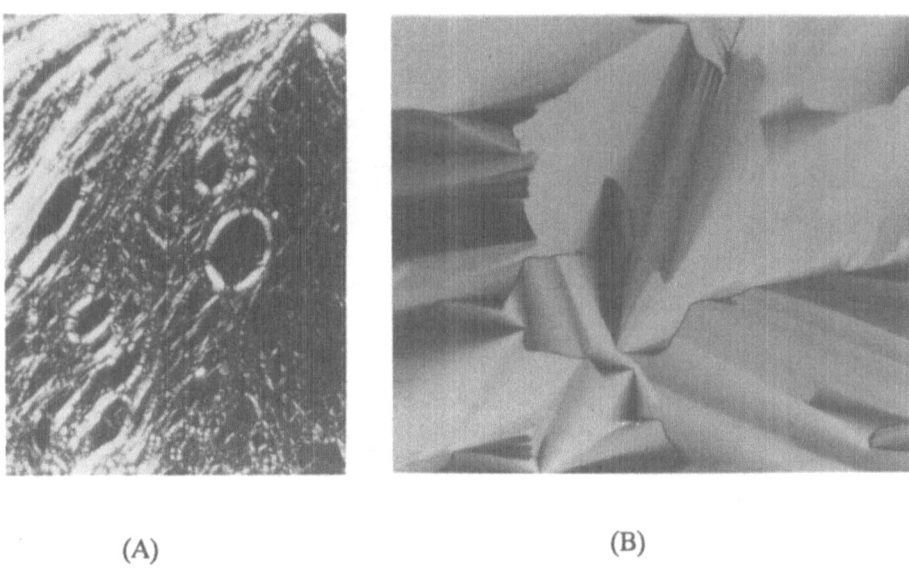

(A)   (B)

Figure 13. Optical patterns of a lamellar (a) and hexagonal (b) liquid crystal.

Small x-ray diffraction patterns also distinguish between these two varieties. The lamellar phase has a ratio of $1:1/\sqrt{2}:1/\sqrt{3}:1/\sqrt{4},...$ between the interlayer spacings, while the hexagonal array of cylinders has ratios $1:1/\sqrt{3}:1/\sqrt{4}:1/\sqrt{5}...$ (20).

The presence of the liquid crystals may also be detected directly in the emulsion using optical microscopy. An emulsion without a liquid crystal shows only the two phases while the liquid crystalline cover on the droplets is conspicuous from its radiance in polarized light, Fig.14.

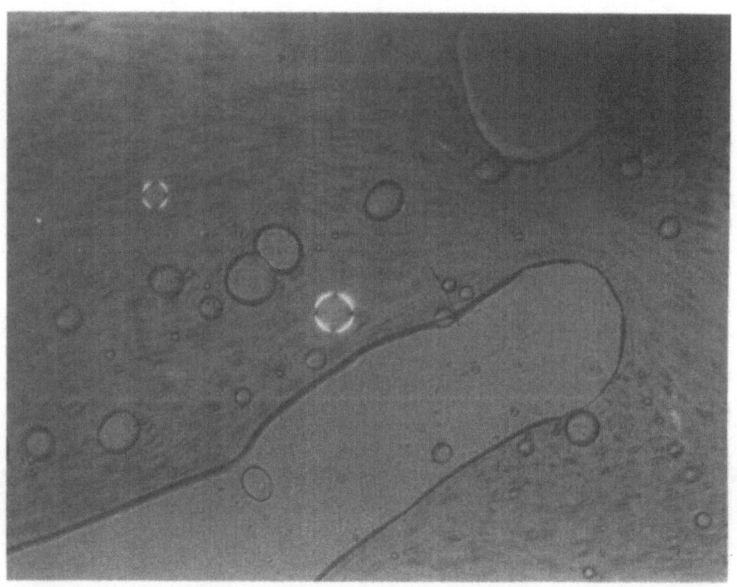

Figure 14. Droplets with a liquid crystalline cover are easily observed from their radiant halo with a Maltese cross when viewed between crossed polarizers in the microscope. The droplets without liquid crystal do not show these phenomenon.

The stabilizing action of the liquid crystals is limited to protection against coalescence (e.g., short distance action). It serves in three manners.

1. The lamellar structure leads to a strong reduction of the van der Waals forces during the coalescence step. The mathematical expression is tedious (19) but the van der Waals potential variation with interdroplet distance, Fig.15, illustrates the phenomenon (21).

Without the liquid crystalline phase, coalescence takes place over a thin liquid film in a distance range, where the slope of the van der Waals potential is steep (= huge van der Waals force). The force monitoring the coalescence over the liquid crystalline layers, Fig. 15, on the other hand, is extremely small and, in addition, the liquid crystal is highly viscous. With the combination of small force and large viscosity of the intervening layer the emulsion should be expected to exhibit enhanced stability against coalescence in comparison with the two-phase system with its thin film of low viscosity and huge compressive force.

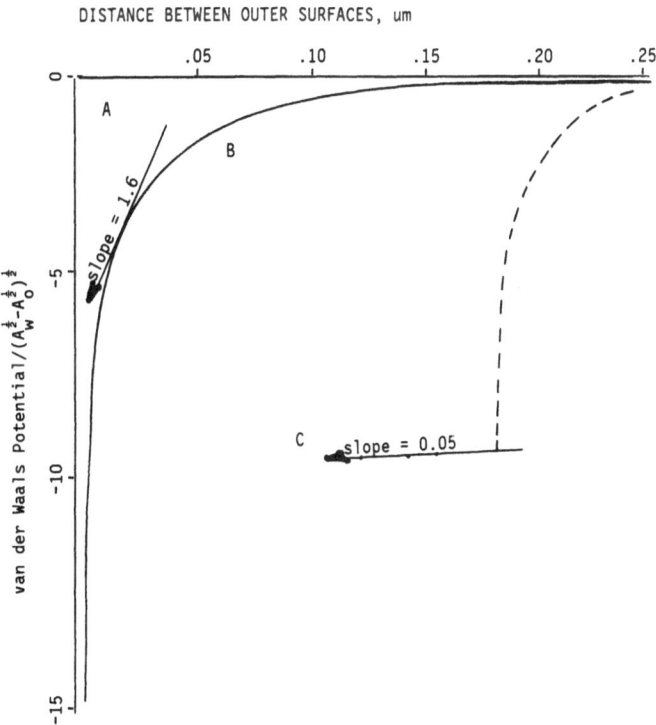

Figure 15. During coalescence, the van der Waals force in a system with liquid crystals (slope = 0.05) is much smaller than in the system with monomolecular layers of surfactant (slope = 1.6).

2.   In addition, the network of liquid crystalline leaflets (22) hinders the free mobility of the emulsion droplets and makes a stabilization contribution similar to that of a rigid continuous phase.

3.   The final consideration of the stability of these three–phase emulsions is probably the most important one. Small changes in emulsifier concentration lead to drastic changes in the amounts of the stabilizing phase. As an example, consider the points 1 to 3 in Figure 16. At point 1, with 2% emulsifier, 49% water and 49% aqueous phase, 50% oil and 50% aqueous phase are the only phases present. At point 2 the emulsifier concentration has been increased to 4%. Now the oil phase constitutes 47% of the total, while the aqueous phase is reduced to 29%; the remaining 24% being a liquid crystalline phase.

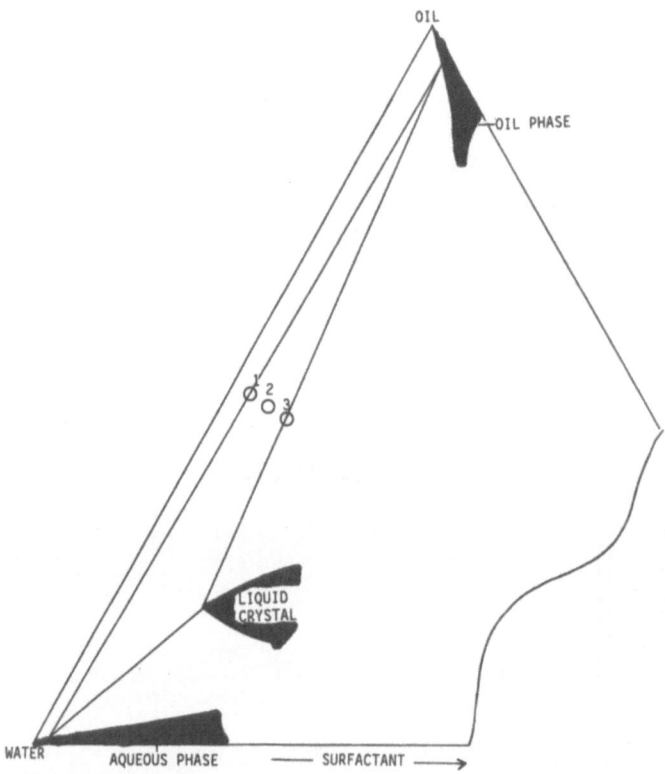

Figure 16. The nature of the emulsion is completely changed from an oil and water system at 2% emulsifier to an oil and liquid crystal system at 7% emulsifier.

At point 3, the aqueous phase has disappeared, and the entire emulsion consists of 42.3% oil and 57.5% liquid crystalline phase. The stabilizing phase is now the major part of the emulsion.

The application of these systems in high internal ratio emulsions (O/W emulsions with high content of oil) is obvious in light of the fact that the liquid crystals stabilize against coalescence but offer no protection against flocculation. In a high internal ratio emulsion, the dispersed phase occupies a large share of the total volume. An emulsion of that kind consists of very close-packed droplets, and flocculation is unavoidable. Protection against coalescence now becomes the essential characteristic of the emulsion, and the structures involving liquid crystals are useful.

The second application is concerned with spontaneous emulsification, a phenomenon treated by Groves in a series of articles (23).

The third application, which has been investigated by Frank (24), is in systems for delayed drug delivery. The relevance of using a lamellar liquid crystal to delay drug delivery should be seen against the diffusion rates in a lamellar liquid crystal (25–27). The diffusion rate parallel to the layers is as fast as in a liquid while perpendicular to the layers a value 3 orders of magnitude smaller is found. Hence, a lamellar liquid crystal covering an emulsion droplet could be expected to have a decisive influence on the mass transfer across the interface.

In summary, the three-phase emulsions with liquid crystals have been investigated to some extent. There is no doubt that in the future, when their properties are better known, they will be used extensively for drug delivery systems. Not only may a liquid crystal dissolve greater amounts of some substances that are only poorly soluble in liquid solvents (28), but the delivery rate of drugs from them may be changed.

The first developments to use liquid crystals as a matrix for solvents to enhance transdermal transport have already been made.

ACKNOWLEDGEMENT

Partial funding by the New York State Commission for Science and Technology through its Center for Advanced Material Processing at Clarkson University is gratefully acknowledged.

TABLE II
Symbols used in the article

A = Hamaker constant

D = diffusion coefficient for one droplet

ΔE = the energy to expel a spherical particle with radius r from the interface into the phase by which it is predominantly wet and toward which its contact angle is Θ

M = moles per liter

T = absolute temperature

V_{Eldl} = Electric double layer potential in $\kappa T/\mu m^2$

$V(kT)$ = potential energy expressed in kT

V_{vdw} = Van der Waals attraction potential in $kT/\mu m^2$

c = concentration of counter ions in moles/liter

d = one half the distance between two parallel plates

g = gravity acceleration

h = the distance perpendicular to the interface that the particle has moved form its equilibrium position.

k = Boltzmann constant

ℓ = distance between droplet centers

n = droplets per volume

p = distance in μm

Δp = difference in density

r = droplet radius

r_p = radius of solid particle

t = time

z = charge of counter ions

$\gamma_{O/w}$ = the interfacial tension between the oil and water phases.

η = viscosity

K = $3.1 \cdot 10^7 \sqrt{c \cdot z} (cm^{-1})$

ν = velocity

References

1. Becher, P. (1965), Emulsions: Theory and Practice, Reinhold, New York.

2. Sherman, P. (1969), Emulsion Science, Academic, New York.

3. Friberg, S.E., Mandell, L., and Larsson, M. (1969), J. Coll. and I. Sci. 29, 1.

4. International Union of Pure and Applied Chemistry, Manual on Colloid and Surface Science, Butterworths, London, (1972).

5. Wasan, D. Chapter in this book.

6. Boehm, J.T.C. and Lyklema, J. (1976), in Theory and Practice of Emulsions Technology (A.L. Smith, ed.) Academic Press, New York.

7. Graham, D.E. and Phillips, M.C. (1976), in Theory and Practice of Emulsion Technology (A.L. Smith, ed.) Academic Press, New York.

8. Napper, D.H. (1983), Polymeric Stabilization of Colloidal Dispersion, Academic Press, New York.

9. Clayfield, E.J. and Lumb, E.C. (1968), Macromolecules, 1,133.

10. Schulman, J.H. and Leja, I. (1954), Trans. Faraday Soc., 50, 598.

11. Shinoda, K. and Arai, H. (1964), J. Phys. Chem. 68, 3485.

12. Lin, T.J. (1978), J. Sco. Cosmet. Chem, 29, 117.

13. Saito, H. and Sh9noda, K. (1971), J. Colloid I. Sci., 35, 359.

14. Shinoda, K. and Arai H. (1964), J. Phys. Chem. 68, 3485.

15. Friberg, S.E. and Lapczynaska, I. (1975), Prog. Coll. Polym. Sci., 56, 16.

16. Lin, T.J. (1978), J. Sco. Cosmet. Chem., 29, 117.

17. Flaim, T.D. and Friberg, S.E. (1981), Separation Sci. Technol., 16, 1467.

24

18.   Friberg, S.E., Mandell, L. and Larsson, M. (1969), J. Coll. I. Sci., 29, 155.

19.   Friberg, S.E., Jansson, P.-O., and Cederberg, E. (1976), J. Coll. I. Sci., 55, 614.

20.   Fontell, K (1974), in "Liquid Crystals & Plastic Crystals" Vol. 2 (G.W. Gray and P.A. Winsor, eds.) Ellis Harwood, p.80.

21.   Jansson, P.-O., and Friberg, S.E. (1976), Mol. Cryst. Liquid Cryst., 34,75.

22.   Barry, B.W. (1975), Adv. Coll. I. Sci., 5,37.

23.   Groves, M.J., and Yalabik, H.S. (1977), Pharm. Technol., 2,21.

24.   Frank, S.G. (1978), Acta Pharm. Suec., 15,1.

25.   Webb, W.W. (1977), Science Vol. 195.

26.   Chidichimo, G. De Fazio, D. Ranieri, G.A. and Terenzi, M. (1986), Mol. Cryst. Liq. Cryst., 133.

27.   Gerritsen, H.C. and Caffrey, M. (1990), J. Phys. Chem, 94.

28.   Wahlgren, S., Lindström, A.L., and Friberg, S.E. (1984), J. Pharm. Sci. 73,10.

ADSORBED PROTEIN LAYERS IN FOOD EMULSIONS

ERIC DICKINSON
Procter Department of Food Science,
University of Leeds,
Leeds LS2 9JT, U. K.

ABSTRACT. Factors affecting the structure, composition and surface rheology of protein layers in oil-in-water emulsions are described. For β-lactoglobulin, direct evidence is presented for time-dependent polymerization of adsorbed protein via sulphydryl—disulphide interchange. Addition of the non-ionic surfactant Tween 20 (polyoxyethylene sorbitan monolaurate) up to a 1:1 molar ratio produces a dramatic reduction in the surface shear viscosity of an adsorbed β-lactoglobulin film, but no change in interfacial polymerization behaviour. The presence of Tween 20 influences the competitive displacement of pure and mixed whey proteins from the emulsion droplet surface. For the case of emulsions made with a 1:1 molar ratio of β-lactoglobulin + α-lactalbumin, addition of emulsifier after homogenization leads to more displacement of β-lactoglobulin than α-lactalbumin. It appears that the kinetics of competitive protein exchange between the aqueous phase and the oil—water interface is facilitated by the presence of the surfactant. Related aspects of emulsion stability are discussed.

1. INTRODUCTION

An adsorbed layer of protein at the oil—water interface is responsible for the primary stabilization of food emulsions such as homogenized milk, mayonnaise or ice-cream. This means that, if we are properly to understand the fundamental physico-chemical factors affecting the formation, rheology and stability of food emulsions, we must first have a good knowledge of the composition, structure and properties of protein films. The problem is complicated for food systems by the fact that the adsorbed layer is not a single pure protein, but is a mixture of several different protein species, together with various small-molecule surfactants (polar lipids and added emulsifiers) and other small molecules and ions.

The composition and structure of the stabilizing layer in a food emulsion is determined by competitive adsorption between proteins and surfactants at the oil—water interface [1,2], and by the nature of surfactant—protein interactions at the interface and in solution [3,4].

J. Sjöblom (ed.), Emulsions – A Fundamental and Practical Approach, 25–40.

The competitive displacement of protein from the interface by surfactant has been observed experimentally [5-9] and predicted by analytic theory [10] and computer simulation [11-13]. Competition between proteins depends on many factors—most notably the sequence and timing of the exposure to the interface, and the degree of configurational disorder of the native structure [14-18].

Proteins adsorb on newly formed oil droplets during the preparation of emulsions. The main thermodynamic driving force is the removal of non-polar side-chains away from the unfavourable aqueous environment of the bulk solution by displacing ordered water molecules from the droplet surface. An additional driving force is associated with the partial unfolding of the protein on adsorption.

This strong affinity of proteins for the oil—water interface means that they generally do not become desorbed from the droplet surface however much the emulsion is diluted by. To this extent, protein adsorption is effectively irreversible, though in other ways the term 'irreversible' is not appropriate as proteins can certainly be partially displaced from the oil—water interface by change in solvent conditions (ionic strength, pH, etc.) and completely displaced by competition with small-molecule surfactants. The equilibrium constant between adsorbed and unadsorbed states is proportional to the Boltzmann factor $\exp(E/kT)$, where E is the total molecular binding energy, T is temperature, and k is Boltzmann's constant. Suppose, for example, we consider the highly disordered protein β-casein composed of over 200 residues, one third of which are directly bound to the interface. While the mean binding energy per residue may be small (say 0.5 kT), the energy per molecule is sufficiently large that the Boltzmann factor ($> 10^{13}$) is weighted over-whelmingly towards the adsorbed state. This is why desorption cannot readily occur by dilution, though it may be induced by changing the solvent conditions to reduce the mean segment binding energy to a value near zero. One way to do this is to add a surfactant 'displacer' to the system. The small molecules compete strongly with protein residues for sites at the interface, thereby reducing the effective average binding energy per segment to below that required for protein displacement. Similar general behaviour to the above occurs at solid and liquid surfaces with synthetic polymers that are chemically and structurally much less complicated than proteins. Hence, there seems to be nothing unique about proteins in terms of the thermodynamics of adsorption.

This paper is concerned with systems containing milk proteins (β-casein, β-lactoglobulin and α-lactalbumin) and a low-molecular-weight surfactant (e.g. Tween 20). A protein like β-casein is a much more efficient adsorber than Tween 20 because it saturates the oil—water interface at a much lower bulk concentration. An effective monolayer saturation coverage is reached with the protein at a weight concentration some 100 times lower than for the surfactant (10^3 to 10^4 times lower in terms of molar concentration). This greater affinity of proteins for the adsorbed state leads to a greater lowering of the interfacial tension than with surfactants at low bulk concentrations. On the other hand, small amphiphilic molecules give a lower interfacial tension at high bulk concentrations. This difference is illustrated in Figure 1 by some surface pressure data [19] for lysozyme and three surfactants

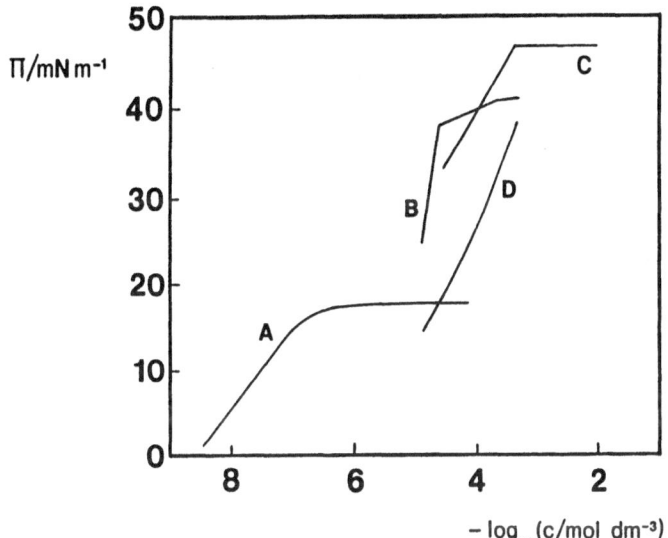

Figure 1. Surface pressure Π as a function of logarithm of bulk
concentration c for (A) lysozyme at the toluene—water interface,
(B) sorbitan monooleate (Span 80) at the octane—water interface,
(C) octylphenoxyethoxyethanol at the isooctane—water interface,
and (D) sodium dodecyl sulphate (SDS) at the heptane—water interface.
The surface pressure Π is defined by $\Pi = \gamma_0 - \gamma$, where γ and γ_0 are
the tensions in the presence and absence of surfactant.

adsorbing at oil—water interfaces. The reason why the small amphiphiles
reach a higher surface pressure is that they can pack more effectively
in the vicinity of the Gibbs dividing plane, and so can lower the
surface free energy to a greater extent than the protein molecules,
whose orientation of hydrophobic groups towards the oil phase and
hydrophilic groups towards the aqueous phase is severely restricted by
entanglements caused by the topological constraints of the chains. So,
even though it is a more efficient adsorber (at low concentrations), we
shall say that the protein is less surface—active than the surfactant
since it is readily displaced from the interface in the presence of high
bulk concentrations of the latter. In the same way, we shall say that
one protein is more surface—active than another if it gives a lower
interfacial tension (under steady—state conditions) at (similar) high
bulk concentrations. At equilibrium, the more surface—active protein is
expected to predominate at the interface. This is indeed what has been
observed in the competitive adsorption of the two disordered proteins,
α_{s1}—casein and β—casein, at the emulsion droplet surface and at the
macroscopic oil—water interface [15]. With proteins other than the
highly disordered caseins, however, deviations from this simple picture
occur because of interactions in the adsorbed layer.

2. INTERACTIONS IN PURE PROTEIN FILMS

An adsorbed protein film is a dense layer of polymer molecules. Based on experimental evidence and theoretical modelling, it is known that, on average, only a fraction of the segments of an adsorbing polymer is in direct contact with the surface. The remaining segments protrude into the bulk phase(s) forming an interfacial region that is much thicker than the width of the chain. Segments of a flexible polymer lying in direct contact with the surface are called 'trains', whereas 'loops' and 'tails' dangle into the bulk phase(s). At saturation coverage, only about one-third of the available surface area is occupied by train segments, as compared with nearly 100 % for a small-molecule surfactant monolayer. This rather low coverage of sites at the surface by flexible polymers arises from the configurational constraints of chain linkages in the presence of other polymer chains. It means that there are many small gaps in the adsorbed layer structure which are readily accessible to other (small) molecules in the system.

Proteins, of course, are much more complicated than the flexible random-chain homopolymers for which the idealized train—loop—tail model was originally devised. Nevertheless, it is still useful to regard a highly disordered protein such as β-casein, with little secondary structure and no intramolecular disulphide bonds, to be described by this model to a first approximation. Indeed, based on a simple statistical analysis of surface pressure data, it is estimated [20] that roughly one-third of the amino-acid residues in β-casein do reside in trains at the oil—water interface, with the rest in loops and tails. In addition, it appears [21] that the most hydrophilic part of the β-casein molecule (residues 1—48 out of a total of 209) protrudes substantially into the aqueous phase, and makes a major contribution to the effective hydrodynamic thickness of an adsorbed β-casein layer at the emulsion droplet surface. This contrasts with the globular protein β-lacto-globulin, which gives a thin compact adsorbed layer, i.e., thickness ca. 2 nm, as compared with 10—15 nm for β-casein [21].

Adsorption of a protein produces a change in its environment. This, in turn, may induce configurational rearrangements, since the Gibbs free energy of stabilization of the native structure in bulk solution is usually much less than 1 kT per amino-acid residue [22]. Hydrophobic interaction between non-polar residues, which stabilizes organization into α-helices and β-sheets in the native structure, is partly replaced by hydrophobic interaction with non-polar molecules at the surface of the oil. A loss of intramolecular hydrophobic interactions therefore leads to less secondary structure and increased conformational entropy of the molecule. Nonetheless, with globular proteins such as lysozyme or β-lactoglobulin, much of the secondary structure—and even some of the tertiary structure—is retained on adsorption. In these cases it is certainly inappropriate to describe the adsorbed protein layer by the train—loop—tail model, since much of the original tightly coiled globular shape is retained in the adsorbed state by the covalent strong disulphide cross-links. Thus, it may be more realistic to regard an adsorbed monolayer of globular protein molecules, not as an entangled layer of flexible chains, but rather as a dense two-dimensional system of interacting deformable particles of similar size to the native protein molecule in bulk solution [23]. A computer simulation of such deformable particles modelled as cyclic lattice chains has recently been described [24,25].

The equilibrium structure of an adsorbed protein film may not be reached over the normal experimental time-scale. This is especially so when fast adsorption takes place from a concentrated solution during emulsion formation. A protein film may contain molecules in many different states of unfolding with only very slow exchange between the states due to the high density of close-range interactions and slow diffusion in the layer. Protein molecules adsorbing from concentrated solution at a fresh oil—water interface may not have time to unfold properly before they are rapidly surrounded by others, producing a congested close-packed arrangement in which the possibility for modest configurational readjustment is severely hampered. The high protein concentration in the adsorbed film means that its physical state is more like that of a viscoelastic solid (a glass or a gel) than that of a protein solution.

Surface rheology of protein films at the oil—water interface is extremely sensitive to protein structure and to the nature of protein—protein interactions in the adsorbed layer [26-28]. Adsorbed films of the disordered caseins have a low surface shear viscosity, whereas films of the globular whey proteins are highly viscoelastic. Figure 2 shows time-dependent surface viscosity data [28] for adsorbed films of pure α-lactalbumin and pure β-lactoglobulin at the hydrocarbon—water surface (0.001 wt %, pH 7, 25 °C). The surface viscosity of α-lactalbumin attains an effectively constant value of 550 mN m^{-1} s after ca. 40 h, whereas that for β-lactoglobulin is substantially higher (1300 mN m^{-1} s) and is still slowly increasing with aging time at the interface. Under similar experimental conditions, the surface viscosity of β-casein (< 1 mN m^{-1} s) is too small to be plotted on the same graph. In contrast to surface pressure measurements, which are related to free energy changes occurring in the immediate vicinity of the Gibbs dividing surface (train segments), the determination of surface rheological properties gives information about connected regions of the protein film further away from the interface (loops and tails, and secondary layers). The origin of the slow steady increase in surface viscosity with time appears not to be due to any single type of molecular interaction—as it occurs with proteins having a wide range of structures. However, the pronounced increase in surface viscosity with time for β-lactoglobulin appears to be associated with intermolecular disulphide bond formation in the adsorbed layer, as this extended increase with time is eliminated in experiments with chemically modified β-lactoglobulin [28] for which the disulphide bonds have first been cleaved and blocked (see Figure 3). The surface viscosity of κ-casein is much higher than that of β-casein or α$_{s1}$-casein [14], possibly also as a result of intermolecular disulphide bond formation at the interface.

Direct evidence for time-dependent intermolecular sulphydryl—disulphide interchange involving β-lactoglobulin at the emulsion droplet surface has recently been obtained [29] from an analysis of protein displaced from the interface by sodium dodecyl sulphate (SDS). Samples

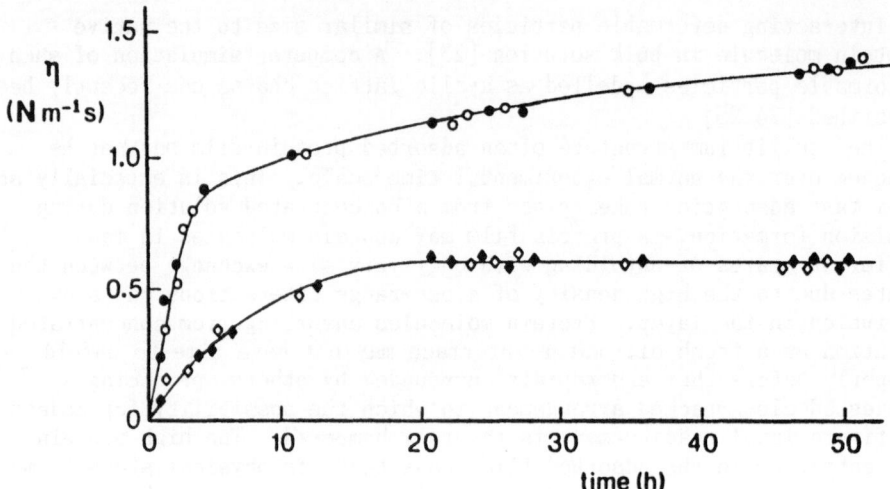

Figure 2. Surface shear viscosity η of pure whey proteins at the n-tetradecane—water interface plotted against adsorption time t: ♦, ◊, α-lactalbumin; •, o, β-lactoglobulin.

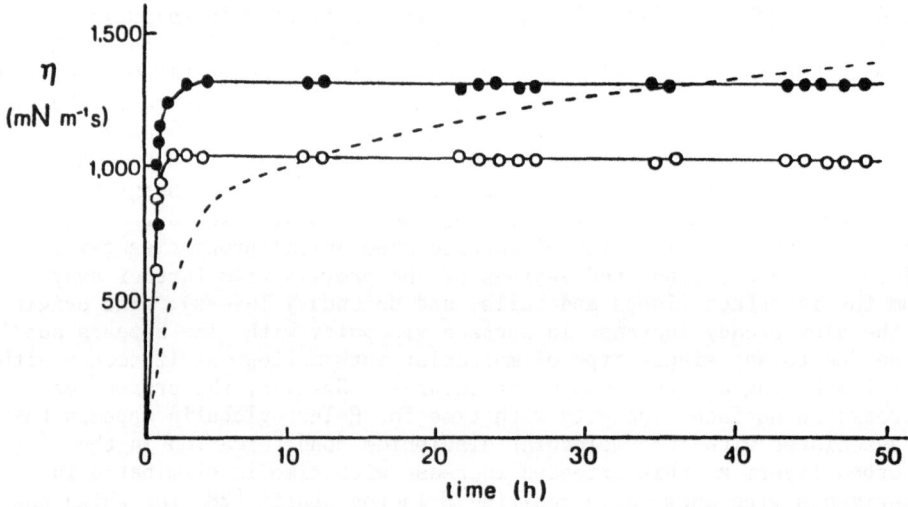

Figure 3. Surface shear viscosity η as a function of adsorption time t: o, S-carboxymethyl-β-lactoglobulin; •, amidoethyl-β-lactoglobulin; ---, native β-lactoglobulin.

of β-lactoglobulin-stabilized oil-in-water emulsion (10 wt % oil, 0.5 wt % protein, pH 7) were aged for various times prior to complete protein displacement by SDS and subsequent analysis by SDS—polyacrylamide gel electrophoresis. Table 1 gives the relative amounts of various oligomers of β-lactoglobulin existing at the interface at various times after emulsion formation. Just monomers are detected in the adsorbed layer immediately after emulsification, but on aging there is a gradual growth in the amount of polymerized β-lactoglobulin and in the average size of oligomers. In contrast, over the same 72-hour observation time,

TABLE 1. Percentages of different oligomers of β-lactoglobulin at the oil—water interface for emulsions aged for various periods of time [29]

Oligomer	Time (h)			
	0	2	24	72
Monomer	100.0	83.1	60.9	38.2
Dimer	0.0	14.6	25.2	26.1
Trimer	0.0	2.0	7.8	8.2
> Trimer	0.0	0.3	6.5	27.5

there is no discernible polymerization of adsorbed protein in emulsions made with pure α-lactalbumin, or in emulsions made with β-lactoglobulin in the presence of a reagent for modifying sulphydryl groups. Monomeric β-lactoglobulin (18 350 daltons) has a free sulphydryl group which can participate in sulphydryl—disulphide interchange at the oil—water interface. Native α-lactalbumin has no free sulphydryl group, and so pure α-lactalbumin does not polymerize at the interface (though in mixed films of the two whey proteins, α-lactalbumin may participate in hetero-polymer formation with β-lactoglobulin, again presumably by a sulphydryl —disulphide interchange reaction). These results are broadly in line with the surface rheology data for the pure and mixed whey proteins adsorbed at the macroscopic oil—water interface [28].

3. COMPETITIVE ADSORPTION OF MIXED WHEY PROTEINS + TWEEN 20

We have performed experiments [30] to investigate the influence of the water-soluble surfactant Tween 20 (polyoxyethylene sorbitan monolaurate) on the competitive adsorption of α-lactalbumin + β-lactoglobulin at the oil—water interface in a hydrocarbon oil-in-water emulsion. The aim of the study is to learn how the presence of a food-grade emulsifier can affect protein surface composition in a model dairy emulsion system.

Let us start by considering the binary whey protein system in the absence of emulsifier. In contrast to α_{s1}-casein + β-casein, for which the competitive adsorption is rapid and reversible [15], the process of protein interchange at the oil—water interface in the α-lactalbumin +

β-lactoglobulin system is slow and irreversible [16,17]. In emulsions made with various ratios of the two whey proteins, we have found [17] that the two proteins adsorb together at the fresh interface during emulsification roughly in proportion to their bulk concentrations. When β-lactoglobulin in solution is added to an emulsion prepared with an equal amount of pure α-lactalbumin, only ca. 10 % of the originally adsorbed α-lactalbumin is displaced over a 30-hour period; over the same period approximately 3 times as much β-lactoglobulin becomes adsorbed [16]. This increase in the total amount of protein at the emulsion droplet surface on addition of the second component seems to be an example of what may be a rather general phenomenon, since it has been observed also in the binary casein system [15] and in binary mixtures of synthetic polymers adsorbing competitively at solid surfaces [31,32]. When the same experiment was carried out the other way round (i.e. with α-lactalbumin added to a β-lactoglobulin emulsion that had been made for about 30 minutes), it was observed [16] that a negligible amount of β-lactoglobulin is displaced and only a small amount of α-lactalbumin is adsorbed. The fact that β-lactoglobulin is more difficult to displace than α-lactalbumin, even though the latter is apparently more surface-active (see later), may have something to do with its polymerization at the interface via disulphide bonds. What is confirmed from these whey protein emulsion experiments (and complementary interfacial viscosity experiments [2,28]) is that the adsorption is kinetically controlled. In mixed globular protein emulsion systems, the general rule seems to be that the protein component that gets to the interface first (e.g. during homogenization) will be the one that predominates there thereafter.

We turn now to emulsion systems containing a mixture of Tween 20 and a single pure whey protein. Figure 4 shows how the presence of an increasing concentration of emulsifier affects the protein concentration at the interface in a β-lactoglobulin emulsion system. The interfacial coverage of β-lactoglobulin is plotted both as a function of the Tween 20 concentration c and the emulsifier-to-protein molar ratio R. Full experimental details are given elsewhere [30]. We see that, at low Tween 20 concentration ($c \lesssim 0.01$ wt %, $R \lesssim 0.5$), there is no real change in protein surface coverage. On the other hand, there is effectively complete displacement of protein from the droplet surface for $c \gtrsim 0.3$ wt % ($R \gtrsim 10$). In confirmation of earlier work [6], we see that the general form of the displacement curve in Figure 4 is qualitatively the same irrespective of whether the emulsifier is present during homogenization or is added afterwards. That being said, there is a discernible trend towards lower protein surface coverages in the systems with Tween 20 present during emulsification at concentrations corresponding to R values from 1 to 4. This effect is probably attributable [30] to the average droplet size of the latter emulsions being slightly smaller (by 10—20 %).

At emulsifier concentrations below those required for significant protein displacement, Tween 20 forms a 1:1 molar complex ($R = 1$) with β-lactoglobulin [33,34]. It is noteworthy, therefore, that adsorbed films of β-lactoglobulin at the oil—water (or air—water) interface become highly mobile at $R \simeq 1.0$, in contrast to the highly viscoelastic behaviour in the absence of surfactant (see Figure 2). Figure 5 shows

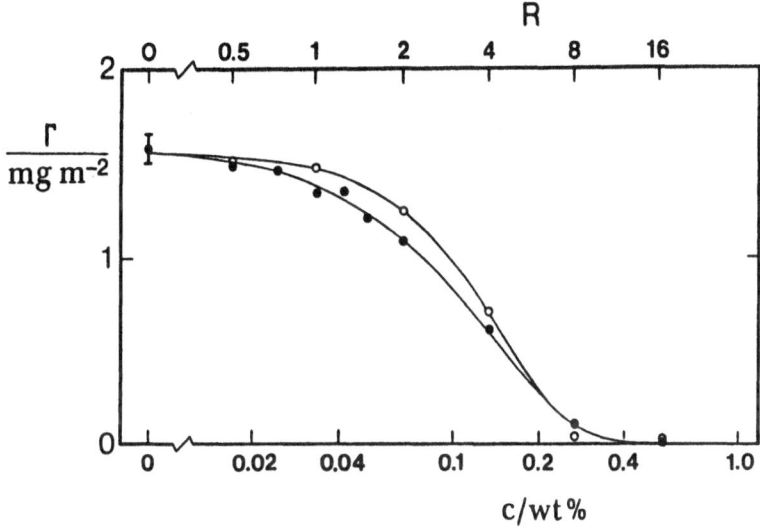

Figure 4. Effect of Tween 20 on surface coverage in emulsions made with pure β-lactoglobulin (10 wt % oil, 0.5 wt % protein). Protein surface concentration Γ is plotted against emulsifier concentration c and molar ratio R: ●, Tween 20 present during homogenization; o, Tween 20 added after homogenization.

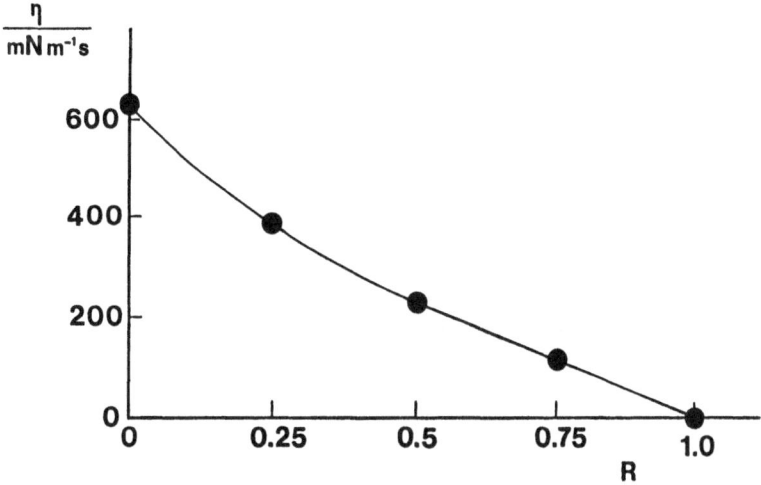

Figure 5. Surface shear viscosity η of β-lactoglobulin film with Tween 20 present after 5 hours at n-tetradecane–water interface as a function of the emulsifier-to-protein molar ratio R.

surface viscosity values after 5 hours adsorption at the oil—water interface [8]. The surface viscosity at R = 1.0 is below 10 mN m^{-1} s as compared with over 1000 mN m^{-1} s for R = 0 (see Figure 2). A similar effect of a very low surfactant concentration in drastically reducing the surface viscosity was found [13] with sodium caseinate + non-ionic surfactant $C_{12}E_8$ (octaoxyethylene dodecyl ether); in this latter case, there is no evidence for complex formation. We infer, therefore, that the phenomenon of the sharp drop in surface rheological parameters on addition of surfactant at concentrations well below that required for protein displacement is not directly related to complex formation. Nor does the interaction between Tween 20 and β-lactoglobulin have any significant effect on the protein polymerization at the interface [30]. Whatever is the reason for the behaviour represented by the data in Figure 5, it cannot be interpreted in terms of an inhibition of sulphydryl—disulphide interchange due to the presence of the emulsifier.

The general trend of competitive displacement behaviour for pure α-lactalbumin is similar to that described above for β-lactoglobulin, though it is slightly more difficult to displace the former from the emulsion droplet surface than the latter. That is, it takes more Tween 20 to displace the same amount of protein with pure α-lactalbumin than with pure β-lactoglobulin. Whereas all of the pure β-lactoglobulin is displaced at R ≈ 10, the pure α-lactalbumin is not even fully displaced at R = 32.

Let us now consider the system α-lactalbumin + β-lactoglobulin + Tween 20. Figure 6 shows the effect of adding emulsifier following homogenization to an emulsion made with a 1:1 molar ratio of the two proteins [30]. The total protein surface concentration and the surface concentrations of the separate whey proteins are plotted as a function of the emulsifier-to-protein molar ratio R. In the absence of Tween 20 (R = 0), the surface concentrations of the two proteins are similar. In the range 2 ≤ R ≤ 4, under conditions of substantial partial displacement of protein from the interface, there is almost twice as much of the α-lactalbumin at the surface as β-lactoglobulin. At R = 16, we see that the β-lactoglobulin is completely displaced, whereas the α-lactalbumin is not. This mirrors the behaviour in the pure protein systems.

The tendency of the more surface-active α-lactalbumin to predominate over β-lactoglobulin at the hydrocarbon oil—water interface is even more evident when the Tween 20 is present during homogenization. This is illustrated in Figure 7. On introduction of emulsifier, the surface coverage of α-lactalbumin remains essentially constant, but there is a reduction in the surface coverage of β-lactoglobulin, and hence in the total protein surface coverage. At R = 2, there is about three times as much α-lactalbumin at the emulsion droplet surface as β-lactoglobulin. So we see that the presence of surfactant during emulsification allows preferential adsorption of α-lactalbumin from the binary mixture of whey proteins, whereas there is no such competitive effect with the whey proteins in the absence of surfactant. Its preferential adsorption at the oil—water interface may be related to the fact that α-lactalbumin does not form a complex with Tween 20 whereas β-lactoglobulin does. If the latter complex is less surface-active than native β-lactoglobulin, then it will compete less successfully for space at the interface.

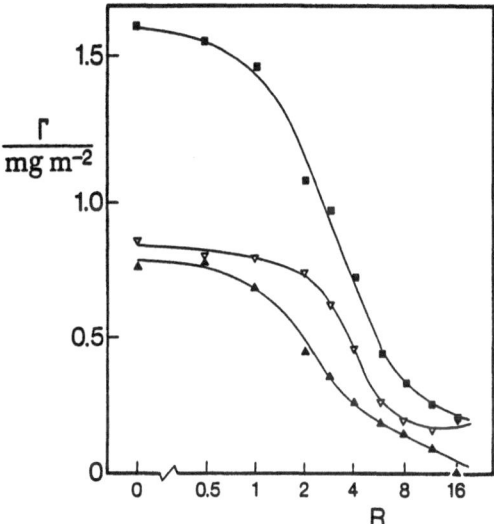

Figure 6. Effect of Tween 20 added after homogenization on protein surface concentration Γ in emulsions made with 1:1 molar ratio of α-lactalbumin + β-lactoglobulin: ■, total protein; ▽, α-lactalbumin; ▲, β-lactoglobulin.

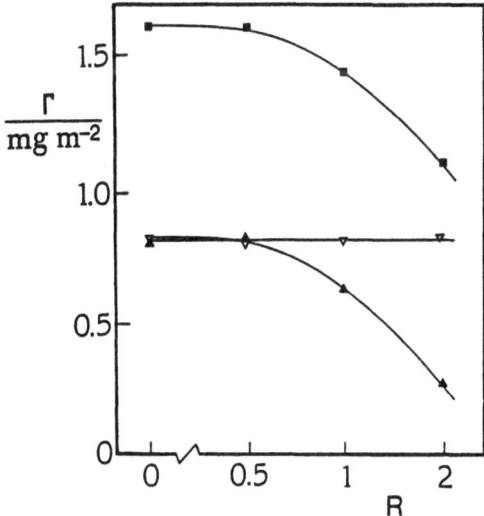

Figure 7. As Figure 6, except that the emulsifier is present during homogenization.

4. ASPECTS OF EMULSION BEHAVIOUR

At high concentrations, a water-soluble surfactant like Tween 20 can act
as a vehicle for mass transport of oil between emulsion droplets via a
mechanism involving solubilization of oil into surfactant micelles in
the aqueous phase between the droplets. This phenomenon is illustrated
by recent experiments [35] on binary emulsions made up from equal
volumes of n-hexadecane droplets + n-octadecane droplets, prepared
separately as emulsions having the same average droplet size, and then
mixed together after homogenization. The oil crystallization in these
binary emulsions (20 wt % oil) was studied as a function of temperature
by low-intensity ultrasonic velocity measurement [36], enabling direct
comparison to be made with the crystallization behaviour in the entirely
equivalent emulsions containing just n-hexadecane or n-octadecane
droplets alone [37]. With sodium caseinate (0.5 wt %) as the emulsion
stabilizer (no Tween 20 present), it was observed [35] that the two
sorts of oil droplets in the binary emulsion exhibited independent
supercooling and crystallization behaviour similar to what was found in
the corresponding single oil emulsions. That is, on cooling at a slow
steady rate, the n-octadecane droplets crystallize first over a narrow
temperature range at ca. 15 °C, and then the n-hexadecane droplets
crystallize over a narrow temperature range at ca. 3.5 °C. However,
when the same experiment was performed with Tween 20 as emulsifier (2
wt % surfactant, no protein, same average droplet size), there was found
to be a different crystallization behaviour in the binary emulsion from
in the corresponding single oil emulsions. In particular, it was found
that, on cooling, the onset of n-octadecane freezing is depressed by a
few degrees, and that the freezing process, once it does occur, is not
sharp (as with the protein-stabilized emulsions) but takes place over a
range of several degrees. This behaviour is indicative of some inter-
mixing of the oils in the n-hexadecane and n-octadecane droplets in the
time period (a few hours) between making up the binary emulsion and the
taking of the ultrasonic velocity measurement.

In an emulsion stabilized solely by Tween 20, we can assume that the
excess surfactant exists as a micellar solution in the aqueous medium,
and that these surfactant aggregates are able to solubilize hydrocarbon
oil into the continuous phase. Swollen micelles can therefore act as a
mode of transport for the diffusion of n-hexadecane and n-octadecane
molecules between emulsion droplets. In addition, some direct exchange
of oil molecules may occur during droplet collision. (Evidence for a
direct interaction between the contents of droplets is provided by
experiments on emulsions containing a mixture of solid and supercooled
liquid droplets: we have found [38] that crystallization is induced in
liquid n-hexadecane droplets when crystalline n-hexadecane droplets are
mixed with them.) In contrast to water-soluble surfactants, however,
proteins are not able to solubilize individual hydrocarbon molecules in
solution to any significant extent. In addition, the thick adsorbed
protein layer inhibits direct oil transport between droplets by not
allowing the colliding droplet surfaces to get very close together. The
large thermodynamic driving force for the mass transport process is the
entropy of mixing of the two pure oils; this would be expected to be

close to the ideal entropy of mixing for the mixing of two saturated hydrocarbons of similar chain length. This driving force for the inter-mixing of different liquids is several orders of magnitude larger than the driving force for conventional Ostwald ripening in pure oil systems. The inter-mixing is not accompanied by any change in average emulsion droplet size.

The state of aggregation of an emulsion may be affected by the competitive adsorption of proteins during emulsification [39-41]. In a recent investigation of hydrocarbon oil-in-water emulsions made with a mixture of highly surface-active protein (sodium caseinate) + moderately surface-active hydrocolloid (gum arabic), it was found [42] not to be possible to make stable unflocculated emulsions with certain (low) emulsifier concentrations and compositions. This behaviour is explained in terms of partial displacement of the hydrophilic polysaccharide molecules from the oil—water interface by the more surface-active casein molecules, leading to bridging flocculation of emulsion droplets by the preferentially adsorbing protein. Replacement of gum arabic by the hydrophilic protein gelatin leads to behaviour of a qualitatively similar nature [39,40]. A recent Monte Carlo simulation has shown [41] how bridging flocculation can occur in the two-component emulsifier system when it is absent in both the equivalent one-component systems.

The results reported in this article show how a study of model systems containing pure proteins and surfactants can give information about the structural and compositional properties of protein films in protein-stabilized emulsions. The presence of surfactants during the emulsification generally leads to a reduction in mean droplet size and an improvement in creaming stability [2,43]. Small quantities of surfactants, well below what is required for protein displacement, can produce large changes in the surface rheology of adsorbed protein films. These surface rheological changes seem likely to have very important implications for the aggregation behaviour of emulsion droplets and the drainage and rupture of thin films between droplets. An understanding of these physico-chemical effects will assist in controlling the bulk manufacture and processing of food emulsions and in extending the shelf-life of emulsion products.

ACKNOWLEDGEMENT

Continued financial support from the Agricultural and Food Research Council is gratefully acknowledged.

REFERENCES

1 Dickinson, E. (1986) 'Mixed proteinaceous emulsifiers: review of competitive protein adsorption and the relationship to food colloid stabilization', Food Hydrocolloids, 1, 3-23.
2 Dickinson, E. (1991) 'Competitive adsorption and protein—surfactant interactions in oil-in-water emulsions', American Chemical Society Symp. Ser., 448, 114-129.

38

3 Dickinson, E. and Woskett, C. M. (1989) 'Competitive adsorption between proteins and small-molecule surfactants in food emulsions', in R. D. Bee, P. Richmond and J. Mingins (eds.), Food Colloids, Royal Society of Chemistry, Cambridge, pp. 74-96.

4 Jones, M. N. and Brass, A. (1991) 'Interactions between small amphipathic molecules and proteins', in E. Dickinson (ed.), Food Polymers, Gels and Colloids, Royal Society of Chemistry, Cambridge, pp. 65-80.

5 Oortwijn, H. and Walstra, P. (1979) 'The membranes of recombined fat globules. 2. Composition', Neth. Milk Dairy J., $\underline{33}$, 134-154.

6 de Feijter, J. A., Benjamins, J. and Tamboer, M. (1987) 'Adsorption displacement of proteins by surfactants in oil-in-water emulsions', Colloids Surf., $\underline{27}$, 243-266.

7 Dickinson, E., Narhan, S. K. and Stainsby, G. (1989) 'Stability of cream liqueurs containing low-molecular-weight surfactants', J. Food Sci., $\underline{54}$, 77-81.

8 Courthaudon, J.-L., Dickinson, E., Matsumura, Y. and Clark, D. C. (1991) 'Competitive adsorption of β-lactoglobulin + Tween 20 at the oil—water interface', Colloids Surf., $\underline{56}$, 293-300.

9 Courthaudon, J.-L., Dickinson, E. and Dalgleish, D. G. (1991) 'Competitive adsorption of β-casein and nonionic surfactants in oil-in-water emulsions', J. Colloid Interface Sci., in press.

10 Cohen Stuart, M. A., Fleer, G. J. and Scheutjens, J. M. H. M. (1984) 'Displacement of polymers. 1. Theory. Segmental adsorption energy from polymer desorption in binary solvents', J. Colloid Interface Sci., $\underline{97}$, 515-525.

11 Dickinson, E. (1988) 'Monte Carlo model of competitive adsorption between interacting macromolecules and surfactants', Molec. Phys., $\underline{65}$, 895-908.

12 Dickinson, E. and Euston, S. R. (1989) 'Computer simulation of the competitive adsorption between polymers and small displacer molecules', Molec. Phys., $\underline{68}$, 407-421.

13 Dickinson, E., Euston, S. R. and Woskett, C. M. (1990) 'Competitive adsorption of food macromolecules and surfactants at the oil—water interface', Prog. Colloid Polym. Sci., $\underline{82}$, 65-75.

14 Dickinson, E., Murray, A., Murray, B. S. and Stainsby, G. (1987) 'Properties of adsorbed layers in emulsions containing a mixture of caseinate and gelatin', in E. Dickinson (ed.), Food Emulsions and Foams, Royal Society of Chemistry, London, pp. 86-99.

15 Dickinson, E., Rolfe, S. E. and Dalgleish, D. G. (1988) 'Competitive adsorption of α_{s1}-casein and β-casein in oil-in-water emulsions', Food Hydrocolloids, $\underline{2}$, 397-405.

16 Dickinson, E., Rolfe, S. E. and Dalgleish, D. G. (1989) 'Competitive adsorption in oil-in-water emulsions containing α-lactalbumin and β-lactoglobulin', Food Hydrocolloids, $\underline{3}$, 193-203.

17 Dalgleish, D. G., Euston, S. E., Hunt, J. A. and Dickinson, E. (1991) 'Competitive adsorption of β-lactoglobulin in mixed protein emulsions', in E. Dickinson (ed.), Food Polymers, Gels and Colloids, Royal Society of Chemistry, Cambridge, pp. 485-489.

18 Dickinson, E., Hunt, J. A. and Dalgleish, D. G. (1991) 'Competitive adsorption of phosvitin with milk proteins in oil-in-water

emulsions', Food Hydrocolloids, **4**, 403-414.

19 Fisher, L. R. and Parker, N. S. (1988) 'Effect of surfactants on the interactions between emulsion droplets', in E. Dickinson and G. Stainsby (eds.), Advances in Food Emulsions and Foams, Elsevier Applied Science, London, pp. 45-90.

20 Graham, D. E. and Phillips, M. C. (1979) 'Protein at liquid interfaces', J. Colloid Interface Sci., **70**, 403-439.

21 Dalgleish, D. G. and Leaver, J. (1991) 'Dimensions and possible structures of milk proteins at oil—water interfaces', in E. Dickinson (ed.), Food Polymers, Gels and Colloids, Royal Society of Chemistry, Cambridge, pp. 113-122.

22 Cohen Stuart, M. A., Fleer, G. J., Lyklema, J., Norde. W. and Scheutjens, J. M. H. M. (1991) 'Adsorption of ions, polyelectrolytes and proteins', Adv. Colloid Interface Sci., **34**, 477-535.

23 de Feijter, J. A. and Benjamins, J. (1982) 'Soft-particle model of compact macromolecules at interfaces', J. Colloid Interface Sci., **90**, 289-292.

24 Dickinson, E. and Euston, S. R. (1989) 'Statistical study of a concentrated dispersion of deformable particles modelled as an assembly of cyclic lattice chains', Molec. Phys., **66**, 865-886.

25 Dickinson, E. and Euston, S. R. (1990) 'Simulation of adsorption of deformable particles modelled as cyclic lattice chains: a simple statistical model of protein adsorption', J. Chem. Soc., Faraday Trans., **86**, 805-809.

26 Castle, J., Dickinson, E., Murray, B. S. and Stainsby, G. (1987) 'Mixed protein films adsorbed at the oil—water interface', American Chemical Society Symp. Ser., **343**, 118-134.

27 Dickinson, E., Murray, B. S. and Stainsby, G. (1988) 'Protein adsorption at air—water and oil—water interfaces', in E. Dickinson and G. Stainsby (eds.), Advances in Food Emulsions and Foams, Elsevier Applied Science, London, pp. 123-162.

28 Dickinson, E., Rolfe, S. E. and Dalgleish, D. G. (1990) 'Surface shear viscosity as a probe of protein—protein interactions in mixed protein films adsorbed at the oil—water interface', Int. J. Biol. Macromol., **12**, 189-194.

29 Dickinson, E. and Matsumura, Y. (1991) 'Time-dependent polymerization of β-lactoglobulin through disulphide bonds at the oil—water interface in emulsions', Int. J. Biol. Macromol., **13**, 26-30.

30 Courthaudon, J.-I., Dickinson, E., Matsumura, Y. and Williams, A. (1991) 'Influence of emulsifier on the competitive adsorption of whey proteins in emulsions', Food Structure, in press.

31 Huldén, M. and Sjöblom, E. (1990) 'Adsorption of some common surfactants and polymers on TiO_2-pigments', Prog. Colloid Polym. Sci., **82**, 28-37.

32 Frantz, P. and Granick, S. (1991) 'Kinetics of polymer adsorption and desorption', Phys. Rev. Lett., **66**, 899-902.

33 Coke, M., Wilde, P. J., Russell, E. J. and Clark, D. C. (1990) 'The influence of surface composition and molecular diffusion on the stability of foams formed from protein/surfactant mixtures', J. Colloid Interface Sci., **138**, 489-504.

34 Clark, D. C., Coke, M., Wilde, P. J. and Wilson, D. R. (1991)

'Molecular diffusion at interfaces and its relation to disperse phase stability', in E. Dickinson (ed.), Food Polymers, Gels and Colloids, Royal Society of Chemistry, Cambridge, pp. 272-276.

35 Dickinson, E., Goller, M. I., McClements, D. J. and Povey, M. J. W. (1991) 'Monitoring crystallization in simple and mixed oil-in-water emulsions using ultrasonic velocity measurement', in E. Dickinson (ed.), Food Polymers, Gels and Colloids, Royal Society of Chemistry, Cambridge, pp. 171-179.

36 Povey, M. J. W. (1988) 'Ultrasonics as a probe of food emulsions and dispersions', in E. Dickinson and G. Stainsby (eds.), Advances in Food Emulsions and Foams, Elsevier Applied Science, London, pp. 285-327.

37 Dickinson, E., Goller, M. I., McClements, D. J., Peasgood, S. and Povey, M. J. W. (1990) 'Ultrasonic monitoring of crystallization in an oil-in-water emulsion', J. Chem. Soc., Faraday Trans., 86, 1147-1148.

38 McClements, D. J., Dickinson, E. and Povey, M. J. W. (1990) 'Crystallization in hydrocarbon-in-water emulsions containing a mixture of solid and liquid droplets', Chem. Phys. Lett., 172, 449-452.

39 Castle, J., Dickinson, E., Murray, A. and Stainsby, G. (1988) 'Rheological and stability properties of concentrated emulsions made with a binary mixture of proteins', in G. O. Phillips, D. J. Wedlock and P. A. Williams (eds.), Gums and Stabilisers for the Food Industry, IRL Press, Oxford, vol. 4, pp. 473-482.

40 Dickinson, E., Flint, F. O. and Hunt, J. A. (1989) 'Bridging flocculation in binary protein stabilized emulsions', Food Hydrocolloids, 3, 389-397.

41 Dickinson, E. and Galazka, V. B. (1991) 'Bridging flocculation induced by competitive adsorption: implications for emulsion stability', J. Chem. Soc., Faraday Trans., 87, 963-969.

42 Dickinson, E. and Galazka, V. B. (1991) 'Bridging flocculation in emulsions made with a mixture of protein and polysaccharide', in E. Dickinson (ed.), Food Polymers, Gels and Colloids, Royal Society of Chemistry, Cambridge, pp. 494-497.

43 Dickinson, E., Mauffret, A., Rolfe, S. E. and Woskett, C. M. (1989) 'Adsorption at interfaces in dairy systems', J. Soc. Dairy Technol., 42, 18-22.

EMULSIONS IN THE FOOD INDUSTRY

KÅRE LARSSON
Department of Food Technology, University of Lund
Box 124, S-221 00 Lund, Sweden

ABSTRACT. Most food emulsions contain two types of emulsifiers – polar lipids and proteins – at the oil/water interface. Another characteristic feature is the common presence of fat crystals. The paper is focused on the crystallization behaviour of polar lipids together with oils/fats. The most important lipid type of food emulsifiers forms crystals with a polar surface exposed towards water and non-polar surfaces in the oil (surface-active crystals). When a triglyceride oil is crystallized at the oil/water interface, a remarkable alpha-like monolayer is formed, which also can form multilayers. There are close structural relations between the monolayer phases and the bulk polymorphic forms.

The effect of emulsion structure has also a strong effect on fat oxidation. Recent work in our laboratory has shown that marine oils can be well stabilized against oxidation if a microemulsion of glycerol with ascorbic acid is formed with tocopherol in the oil.

1. Introduction

The complex character of food emulsions is illustrated in Table 1. As both proteins and lipids are amphiphilic they will compete at the interface, and as protein adsorption usually takes place in an irreversible way, the kinetics involved in the emulsification process plays an important role. The simplest of these emulsions is margarine, which in exceptional cases consists of water without any protein present. It should be pointed out that both this water-in-oil (w/o) emulsion as well as butter are not emulsions in strict meaning as the continuous phase is not liquid but plastic. As son as the fat crystals are melted, the water phase separates from the oil. These emulsions are thus stabilized by fat crystals in a similar way as hydrocolloids due to 'immobilization' effects can stabilize oil-in-water emulsions.

41

J. Sjöblom (ed.), Emulsions – A Fundamental and Practical Approach, 41–49.
© 1992 *Kluwer Academic Publishers.*

TABLE 1. Examples of food emulsions [1]

Food product	Emulsion type	Dispersed phase	Continuous phase
Milk, cream	O/W	Butterfat triglycerides partially crystalline and liquid oils Droplet size: 1-10 μm Volume fraction: Milk: 3-4% Cream: 10-30%	Aqueous solution of milk proteins, salts, minerals, etc.
Ice cream	O/W	Butterfat (cream) or vegetable, partially crystallized fat Volume fraction of air phase: approx. 50%	Water and ice crystals, milk proteins, carbohydrates (sucrose, corn syrup) Approx. 85% of the water content is frozen at -20°C
Butter	W/O	Buttermilk: milk proteins, phospholipids, salts Volume fraction: 16%	Butterfat triglycerides, partially crystallized and liquid oils; genuine milk fat globules are also present
Imitation cream (to be aerated)	O/W	Vegetable oils and fats Droplet size: 1-5 μm Volume fraction: 10-30%	Aqueous solution of proteins (casein), sucrose, salts, hydrocolloids
Margarine and low-calorie spread	W/O	Water phase, may contain cultured milk, salts, flavors Droplet size: 1-20 μm Volume fraction: 16-50%	Edible fats and oils, partially hydrogenated, of animal or vegetable origin Colors, flavor, vitamins
Mayonnaise	O/W	Vegetable oil Droplet size: 1-5 μm Volume fraction: Minimum 65% (U.S. food standard)	Aqueous solution of egg yolk, salt, flavors, seasonings, ingredients, etc. pH: 4.0-4.5
Salad dressing	O/W	Vegetable oil Droplet size: 1-5 μm Volume fractions: Minimum 30% (U.S. food standard)	Aqueous solutions of eff yolk, sugar, salt, starch, flavors, seasonings, hydrocolloids, and acidifying ingredients pH: 3.5-4.0

As proteins are considered in other papers in this volume, the discussion below will be concentrated on lipids at the interface and the crystallization of these lipids as well as the oil.

2. Phospholipids at the Oil/Water Interface

It is simple to demonstrate how a liquid-crystalline interfacial layer builds up when a triglyceride oil containing some amount of solved phospholipids comes in contact with water. Using pure phosphatidylcholine (PC) from soy beans and soy bean oil, about 0.1% PC can be solved in the oil, and a visible interfacial film is formed towards water within an hour. This film can be separated, and by X-ray diffraction it shows similar diffraction characteristics as the L_α-phase formed in excess of water [2] (a repeat distance of 44 Å was observed in PC from rapeseed). There are various reports in the literature of the surface tension of triglyceride oils towards water and the effect of phospholipids (cf. [3]). Such measurements, however, are not meaningful as a separate phase builds up at the oil/water interface.

Charged phospholipids exhibit "infinite" swelling in water; even if they are present in just a few percent in a phospholipid mixture. For this reason they are most effective as emulsifiers, used for example in emulsions used in parenteral nutrition – "Intralipid", as only monolayer is formed.

3. Crystals and Gel-Phases of Polar Lipids at the Oil/Water Interface

Most food emulsions based on polar lipids, such as monoglycerides, are prepared by a cooling process, where the crystallization step is crucial for the emulsification effect. Sometimes a gel-phase is formed at the interface but more common is the formation of crystals of the polar lipids; exhibiting polar surfaces towards water and non-polar surfaces towards the oil. This structure at the interface was recently determined in a joint work with Krog, and is further described in his paper in this volume.

4. Crystalline Triglycerides at the Oil/Water Interface

Since Benjamin Franklin's classical experiments on wave damping by a droplet of vegetable oil, it has been known that a triglyceride spreads spontaneously on a water surface as a monomolecular layer. The spreading pressure at the air/water interface is about 15 mN/m, corresponding to a surface tension of about 58 mN/m. The interfacial tension between a pure triglyceride in the liquid state and water is close to 30 mN/m [2]. The fact that the triglyceride molecule itself is surface-active is often neglected, and particularly the effect of crystallization of the triglycerides is a field of knowledge clearly relevant in the understanding of emulsions which has been ignored. It will therefore be considered below.

44

The monolayer behaviour of tripalmitin is shown in Figs 1 and 2 [4]. When a spread monolayer is compressed, the film pressure (Π) shows the existence of two monolayer phases which according to the linear Π-area isotherm are solid condensed phases (C_1 and and C_2). The molecular area at the transition $C_1 \rightarrow C_2$ is 58.3 Å/molecule. From a number of Π-A isotherms the monolayer phase diagram can be constructed, which is shown in Fig. 2. The phase properties and the area and compressibility data indicate that the C_1 phase corresponds to vertical chains arranged as in the α-phase of triglyceride, whereas the C_2 monolayer, with close-packed chains, is metastable. The equilibrium pressure curve given in Fig. 1 shows that the C_1-monolayer phase forms a stable monolayer.

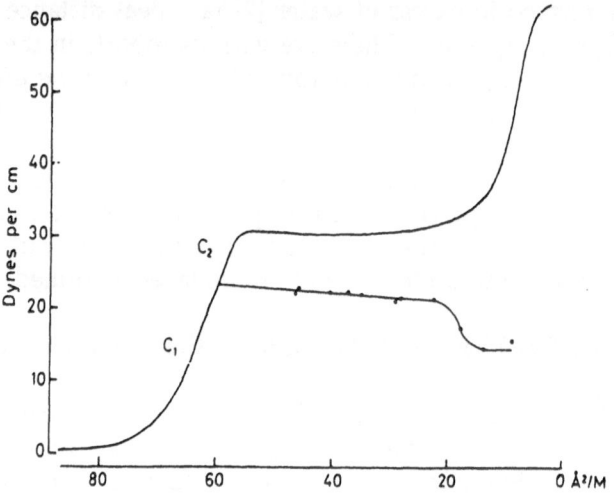

Figure 1. Pressure-area isotherm at 40.7°C of tripalmitin, compressed below monolayer collapse. The points correspond to equilibrium pressure values [4].

There is also a remarkable steep rise in pressure [4], which corresponds to the formation of a triple-chain layer, and the probable structure is indicated in Fig. 3. As we know from the solid state, there is a strong tendency to form a so-called tuning-fork conformation of the molecules (chains in 1- and 3-position pointing in opposite direction to that in 2-position). This conformation can thus coexist with the E-conformation (all three chains pointing in the same direction), which is a necessity on the water surface. In this connection, the α-structure is of interest, as it coexists with the monolayer. From molecular models a probable structure has been derived which is shown in Fig. 4. The dilemma is the accommodation of vertical chains to the end-group layer. We know from polymorphic transitions and the structures of the β- and β'-forms that the dimer is basically the same [5]. The most favour-

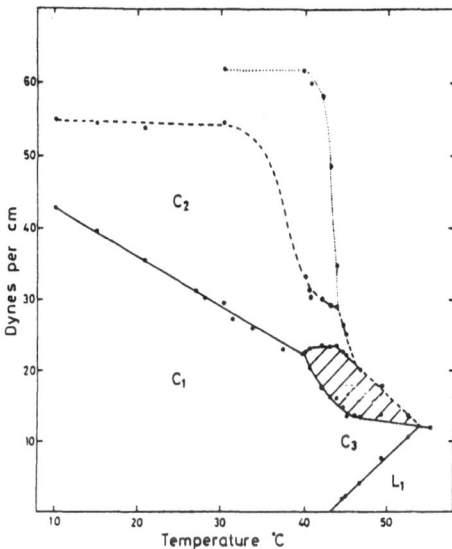

Figure 2. Pressure-temperature phase diagram of tripalmitin. Phase boundaries are denoted by full-drawn lines, collapse by a dashed line and triple layer formation by a dotted line. The transition region is shaded [4].

able way to obtain the vertical chains (obvious from X-ray data) is shown in Fig. 4. The only way to account for the irregular end-group structure is the occurrence of some disorder.

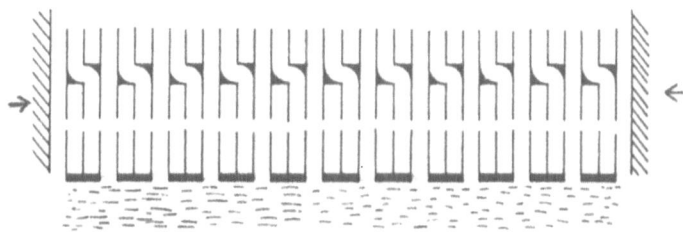

Figure 3. Monolayer structure in coexistence with a bilayer as indicated by pressure-area behaviour ([4].

Equilibrium spreading pressures (ESP) were recorded by putting a few crystals on the water surface in the surface balance. When about 1 mg/100 cm^2 β-crystals of trimyristin were added, the pressure change within 20 minutes was less than 1 mN/m, and after 20 minutes there was a steep change, which gave a plateau value of 6 mN/m. This is obviously the

46

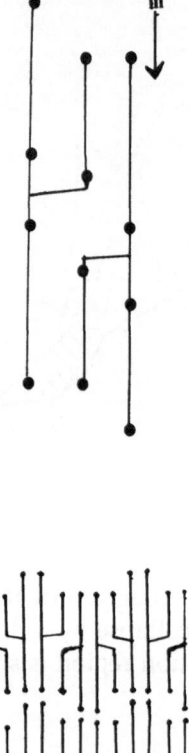

Figure 4. Proposed structure of the α crystal form of triglycerides. The dimer unit of structure is shown above, and *m* indicates the mirror symmetry introduced to obtain the full structure shown below.

ESP of β-crystals, and the time with almost no change in pressure corresponds to very few molecules on the surface in a gas-type of molecular arrangement. This value of the ESP is considerably lower than the film spreading pressure from the liquid state of triglycerides, which is about 15 mN/m. The difference illustrates the problems to arrange triglyceride molecules with close-packed chains in an *E*-conformation, and the preference for the crystalline state, where molecules can adopt the turning-fork structure. In the case of the liquid-type of disordered chains, the *E*-conformation requires less strain at the chain linkages to the polar head. These data demonstrate that crystals of triglycerides should not be expected to be locali-

zed at the oil/water interface, nor should they contribute to interfacial monolayers if liquid triglycerides are present.

The ESP-value of the α-form, however, is higher than that of the liquid
state as indicated from Fig. 1. In order to examine this state further, a droplet
(about 50 mg) of a trimyristin melt at 35°C was spread on a 200 cm^2 water
surface at 30°C, *i.e.* 3°C below the α melting point. The α-form is remarkable
in the respect of not exhibiting supercooling and an ability of two-dimensional nucleation, which may be the explanation of the immediate response we
observed in film pressure. Thus, a value of 19 mN/m was recorded. After 3
minutes of constant pressure, a reduction was observed and a plateau value
of 14 mN/m was observed for about 2 minutes. Then the pressure dropped to
a final plateau value of 6 mN/m, which is that recorded of β-crystals. The
intermediate plateau should be expected to correspond to the β'-type of chain
packing, but as the β'-crystals are so unstable they could not be used in an
experimental verification.

The formation of a two-dimensional analogue to the α-form from α-
crystals is probably a significant factor in emulsification technology. Such a
film due to its high pressure will compete with proteins at the interface, and
even if it only exists as a transitory state, it will influence the final emulsion
structure due to irreversibility involved with proteins at the interface.

5.   Lipase Digestion of Emulsions

It has been known for a long time that the triglyceride molecule is only
digested by lipase if the enzyme can attack at an oil/water interface. This
means that the lipase is adsorbed at an interface with triglyceride molecules
in the E-type of conformation, and the acyls in 1- and 3-position will be preferentially hydrolysed. Any food emulsion, however, will not expose such a
"naked" triglyceride surface towards the aqueous medium containing the
pancreatic (or lingal) lipase. In protein-based emulsions, proteolytic degradation starts already in the gastric region. Furthermore, the bile salts in the
intestine can contribute to the surface exposure of a triglyceride monolayer.
We have a more complicated situation if the emulsion is based on surface-
active polysaccharides, which cannot be digested, and at present we lack
information on the nutritional effects of such emulsion.

Let us consider ethyl-hydroxyethylcellulose (EHEC), which depending
upon the block character of the substituents can be highly surface-active. If
proteins or bile salts cannot displace the EHEC-surface coat, the emulsion will
not be digested. Whether or not such an emulsion will be non-caloric,
however, depends on the fermentation in colon. Even less is known about
this environment from a surface chemistry point of view. (EHEC is used in
foods in Sweden mainly in bread products).

So-called natural food emulsifiers, like phospholipids, will be digested and
in this way there are good possibilities for the lipase to reach the triglyceride

48

oil surface. Synthetic emulsifiers, however, such as polyglycerolesters, lactic acid esters or citric acid ester, are also frequently used in foods. Even if they might be degraded by lipase or phospholipase, there are no studies reported in the literature, known by the present author, showing how much if any of such emulsions contribute nutritionally. In this context it is natural to point out the general ignorance of physico-chemical aspects in the gastrointestinal system. Thus an obvious risk factor of surfactants in foods must be their potential damaging effects on surface-exposed biomembranes (cf. [6]).

In conclusion the properties of the emulsion coat with regard to fat digestion in the intestine is poorly understood. There should be obvious possibilities to find emulsion surface coats which can give considerable caloric reductions, as an alternative to the use of non-caloric oils like polysucrose esters.

6.   Fat Oxidation and Emulsion Structure

In a simple demonstration of the significance of emulsion structure with regard to oxidation stability we solubilized α-tocopherol in the bilayers of liposomes. Then we encapsulated oils in water by such bilayers and compared the rate of oxidation with an emulsion with the same composition and particle size distribution [7]. A much higher antioxidant effect was achieved with tocopherol localized in a bilayer outside the oil globule.

Another example of such structural effects was observed in the case of microemulsions. Our aim was to try to simulate the coupled antioxidant effect in biomembranes of tocopherol in the bilayer, which in its radical state is reactivated by ascorbic acid in the outside water phase. By using soy bean oil with about 5% (w/w) water incorporated as a microemulsion by monoglycerides, a system containing both tocopherol and ascorbic acid was achieved [8]. A drastic reduction in oxidation rate was observed. In a recent work in our laboratory, we used glycerol instead of water in order to protect a fish oil against oxidation. The same type of microemulsion was prepared with the aid of monoglycerides from sunflower oil. The antioxidant effect in this system was even better than with the same concentration of ascorbic acid in water.

7.   References

[1]   Krog, N., Riisom, T. and Larsson, K. (1985) Application in the food industry', in P. Becher (ed.), Encyclopedia of Emulsion Technology, Marcel Dekker, New York, pp. 321-365.

[2]   Söderberg, I. (1990) 'Structural Properties of Monoglycerides, Phospholipids and Fats in Aqueous Systems', Thesis, University of Lund, Sweden.

[3] Ogino, K. and Ogino, M. (1980) 'Interfacial action of natural surfactants in oil/water systems', J. Colloid Interface Sci. 83, 18-25.

[4] Bursh, T., Larsson, K. and Lundquist, M. (1968) 'Polymorphism in monomolecular triglyceride films on water and formation of multi-molecular films', Chem. Phys. Lipids 1, 102-113.

[5] Hernqvist, L. and Larsson, K. (1982) 'On the crystal structure of the β'-form of triglycerides and structural changes at the phase transitions liq. → α → β' → β′, Fette-Seifen-Anstrichmittel 84, 349-354.

[6] Larsson, K. and Johansson, L.-Å. (1978) 'Hemolytic effect of some polar lipids used as food additives', Lebensm.-Wiss. u. -Technol. 11, 206-208.

[7] Ruben, C. and Larsson, K. (1985) 'Relations between antioxidant effect of α-tocopherol and emulsion structure', J. Dispersion Sci. Technol. 6, 213-221.

[8] Moberger, L., Larsson, K., Buchheim, W. and Timmen, H. (1987) 'A study of fat oxidation in a microemulsion system', J. Dispersion Sci. Technol. 8, 207-215.

ADSORPTION STRUCTURES IN FOOD EMULSIONS

B. BERGENSTÅHL, P. FÄLDT AND B. LASSEN
Institute for Surface Chemistry,
P.O.Box 5607
114 86 Stockholm
Sweden

ABSTRACT.Food emulsions in particular, but also most other technical emulsions, are complex mixtures. When surface active components adsorbs from a solution of several different surface active molecules, the formed adsorbed layer can roughly be classified in three different structures i) A monolayer containing one predominant molecule .ii) The formation of one adsorbed monolayer containing a mixture of molecules.iii) Adsorption in layers. A model of an ice cream emulsion is an example in which a layered surface structure is formed.Four different methods were applied to investigate the surface: measurements of the interfacial tension, flocculation rate measurements , electrophoretic mobility measurements and TIRF (Total Internal Reflection Fluorescence). The conclusion we made from this investigation was that the adsorption from solutions with several surface active components might lead to the formation of complex layered structures

1. Introduction

The chemical composition of the emulsion droplet surface is a key factor to determine the surface interactions . Food emulsions in particular, but also most other technical emulsions, are complex mixtures. In addition to oil and water they usually contain low molecular emulsifiers, polymers and inorganic salts. The reason why emulsions are stable is the adsorption of surface active molecules. Thereby the hydrophobic attraction between the droplets is eliminated and repulsive forces are generated (1). Without the presence of stabilizing emulsifiers or polymers the oil and the water would phase separate immediately.

When surface active components adsorbs from a solution containing several different surface active molecules, a race to the available interface occur. In our opinion, the formed layer can roughly be classified in three different structures (1).

i) A monolayer containing one predominant type of molecule at the interface. The layer builds up through competitive adsorption which creates a chemically homogeneous surface.

ii) The formation of one adsorbed monolayer containing a mixture of several different types of molecules. The cooperative character of the adsorption leads to a larger amount of adsorbed material than if just one of the components were present. This adsorption gives rise to a chemically heterogeneous surface.

iii) A multilayer forms when one type of molecule adsorbs on top of an another molecular layer with sequential adsorption.

J. Sjöblom (ed.), Emulsions – A Fundamental and Practical Approach, 51–60.
© 1992 *Kluwer Academic Publishers.*

In the case of competitive adsorption, as defined above, the most surface active component is able to replace a less surface active one at the interface. Several examples might be found For instance, albumin has been observed to displace dextran from silver iodine particles (2), and casein displaces gelatin from the interface between water and hexadecane (3). The sequential change of the protein composition on an artificial surface in contact with blood, the Vroman effect (4), is also an example of a truly competitive adsorption To emulsion droplets proteins usually adsorbs more or less irreversible. Hence, the composition of the interfacial layer depends strongly on the order of mixing. For instance, Dickinson and coworkers (18) found in an experiment with adsorption of

$\alpha-$ lactoglobulin and $\beta-$ lactoglobulin that the first added protein dominated the interface. During the homogenization the oil water interface is fresh and the character of the adsorption might be more competitive. It was found that in a homogenized emulsion

of $\beta-$ casein and phosvitin (an egg protein) that the interface was dominated with casein when the casein fraction of the protein content was above 0.25.(19).

Low molecular emulsifiers, for instance unsaturated monoglycerides, has been shown to completely remove even strongly adsorbed protein layers when present in significant concentration (20). However, these experiments were performed on a macroscopic oil water interface . The experience from emulsions shows that a normal use level of monoglycerides or sorbitan esters give a reduction, but not a complete removal, of the protein layer (21).

When cooperative adsorption occurs a mixed surface is formed. A typical example of such a system might be a long alcohol (for instance decanol) and charged surfactants (e.g. soaps). The alcohol acts as a spacer between the charged groups, decrease the headgroup repulsion within the layer and reduces the surface energy. This increases the adsorption, enhances the surface activity. Similarly, a lamellar phase is formed in the corresponding three component phase diagram water-sodium caprylate-decanol (5). Mixed layers due to a cooperative adsorption is commonly formed with natural and technical emulsifier blends. This is the reason why average HLB numbers (6) can be used to characterize the properties of emulsifier blends. Mixed layers are also formed by several natural protein mixtures, such as sodium caseinate (7), or lipid-protein mixtures as egg yolk (8).

Sequential adsorption with different compounds in different layers is possible when several different classes of surface active components are present in the mixture. For instance, the adsorption of polysaccharides to emulsifier covered oil droplets (9) has been demonstrated. However, adsorption of one type of protein on top of another type of protein is rarely observed (10). Similarly, the adsorption of one type of surfactant on top of a surfactant layer composed of a different type of surfactant is uncommon. But

electrophoretic measurements on a $\beta-$ casein stabilized emulsion in the presence of gelatin showed that gelatin adsorb on top of the casein layer, forming an outer layer changing the mobility (19).

An illustrative example of a layered surface structure in emulsions is found in a model of an ice cream emulsion . Three different methods were applied to investigate the surface of the oil droplets in the emulsion: measurements of the interfacial tension between the oil phase and the water phase, flocculation rate measurements (11,12), and electrophoretic mobility measurements. In addition to the characterization of the microscopic surfaces in the disperse system we have also performed measurements on macroscopic model surfaces with TIRF (Total Internal Reflection Fluorescence (13)) A decreased interfacial tension or a reduced flocculation rate indicated adsorption. A changed electrophoretic mobility indicate adsorption. The surface fluorescence from a marked adsorbing species is a direct measure of the adsorption.

2. Material and methods.

The model consist of a mix of the ingredients shown in table 1. All ingredients are of commercial origin. All other chemicals are of analytical grade.

A special homogenization technique, microfluidization (Microfluidics, Newton, Mass. USA) (14), was employed for making emulsion droplets small enough (d<1mm) to allow for turbidimetric flocculation rate measurements (15-17).

The flocculation experiments were performed at 600 nm with a dilute emulsion (the absorbance below 0.5). The flocculation rate was measured as t he initial increase in turbidity versus time. The flocculation was induced either with $MgCl_2$ or acid buffer.

The interfacial tension was measured with the pendent drop method. The profile of an oil droplet is photographed and the interfacial tension calculated from the profile.

The electrophoretic measurements has been performed with a Zeta Sizer II. (Malvern Ltd U.K.) The electrolyte concentration was 0.015 M. A citrat lactate buffer was used for the pH control and the conductivity was adjusted to receive constant conductivity independent of the pH.

The results from the electrophoretic mobility measurements were corrected for the increased viscosity due to the prescens of polymer at all points with a polymer concentration at 100 ppm or above.

The viscosity of the polymers were measured with an Ostwald viscometer in a thermostated bath. Three measurements were done at each concentration. The ionic strength and pH were kept similar to the one used in the mobility measurement

The TIRF measurements has previously been described in (13). As a model surface a hydrophobized silica surface was used. Dichlorodimethylsilane (DDS) was used as a hydrophobizing agent. The protein and the alginate were marked with fluoroscein. The surplus of the probe in the protein was eliminated with chromatographic cleaning through a Sephadex G 25 (Pharmacia, Uppsala, Sweden) column. The surplus in the alginate sample was removed by a repeated precipitation with ethanol and dissolution. The degree of labeling of the alginate was about 0.3 counted on a molecular weight of about 100000
The degree of labeling of the protein was about 1 counted on a molecular weight of about 100 000. The concentration of the protein was 10 ppm the alginate 80 ppm.

Table 1. The recipe to the ice cream model emulsion

Component	Amount in % (w/w)
Coconut butter	10.0
Milk powder	11.5
Sugar	14.0
Mono/diglyceride	0.3
Stabilizer[1]	0.1-0.25

1.Sodium alginate, carboxymethylcellulose, carrageenan, xanthan, guar gum or locoust bean gum.

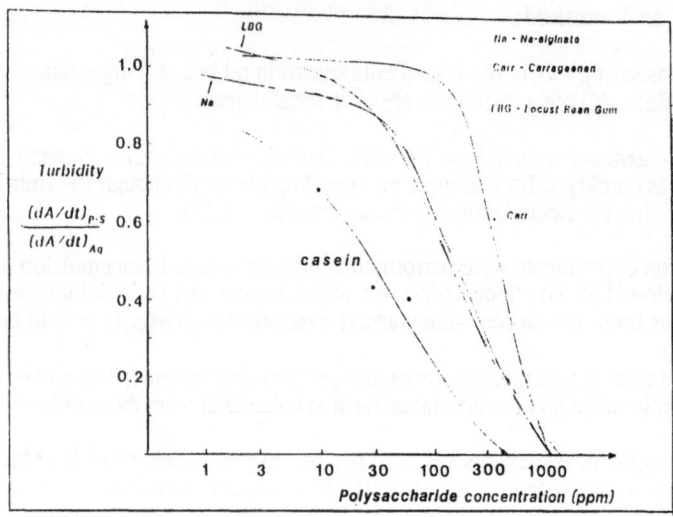

Fig 1. The flocculation rate of a monodiglyceride coated emulsion stabilized by different gums or casein. The flocculation is induced by the addition of salt. Modified from ref(1)

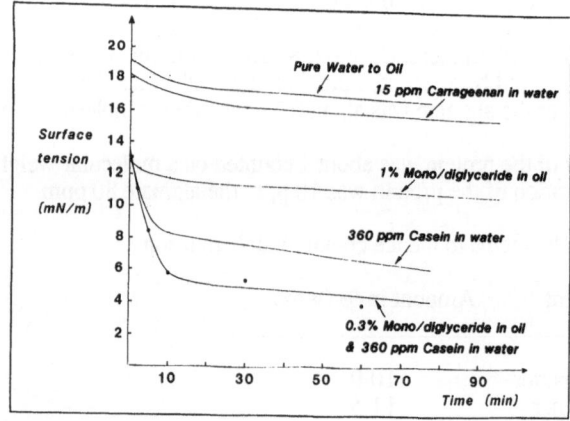

Fig 2. The interfacial tension between oil and water when carrageenan, the monodiglyceride mixture, milk protein or monodiglyceride and casein is present. From ref (1)

3. Results.

3.1 Adsorption from rapid flocculation measurements.

The aim of the experiment was to study the formation of adsorbed structures on the molecular level in an ice-cream emulsion.

From the list of ingredients, table 1, it was assumed that the mono/diglyceride emulsifier was the most surface active component of the mixture. Therefore, we started the investigation by emulsifying a fat phase with the monoglyceride emulsifier in a pure aqueous phase

The adsorption of the different polysaccharides and protein to this surface was studied. The emulsion was diluted until an absorbance o.4 units where received. To induce flocculation we destabilized the emulsion by the addition of salt. The flocculation rate was followed as the turbidity increase versus time. Samples with water and with polymer was investigated. The adsorption of a polymer was then observed as a reduction of the rate of flocculation . The influence of different gums (hydrocolloids) and casein on the flocculation rate of the emulsion is shown in figure 1.

It is obvious that most gums do adsorb to the mono/diglyceride interface. However, it is also clear that the adsorption is rather weak for most of the polysaccharides. The concentration of the different gums in the actual ice cream (table 1) is too low to give rise to any adsorption. However, the milk protein concentration in the ice cream is high, considerably higher than necessary for adsorption ! Hence, it is clear that the oil-water interface partly consists of adsorbed milk proteins The interfacial tension measurements in figure 2 shows that the oil/water interfacial tension is significantly lower in the presence of milk proteins and monoglyceride than in the presence of either the protein solution or the monoglyceride alone. Hence, it is clear that both the mono/diglyceride emulsifier and the casein adsorb to the interface, and that they form a mixed layer.

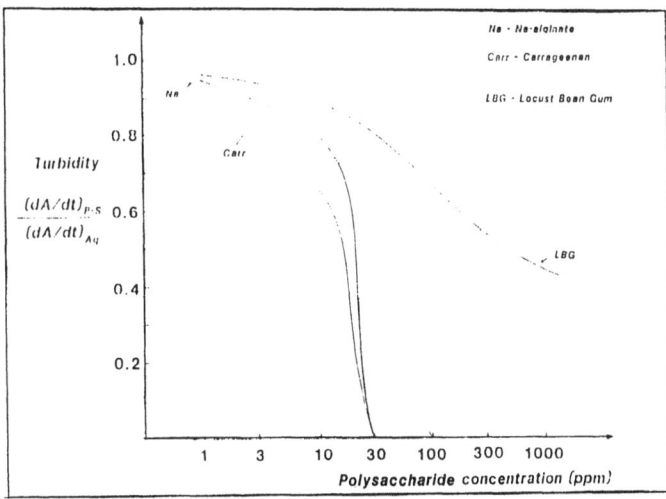

Fig 3. The flocculation rate of an emulsion coated with casein and the monodiglyceride emulsifier when stabilized by different gums. The flocculation was induced by an addition of an acid buffer. Modified from ref (1).

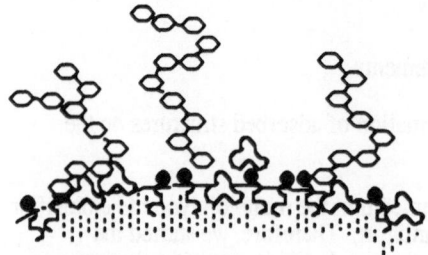

Fig 4 The suggested structure of the adsorbed layer.

To investigate an emulsion coated with both mono/diglyceride and casein, an acidic buffer was added to induce flocculation. The flocculation rate when the different gums were present in the emulsion is shown in figure 4. The results show that several anionic gums reduce the flocculation rate, which is an indication of adsorption to the emulsion droplet surface. The adsorption is enhanced considerably by the presence of casein on the surface. The surface activity of the gums is much lower than that of the casein (Fig. 3). Hence, the polysaccharide should not be able to displace the casein or the mono/diglyceride at the interface. Instead the observed stabilization is due to the formation of a layered structure. The inner layer is composed of mono/diglyceride and casein, whereas the outer layer is made up by polysaccharides. The suggested structure of this emulsion droplet surface is illustrated in figure 5.

However, the flocculation method has a strong limitation in the undefined pH, in which the adsorption is studied. Calculations can show that if every collision between an emulsion droplet and a polysaccharide molecule would lead to adsorption the adsorption kinetics would be rather rapid. But, it is also well-known that there is a kinetic delay in polymer adsorption. Therefore it might not be likely, but also, it might not be excluded, that the adsorption occurs simultaneously to the flocculation process when the acid is added. Hence, it is highly interesting to study the adsorption with alternative methods.

3.2 Adsorption from electrophoretic mobility measurements.

The electrophoretic mobility of an emulsion droplet depends on the charge at the plane of shear. The charge is created by the outer adsorbed layer on the droplet. If a charged species adsorbs to the droplet the mobility will change towards the charge density of the adsorbed material. If an uncharged material adsorbs to the surface the shear plane will be moved outward in the solution and thereby the mobility will be reduced towards zero.

Figure 5,below ,shows the mobility of milkprotein / monoglyceride covered emulsion droplets shown versus the concentration of carrageenan and locust bean gum added.

The results in figure 5 clearly shows that carrageenan adsorbs and that locust bean gum not is adsorbing to the emulsion droplets. Thereby the results confirm the observations in the flocculation experiment. The magnitudes of the changes at the different pHs strongly depends on the variation in charge density at the different pHs and should not necessarily be interpreted as different amount of adsorbed material. However, the adsorption of carrageenan starts at about the same concentration independent of pH. This indicates that the adsorption is not strongly depending on pH.

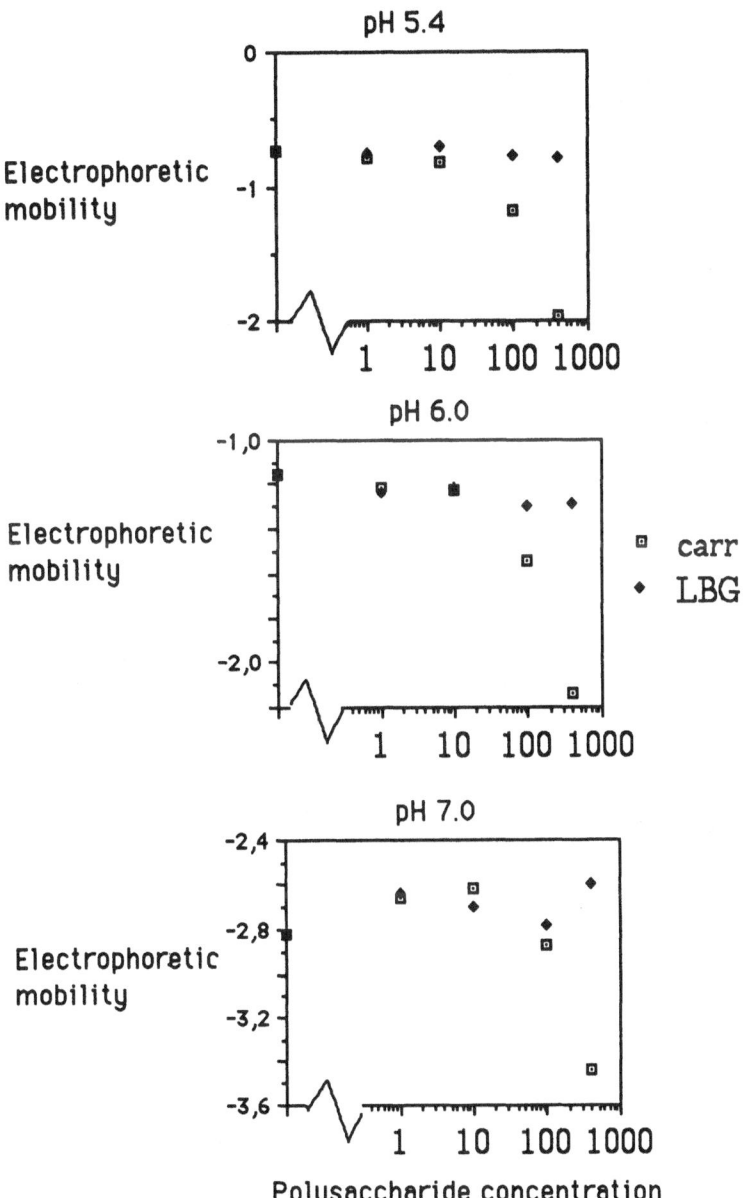

Figure 5. The electrophoretic mobility versus the concentration of added polymer. The mobility results are corrected for the influence of the increased viscosity.

3.3 Adsorption characterized with TIRF

The two previous experiments shows simplified adsorption experiments on emulsion droplets. However, both methods only describe the adsorption phenomena qualitatively. To find a simple and rapid quantitative method we have tried to measure the adsorption in situ on macroscopic model surfaces with surface fluorescence (TIRF). The measurements with this method allows us to follow the adsorption directly. Figure 7 shows the experimental set up.

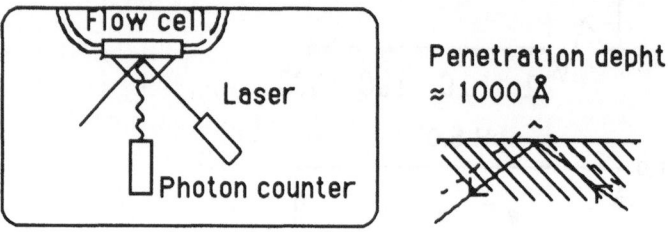

Figure 6. A schematic drawing of the TIRF measurements.

The main purpose with this experiment is to follow the adsorption of polysaccharides to a protein surface and to check if any of the protein is desorbed. In the experiment we used a hydrophobized silica surface as a model surface of the oil. In the first part (figure 7a) we used labeled protein and followed the adsorption. The results showed a quite rapid adsorption. No significant desorption of the milk protein was observed when alginate was introduced. In the second experiment (7b) the alginate was labeled and the adsorption of alginate to the protein surface was followed. We found that an alginate layer was developed within a couple of minutes, even if the adsorption was slower compared to the milk protein adsorption. The pH was kept at 7 with a PBS buffer.

4 Discussion

In an applied project, a couple of years ago, we tried to emulsify a lipid in a proteinaceous water phase. We tried with a number of different emulsifiers but all failed. We received a finely dispersed emulsion that rapidly flocculated. The interfacial tension was measured to control if the emulsifier was removed through competitive adsorption, but that was not the case. It was quite clear that that the emulsifier rather was covered by the adsorbed proteinaceous material than squeezed out of the interface. The final solution we found was a polysaccharide that was able to adsorb on the proteinaceous surface creating a second adsorbed layer.

The conclusion we made from this investigation was that the adsorption from solutions with several surface active components might lead to the formation of complex layered structures and that these structures might be used as a method to overcome problems with destabilizing components in technical emulsions.

We have applied the concept of three different types of adsorption structures from solutions with several surface active components on several problems and we have found it very fruitful. The example given in this paper, the ice cream emulsion, is mainly aimed as an illustration of how the concept can be applied on complex technical emulsions.

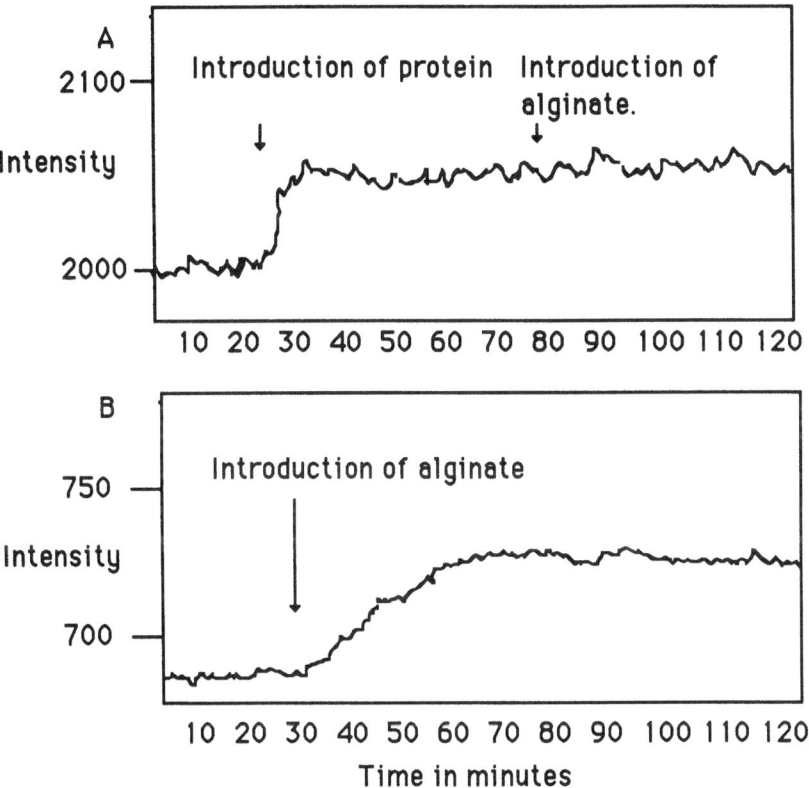

Figure 7. Adsorption of protein to a hydrophobic surface followed by adsorption of alginate to the protein layer. In figure A the protein is labeled. In figure B the alginate is labeled. The protein was introduced into the cell in B about 1 h before the alginate. The curves are reconstructed from original data.

5. References

1 Bergenståhl, B. and Claesson P. Surface forces in emulsions in "Food Emulsions" by Larsson K. and Friberg S. Ed. Marcel and Dekker New York 1990.

2. Matuszewska B., Norde, W. and Lyklema, J., Competitive adsorption of human plasma albumin and dextran on silver iodine particles., *J. Colloid Interface Sci.*, 84, 403 (1983)

3. Dickinson, E., Pogson, D. J., Robson, E. W. and Stainsby, G., Time dependent surface pressures of adsorbed films of caseinate + gelatin at oil water interfaces., *Colloids Surfaces*, 14, 135 (1985)

4. Vroman, L. and Adams, A. J., *J. of Biomed. Mat. Res.*, 3, 43 (1969)

5. Fontell, K., Mandell, L., Lehtinen, H. and Ekwall, P., *Acta polytechnica Scandinavica*, Chap 2, Chem Ser 74 III, 2 (1968)

6. Griffin, W. C. in *Kirk- Othmer Encyclopedia of Chemical Technology*, Vol 8, 1979

7 Dickinson, E.,Robson, E. W. and Stainsby, G., Colloidal stability of casein coated polystyrene particles. J. Chem. Soc. Far. Trans. I, 79, 2937 (1983)

8. Phillips, M. C., Evans, M. T. A. and Mauser, H. *ACS Adv Chem Ser* 144,217 (1978)

9 Bergenståhl, B., Gums as stabilisers of emulsifier covered emulsion droplets, in Phillips G. O., Williams P. A. and Wedlock D. J. (eds) *"Gums and Stabilisers for the food industry"* , *Vol 4*, IRL Press,Oxford, 1988

10. Dickinson,E. Mixed proteinaceous emulsifiers:review of competitive protein adsorption and the relationship to food colloid stabilization., *Food Hydrocolloids*, 1, 3 (1986)

11. Freundlich, H. "Kappilarchemie" Vol 2, p 447-455, Leipzig Akademische Verlag 1932.

12. Ottewill, R. H. and Wilkins, Stability of arachidic acid sols., *Trans. Faraday Soc.*, 58, 608 (1962)

13 Gölander, C. G., Lin, Y. S., Hlady, V. and Andrade, J. D., *Colloids and Surfaces*, 49, 289-302 (1990)

14. E. Cook and A. P. Lagace, US pat. No 4 533 254, ,1985.

15. Kerker, M., "The scattering of light and other electromagnetic radiation", Academic Press, New York 1969.

16. Lichtenbelt, J. W. T., Ros, J. M. C. and Wiersema H., Turbidity of coagulating lyophobic sols, *J. Colloid Interface Sci.*, 46, 522 (1974)

17. Egusa S., Stopped-flow spectroscopy as a simple method for particle size determination., *J. Colloid Interface Sci.*, 86 ,135 (1982)

18. Dickinson, E., Rolfe, S. E. and Dalgleish, D. G., Competetive adsorption in oil in water emulsions containing α-lactoalbumin and β-lactoalbumin, *Food hydrocolloids*, 3, 193-203 (1991)

19. Dickinson, E., Hunt, J. A. and Dalgleish, D. G., Competetive adsorption of phosvitin with milk proteins in oil in water emulsions *Food hydrocolloids*, 4, 403-414 (1991)

20 Hertje, I., Nederlof, J., Hendrickx, H. A. C. M. and Lucassen-Reynders. E. H.. The observation of displacement of emulsifiers by confocal scanning laser microscopy. *Food Structure*, 9, 305-316 (1990)

21 Krog, N., Barfod, N. M. and Sanchez, R. M., Interfacial phenomena in food emulsions, *J. Dispersion Sci Tech.*, 10 , 483-504 (1989)

THE ROLE OF LOW-POLAR EMULSIFIERS IN PROTEIN-STABILIZED FOOD EMULSIONS

Niels Krog
Grindsted Products
Edwin Rahrs Vej 38
8220 Brabrand
Denmark

ABSTRACT. Low polar emulsifiers are used in whippable emulsions to facilitate aeration and improve foam texture. The function is related to interfacial monolayer properties in ice cream emulsions and liquid imitation creams and results in a protein desorption from the fat globule surface at low temperatures (e.g. 5°C). This destabilizes and enhances the whippability and foam stiffness. In spray-dried powdered toppings the low-polar emulsifiers are used in a higher concentration compared to imitation creams and consequently a more pronounced destabilization of the reconstituted topping emulsion takes place, making it possible to create a foam with good stability and stiffness combined with a lower fat content than conventional whipped cream.

1. Introduction

The industrialization of food production, which has taken place in this century, has created a demand for food processing aids such as food emulsifiers or surfactants. The function of these minor ingredients in foods is to facilitate large-scale, highly automated production of uniform products with improved shelf-life which can stand transport and storage until consumption takes place. A modern supermarket has a large variety of food products which are produced in one location and then distributed over long distances before they reach the consumer. Typical examples are a) margarine, spreads, dressings, b) bakery products, and c) dairy-based foods such as ice creams, whipping creams and powdered toppings. The functional properties of emulsifiers in category a) provide emulsification and longer stability (shelf-life). In bakery products b) the starch complexing ability of emulsifiers play a major role together with improvement in texture and volume resulting in good quality and long shelf-life. In the dairy-based foods c), aeration and formation of stable foam, which gives a creamy mouthfeel are, the essential requirements to the emulsifiers used. The physical properties of the low-polar emulsifiers and their use in aerated dairy-based foods will be discussed in the following.

2. Low-Polar Emulsifiers

Food-grade emulsifiers vary in polarity over the entire HLB range depending on their chemical composition as shown in Table 1.

J. Sjöblom (ed.), Emulsions – A Fundamental and Practical Approach, 61–74.

62

TABLE 1. Polarity of Emulsifiers

Emulsifier type	Polarity	HLB Range
Monoglycerides (MG)		
Acetylated MG (AMG)	Low	1-4
Lactylated MG (LMG)		
Propylene glycol mono-stearate (PGMS)		
Stearyl lactylates		
Diacetyl tartaric acid esters of MG (DATEM)		
Citric acid esters of MG (CITREM)	Medium	5-11
Lecithin		
Sorbitan monostearate		
Polyglycerol monostearate		
Polysorbates		
Sucrose monostearate	High	12-20
Lyso-lecithin		
Sodium oleate/stearate		

The low polar emulsifiers dealt with here are monoglycerides together with their esters of acetic or lactic acid, and propylene glycol monostearate.

2.1 MONOGLYCERIDES

Industrial production of monoglycerides began in the middle of the 1930s. Production methods are interesterification of fats with glycerol, resulting in an equilibrium blend of mono-, di- and triglycerides. The monoglyceride content is usually 40-60%. Products containing min. 90% monoglycerides are made by using a high-vacuum, thin film, molecular distillation technique. Figure 1 shows the chemical formular and molecular structure of glycerol monostearate. Fully hydrogenated fats are normally used in the manufacture of monoglycerides, which can be produced from a variety of edible fats and oils obtained from vegetable and animal sources, such as soybean oil, cotton seed oil, safflower oil, lard or tallow.

Distilled monoglycerides are widely used in many different types of foods. A major application area is the baking industry. Mono-diglycerides are used as the major emulsifier in the ice cream industry.

2.2 ORGANIC ACID ESTERS OF MONOGLYCERIDES

Monoglycerides can be esterified with organic acids such as acetic, lactic, succinic, citric and diacetyl-tartaric acids, which form corresponding derivatives of monoglycerides.The esters are produced by reacting the monoglycerides with the organic acid either directly or with its anhydride. The esters of mono-carboxyl organic acids, such as acetylated or

63

lactylated monoglycerides have lower melting points and greater stability in their alpha-crystalline forms, and display more lipophilic properties than their corresponding monogly-cerides. These esters are often referred to as alpha-tending emulsifiers, and they are mainly used in aerated emulsions such as toppings and whippable dairy emulsions (desserts) or in cake mixes. Figures 2 and 3 show the chemical formulas and molecular structure models of acetylated and lactylated monoglycerides.

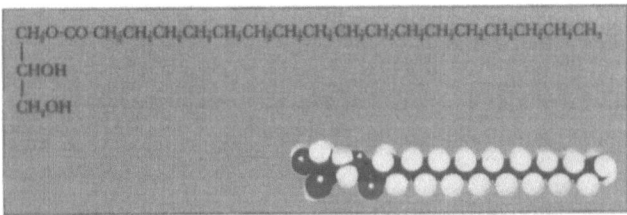

Figure 1. Glycerol-1-monostearate

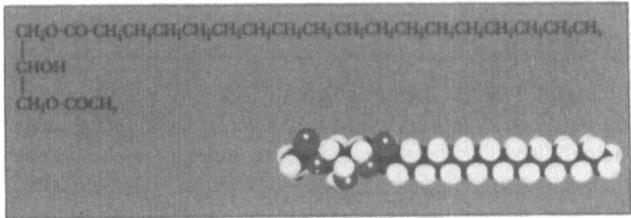

Figure 2. Acetylated monoglyceride

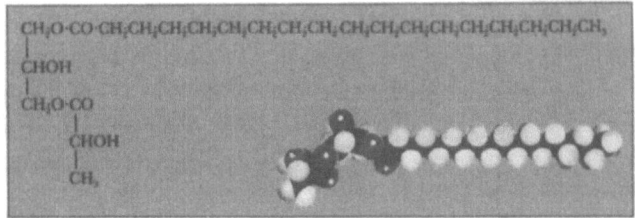

Figure 3. Lactylated monoglyceride

2.3 PROPYLENE GLYCOL ESTERS

Propylene glycol monostearate (PGMS) is made by esterifying propylene glycol with edible fatty acids, usually palmitic and stearic acid. The esterification yields a blend of 45% di-ester and 55% mono-ester from which the mono-ester is separated by molecular distillation, resulting in a high-purity propylene glycol ester. Figure 4 shows the chemical formula and molecular structure of PGMS.

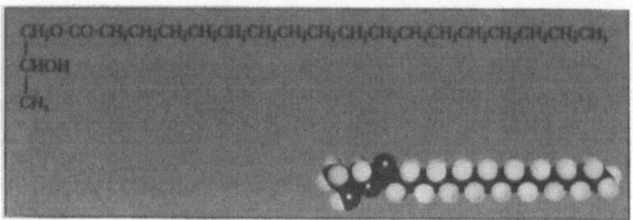

Figure 4. Propylene glycol monostearate.

2.4 PHYSICAL PROPERTIES OF LOW POLAR EMULSIFIERS

Monoglycerides are polymorphic and crystallize when cooled from melt in an unstable alpha-crystal form. If cooled further, a sub-alpha form appears. When stored at room temperature, a transition to a stable beta-crystal form takes place. Table 2 shows melting points of pure mono-palmitin and distilled saturated monoglycerides from hydrogenated palm oil, acetylated monoglycerides, lactylated monoglycerides and propylene glycol monostearate.

Distilled saturated monoglycerides form liquid crystals in water of lamellar or cubic structure above their Kraft temperature (45°-50°C). In unsaturated monoglyceride water systems the cubic and reversed hexagonal structures dominate, and the cubic phase is formed below room temperature (1).

The alpha-stable acetylated or lactylated monoglycerides and propylene glycol monostearate do not form liquid crystals in water, but they can form alpha-gel phases below their melting point due to penetration of water through polar regions in the alpha-crystals (1).

TABLE 2. MELTING AND CRYSTALLIZATION POINTS OF MONOGLYCERIDES

Transition Scheme	Temperature °C	Enthalphy J/g
1-Monopalmitin:		
Stable beta-form -> melt	75.9	210
Melt -> alpha-crystal	65.2	- 113
Alpha -> sub-alpha	37.7	- 43
Distilled monoglycerides:		
Stable beta-form -> melt	71	171
Melt -> alpha-crystal	65	- 97
Alpha -> sub-alpha	16	- 16
Acetylated monoglycerides:		
Stable alpha-form -> melt	38	81
Melt -> alpha-crystal	36	81
Lactylated monoglycerides:		
Stable alpha-form -> melt	38	85
Melt -> alpha-crystal	38	80
Propylene glycol monostearate:		
Stable alpha-form -> melt	44	165
Melt -> alpha-crystal	40	161

3. Aerated, Dairy-based Emulsions

The most food emulsions are stabilized by adsorbed layers of proteins in mono-molecular or micellar configurations. The adsorbed proteins provide stability against coalescence of oil droplets during storage and handling of e.g. dairy emulsions (milk, cream, etc.) until they reach the consumer. Controlled destabilization of the fat globules in whippable emulsions such as dairy cream as well as non-dairy creams, toppings, ice cream emulsions or other frozen dessert products is necessary to obtain a good texture and quality. The microstructure of whipped cream (2,3), toppings (4) or ice creams (3,5) is known to consist of air cells covered with layers of adsorbed fat globules and fat crystals. The three-dimensional network of agglomerated fat particles surrounding the air cells provides texture, creaminess and stability to the whipped cream or ice cream.

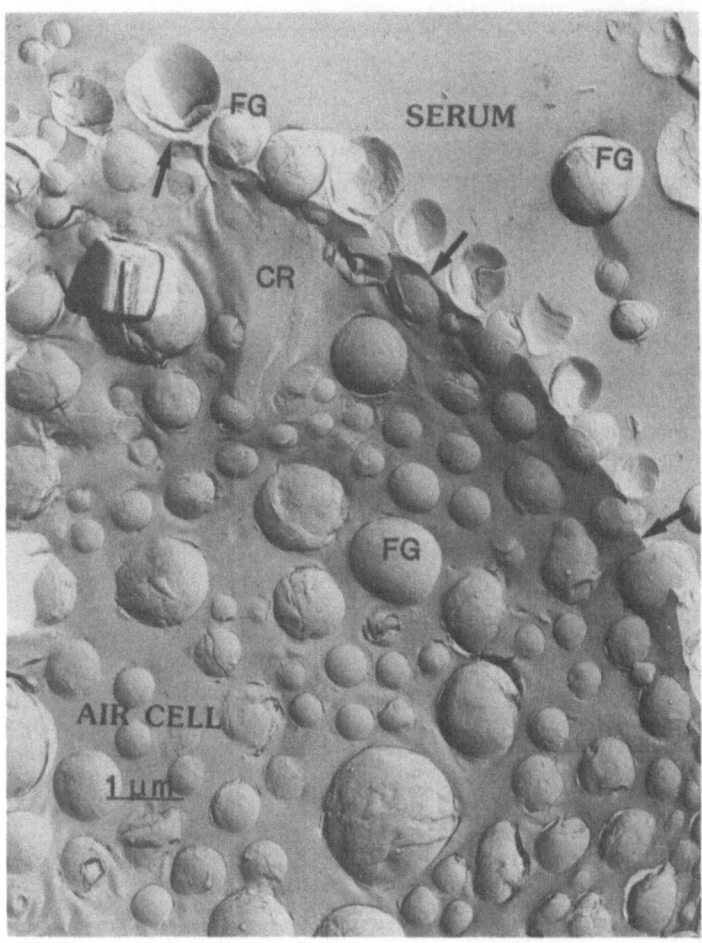

Figure 5. Transmission electron micrograph of whipped imitation cream showing adsorption of fat globules (FG) around an air cell. Fat globules protrude into the air cell. Few fat crystals (CR) are present. Arrows point to the air/water interface. (From ref. 4, courtesy of Dr. W. Buchheim, Kiel).

The driving force for this adsorption of fat globules or crystalline aggregrates is the hydrophobic properties of triglycerides. The functional role of added polar lipids (emulsifiers) is to control the destabilization of the emulsion by enhancing the desorption of the proteins at the fat globule surface and thus to increase the hydrophobicity and flocculation of fat globules. The microstructure of ice cream is in principle very similar to that of whipped cream as shown in Figure 5, although the amount of fat crystals in relation to intact fat globules may be increased due to the mechanical treatment of the emulsion at high shear and low temperature in the ice cream freezer.

3.1   PROTEIN DESORPTION IN ICE CREAM EMULSIONS

Ice cream contains 10-15% fat (dairy cream, butter fat or vegetable fats), milk proteins, sucrose, and 0.5% emulsifier/hydrocolloid blend. The mix is homogenized at 80°C in a high-pressure (160-200 bar) homogenizer. The mix is then cooled to 5°C and stored at 5°C for a period varying from 4-16 hours before it is frozen and whipped into an ice cream with an "over-run" of approx. 100% volume increase. The finished ice cream is then stored at -20°C to -30°C until it is consumed.

It is well-known that the ageing period at 5°C is important to obtain a good-quality product with respect to texture, stability and creaminess.

Electron microscopy studies combined with analytical measurements have shown (6) that proteins are being desorbed from the fat globule surfaces during the ageing period at 5°C. Simultaneously, an increase in fat globule crystallization takes place and this also enhances protein desorption. The removal of adsorbed proteins increases the hydrophobicity of the fat globules and promotes agglomeration and orientation around the air cells in the finished ice cream.

The protein desorption is related to the temperature treatment of the mix. At temperatures above 40°C the amount of protein adsorbed to the fat phase is high, and it decreases when lowering the temperature of the emulsion.

Studies of interfacial tension of oil/water systems with and without emulsifiers dissolved in the oil phase and with milk proteins dissolved in the water phase have shown that temperature has a pronounced effect on the surface activity of monoglycerides (6,7). At high temperatures the solubility of monoglycerides in the oil phase is high, and the reduction in surface tension is minimal. When cooled, the solubility decreases and adsorption of monoglycerides at the o/w interface increases, resulting in lower surface tension. The adsorption of proteins reduces the interfacial tention to about 12 m N/m and is not affected by changes in temperature. With monoglycerides present, the interfacial tension decreases at low temperatures to values below that of the protein film, and consequently the increased adsorption of monoglycerides will displace the protein molecules.

The interrelationship between decrease in interfacial tension, protein desorption, fat crystallization and agglomeration of fat globules due to increased hydrophobicity is shown in Figure 6.

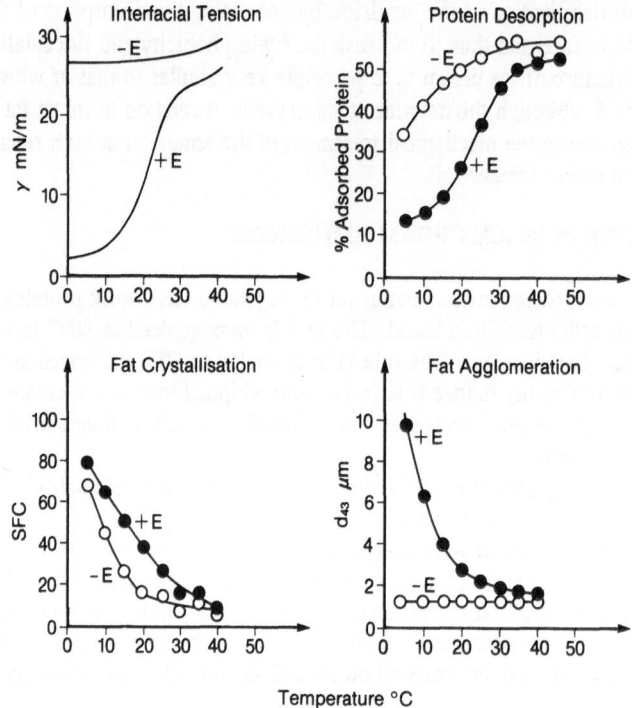

Temperature °C

Figure 6. Changes in interfacial phenomena, such as interfacial tension, protein desorption and fat crystallization related to agglomeration of fat globules in ice cream emulsions as a function of a decrease in temperature with (+E) and without (-E) added emulsifiers (from ref. 6).

The effect of monoglycerides and other surface active lipids on protein desorption is well-documented in the literature (8,9,10). The functionality of monoglycerides in ice cream, is thus undoubtedly related to their increased surface activity at low temperatures promoting protein desorption.

3.2   WHIPPED CREAMS OR TOPPINGS

Dairy cream contains 35-40% fat, and the fat globules are stabilized by the biological membrane formed during secretion of milk. When cooled sufficiently, the fat globule membrane becomes easily disrupted under shear during whipping, and no surface active lipids are needed to enhance the whippability. However, if the fat content is reduced and the cream is homogenized forming new fat droplets covered with adsorbed milk proteins, addition of low-polar emulsifiers may be used to enhance the whippability.

In vegetable fat-based creams acetylated (AMG), lactylated monoglycerides (LMG) or propylene glycol monostearate (PGMS) is used to improve the whipping properties and foam stiffness. The concentration of emulsifiers is critical (1-2% of total formula) and must be optimized by pilot tests. Too low concentrations give poor whippability and too high concentrations increase the viscosity too much due to fat globule agglomeration during the storage and transportation of the liquid cream.

The effect of low-polar emulsifiers used in liquid imitation cream is probably related to a destabilization mechanism as described for ice cream.

Spray-dried, powdered toppings contain approx. 50% fat, 30% sugar or maltodextrin, 10% sodium caseinate and 10% emulsifiers (AMG, LMG or PGMS). When reconstituted in cold milk or water, a highly destabilized "emulsion" containing agglomerated fat particles and crystal platelets is obtained. This gives a high-viscous - often almost paste-like consistency - which makes it easy to aerate into a foam with good texture.

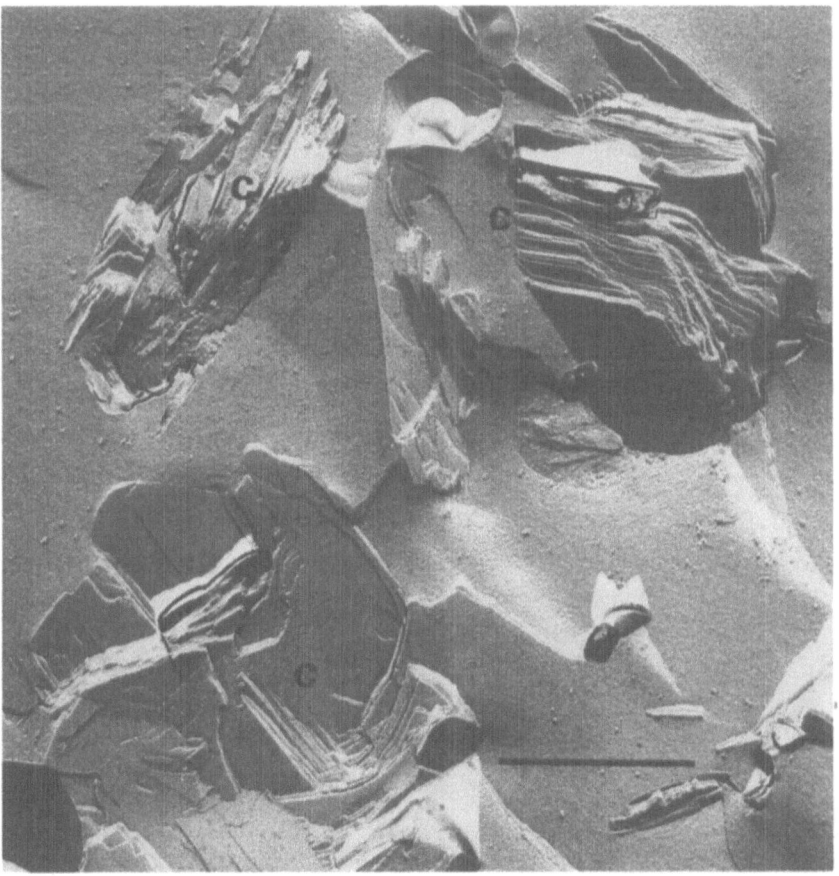

Figure 7. Transmission electron micrograph of a reconstituted topping emulsion with PGMS as emulsifier showing formation of platelet-like fat crystals (C). Bar = 0.5 micron. (Courtesy of Dr. W. Buchheim, Kiel).

The function of the emulsifiers used in toppings is of a different nature than described for ice cream and liquid creams. Studies of model topping systems (4) have shown that a strong destabilization of the reconstituted emulsion (1 part topping powder is mixed with 3 parts water at 5C) takes place forming a high-viscous solution of proteins and carbohydrates containing dispersed fat particles and crystals as shown in Figure 7.

The initial globular structure of the fat globules existing in the topping powder changes drastically. The transformation from the globular state of the fat phase into a matrix of crystal platelets increases the specific surface area of the fat and this has a pronounced effect on the foam stiffness after whipping. A whipped topping contains 15% fat by weight and can give a higher foam stiffness and specific volume than a whipped dairy cream containing 38% fat (11).

The added low polar emulsifiers play a vital role in changing the structure of the fat phase. If the topping powder is made without emulsifiers, it forms a low-viscous emulsion when reconstituted in water. This is demonstrated in Figure 8, which shows globular fat particles with a similar particle size distribution as that of the fat phase in the powder. No crystalline platelets as shown in Figure 7 were found in this case.

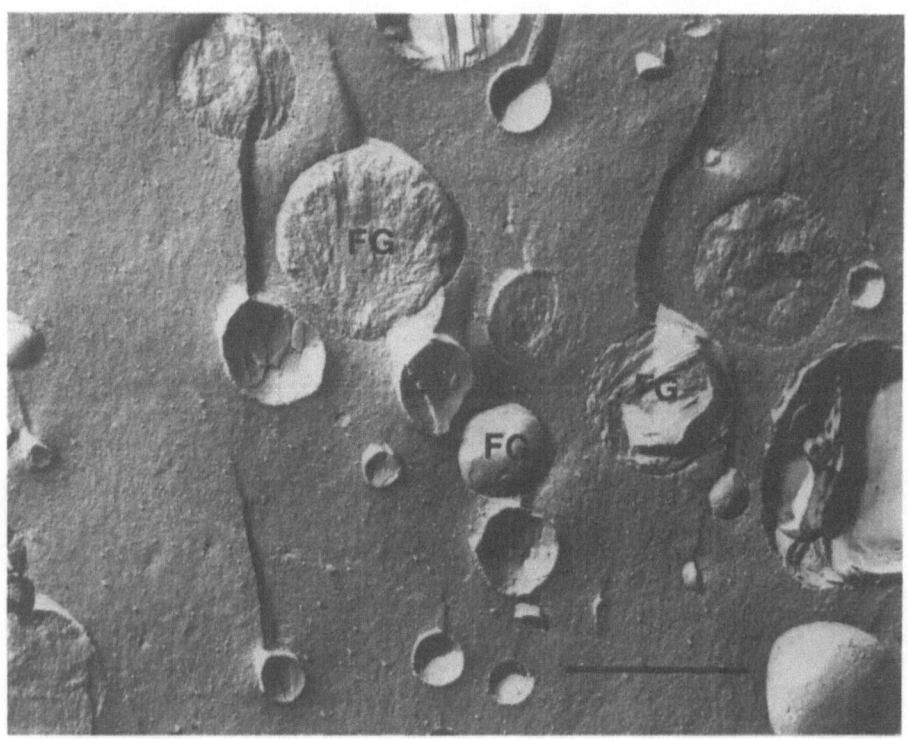

Figure 8. Transmission electron micrograph of a reconstituted topping emulsion without emulsifier added. The figure shows genuine fat globules (FG). Bar = 1 micron (From ref. 4, courtesy of Dr. W. Buchheim, Kiel).

The destabilization of toppings thus depends on two factors: a) relatively high concentrations of low polar emulsifiers in the fat phase, and b) low temperatures (< 20°C) during reconstitution and whipping.

A possible explanation of the function of emulsifiers in toppings is that the amount of emulsifiers present in the fat phase absorbs water into the fat phase and thus disintegrates the structure by a "swelling" process. Studies of topping fat phases by X-ray diffraction analysis have shown that the triglycerides from hydrogenated coconut oil do not co-crystallize with the added PGMS emulsifier (12). The coconut fat crystallizes in a beta-prime form with long spacings of 38 Å, while the PGMS crystallizes in an alpha-form with long spacings of 49 Å. In the blend of coconut oil and PGMS two long spacings of 38 Å and 49 Å are found, showing that the blend contains two separate crystal forms.

The polar region of the PGMS crystals will have a high affinity to water, which can penetrate the crystals and form a gel-like structure within the total lipid phase. A model of such a lipid gel structure is shown in Figure 9. The swelling of polar lipids by water has been described as a hydration force (13), and this mechanism has been referred to in studies of spray-dried emulsions containing lactylated monoglycerides and sodium caseinate (14).

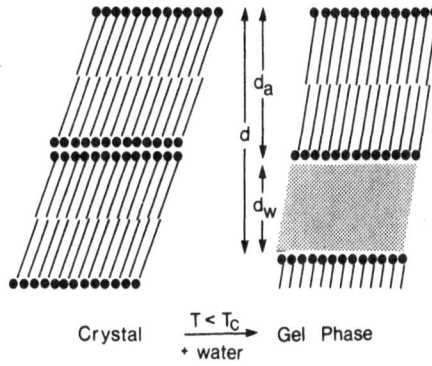

Crystal $\xrightarrow[+ \text{ water}]{T < T_C}$ Gel Phase

Figure 9. Schematic model of the formation of lipid gel-phase by hydration of the polar groups in crystalline regions of emulsifiers. d = interplanar Bragg spacing, da = thickness of lipid bilayer, and dw = thickness of water layer.

The microstructure of a whipped topping is shown in Figure 10. The orientation of crystal platelets at the air/water interface is clearly seen. By comparing the structure of whipped cream shown in Figure 5 to that of whipped topping (Figure 10) it is obvious that the microstructure of the two products is significantly different.

Figure 10. Transmission electron micrograph of a whipped topping foam. The air cell (AC) is surrounded by adsorbed fat crystals (FC) which are also seen at the air/water interface (see arrows) as well as in the serum phase (S). Bar = 1 micron. (From ref. 4, courtesy of Dr. W. Buchheim, Kiel).

The function of low polar emulsifiers in toppings is thus related to interactions between water and crystalline layers of the emulsifiers dispersed within the fat phase in contrast to the function in ice cream or liquid imitation cream where the effect of low polar emulsifiers is linked to the behaviour of interfacial monolayer films at the surface of the fat globules.

In general a lot of research has been devoted to emulsions, in particular to understand the role of emulsifiers for emulsion stability. However, from an industrial and economical point of view the overall destabilizing effect of low-polar emulsifiers in food emulsions is equally important and more research is needed into this area.

References

1.   Krog, N. J. (1990) 'Food Emulsifiers and their Chemical and Physical Properties'
      in K. Larsson and S.E. Friberg (eds) Food Emulsions, Marcel Dekker, Inc., New
      York and Basel, pp. 127-180.

2.   Anderson, M. and Brooker, B.E. (1988) 'Dairy Foams' in E. Dickinson and G.
      Stainsby (eds), Advances in Food Emulsions and Foams, Elsvier Science Publish-
      ing Co. Inc., London and New York, pp. 221-255.

3.   Buchheim, W. and Dejmek, P. (1990) 'Milk and Dairy-type Emulsions' in K.
      Larsson and S.E. Friberg (eds), Food Emulsions, Marcel Dekker, Inc. New York
      and Basel, pp. 203-246.

4.   Buchheim, W., Barfod, N.M. and Krog, N. (1985) 'Relations between
      Microstructure, Destabilization Phenomena and Rheological Properties of Whip-
      pable Emulsions', Food Microstructure, 4, pp. 221-232.

5.   Berger, K.G. (1990) 'Ice Cream', in K. Larsson and S.E. Friberg (eds), Food
      Emulsions, Marcel Dekker, Inc., New York and Basel, pp. 367-444.

6.   Barfod, N. M., Krog, N., Larsen, G. and Buchheim, W (1991) 'Effects of
      Emulsifiers on Protein Fat Interaction in Ice Cream Mix during Ageing I: Quanti-
      tative Analysis', Fat Science Technology, 93, pp. 24-29.

7.   Krog, N. (1991) 'Thermodynamics of Interfacial Films in Food Emulsions' in M.
      El-Nokaly and D. Cornell (eds), 'Microemulsions and Emulsions in Foods,
      American Chemical Society, Washington DC, USA, pp. 138-145.

8.   De Feijter, J.A., Benjamins, J. and Tamboer, M. (1987) 'Adsorption Displacement
      of Proteins by Surfactants in Oil-in-Water Emulsions', Colloids and Surfaces, 27,
      pp. 243-266.

9.   Dickinson, E. and Woskett, C.J. (1989), 'Competitive Adsorption Between
      Proteins and Small Molecule Surfactants in Food Emulsions' in R.D. Bee, P.
      Richmond and J. Minging (eds), Food Colloids, The Royal Society of Chemistry,
      Cambridge, pp. 75-96.

10.   Heertje, I., Nederlof, J., Hendrick, H.A.C.M. and and Lusassen-Reynders, E.H.,
      (1990) 'The Observation of the Displacement of Emulsifiers by Confocal Scanning
      Laser Microscopy', Food Structure, 9, pp. 306-316.

11.   Krog, N., Barfod, N.M., and Buchheim, W., (1987), 'Protein-Fat-Surfactant
      Interactions in Whippable Emulsions' in E. Dickinson (ed.), Food Emulsions and
      Foams, The Royal Society of Chemistry, Cambridge, pp. 144-157.

74

12.   Barfod, N.M., Krog, N., and Buchheim, W., (1989), 'Lipid-Protein-Emulsifier-Water Interactions in Whippable Emulsions' in J.E. Kinsella and W.G. Soucie (eds.), Food Proteins, Amer. Oil Chemist Society, Champaign Ill., pp. 144-158.

13.   Le Neveu, D.M., Rand, R.P., Parsegian, V.A. and Gingell, D., (1977), 'Measurements and Modifications of Forces between Lecithin Bilayers', Biophys. J. 18, p. 209.

14.   Westerbeek, J.M.M. and Prins, A., (1991), 'Function of Alpha-Tending Emulsifiers and Proteins in Whippable Emulsions', in E. Dickinson (ed.) Food Polymers, Gels and Colloids, The Royal Society of Chemistry, Cambridge, pp. 147-158.

SURFACTANT INDUCED FLOCCULATION OF EMULSIONS

Michael P. Aronson
Unilever Research United States
45 River Road
Edgewater, New Jersey 07020
United States of America

ABSTRACT. Previous studies demonstrated that a variety of surfactants induce paraffin oil-in-water emulsions to flocculate when their concentration exceeds a few weight percent. We suggested that this process was driven by micelle exclusion and presented a simple model that was consistent with experimental observations, e.g., effect of droplet size. The study has now been extended to monodisperse nonionics (alcohol ethoxylate) and also a pure cationic (alkyltrimethylammonium bromide) surfactant using highly purified alkanes. The pure nonionics display the same behavior as their commercial counterparts, i.e., accelerated creaming above a critical surfactant concentration. The process is exclusively reversible flocculation and is not accompanied by a change in droplet size. Cationics also display a destabilizing effect: however, the sensitivity of oil-in-water emulsions to surfactant concentration is very dependent on total ionic strength. At low ionic strength, the emulsions are insensitive to cationic surfactant concentration. A few implications of micelle exclusion to emulsion stability and the limitations of the simple model are discussed.

1. Introductions

Many factors control the stability of dispersions stabilized by surface active agents. Emulsions are especially complicated because their interfaces can deform and the phase behavior of the surfactant depends on the nature and composition of both phases. Nevertheless, relatively simple "scaling rules" such as HLB, PIT, equivalent alkane carbon number, etc., provide a way to relate surfactant structure to emulsion stability [1,2].

The stability of emulsions is usually interpreted as an interplay between surfactant adsorption and the accompanying changes it produces in the mechanical properties of the interphases and on the interparticle forces. This notion still persists [1-4]. However, growing evidence demonstrates that surfactant structures formed in solution or at the interface have a dramatic effect on emulsion stability often swamping those arising from monomolecular adsorption [5-8].

J. Sjöblom (ed.), Emulsions – A Fundamental and Practical Approach, 75–96.
© 1992 *Kluwer Academic Publishers.*

In 1983, we noticed that the simple addition of surfactant to a preformed and stable emulsion resulted in an abrupt increase in creaming rate [9] as depicted in Figure 1. A later study showed that this effect was fairly general, and that the transition was accompanied by a significant change in rheology [10]. Furthermore, the behavior was consistent with a simple theory of micelle depletion [10] as first guessed [9]. Recent findings in several laboratories [10-14] support our initial view that depletion by micelles is a general and fairly strong destabilizing force that can influence dispersions stabilized by surfactants. Recently, similar effects have been observed with water-in-oil emulsions further suggesting its generality [15].

The previous studies we carried out were with multicomponent surfactants and oils. Although the conclusions have practical value, there is always a concern that the observed behavior arises from impurities. The present study was undertaken to address this concern as well as to continue to test the generality of the effect with different chemical structures. It will be shown that emulsions prepared from highly purified oils (tetradecane and hexadecane), monodisperse nonionic surfactants (alcohol ethoxylates) and a pure cationic surfactant (tetradecyltrimethylammonium bromide) displays entirely similar behavior to their commercial counterparts. The limitations on the simple model are considered and implications of micelle depletion to emulsions stability are discussed.

2. Experimental

2.1. MATERIALS

Paraffin oil was obtained from Fisher Scientific (laboratory grade light paraffin, 125/135 Saybolt viscosity). Tetradecane and hexadecane were Fluka (Purum > 98% olefin free). The oils were passed through a fluorosil (Fisher Scientific) column and stored in amber jars. They did not spread

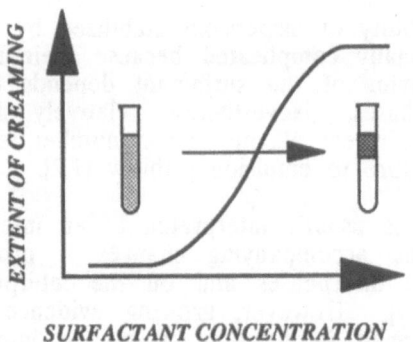

Figure 1. Changes in emulsion stability caused by addition of surfactant.

on the surface of clean water and had no effect on the surface tension (72.3 dyne cm^{-1}) after equilibration for one hour. GC/MS of the hexadecane sample confirmed a purity of greater than 99%.

The monodisperse alcohol ethoxylates were hexaethyleneglycol mono n-dodecyl ether (C12EO6), hexaethyleneglycol mono n-tetradecyl ether (C14EO6), and hexaethyleneglycol mono n-hexadecyl ether (C16EO6). These were obtained from Nikko Chemicals under the Nikkol trade name. Analysis of the C14EO6 sample by super critical fluid chromatography with mass spectrometer detection [16] confirmed the manufacturer's specification of > 99.5% purity. Tetradecyltrimethylammonium bromide was from Fluka (Purum > 98%) and was used as received. Two commercial surfactants not studied previously were also employed. An alkyl polyglycoside, APG 550 (predominantly a mixture of C_{12}-C_{13} alcohols with an average degree of polymerization of about 1.8), was from Henkel. A hydrogenated tallowtrimethylammonium chloride, Adogen 441, was from Sherex. All the remaining surfactants discussed are described in [10].

Sodium bromide was an analytical reagent grade sample from Fisher Scientific. Water was obtained by filtration of deionized water through a Waters Milli-Q ion exchange system. The conductivity was < 1×10^{-6} ohm^{-1} cm^{-1} and bubbles formed by shaking this water in clean glassware had negligible persistence.

2.2. PROCEDURES

Stock Emulsions were prepared by a concentrated emulsion technique [10]. This method involves first making a viscous oil-in-water emulsion by gradually combining about 90 Wt% oil (internal phase) with about 10 Wt% of an aqueous phase (external phase) which contains from about 20 to about 60 Wt% surfactant. Generally a dough mixer is employed. This viscous concentrate is mixed until no change in particle size (PCS) with further mixing is observed. The rheology and visual appearance of these concentrates depend strongly on particle size as is seen in the photographs in Figure 2. They have high conductivity and readily disperse in water confirming that they are oil-in-water emulsions. The concentrates are then simply diluted in water to the desired oil concentration (typically 25-35 Wt%) to form the final stock emulsion. To conserve the pure alcohol ethoxylates, 7-8 g of emulsifier concentrate (30-40 Wt% surfactant) was used and mixing was carried out in a 200 ml tall form beaker using an overhead constant speed mixer (Tekmar model RW20DZM) fitted with a Teflon coated guarded stirrer to minimize foam [17]. The emulsion concentrates were diluted to 35 Wt% oil and stirred overnight before use. The properties of these stock emulsions are recorded in Table 1.

All the droplet subdivision takes place during the formation and mixing of the concentrates. Although they are complex structures, the degree of emulsification in these concentrates appears to follow Taylor's analysis of droplet bursting [18,19]. Thus, the particle size is controlled simply by the ratio of the interfacial tension divided by the viscosity of the external

Table 1. Decription of stock emulsions.

SURFACTANT	STRUCTURE	OIL PHASE	DROP SIZE, μm	
TERGITOL NP-10	NONYLPHENOL ETHOXYLATE	PARAFFIN OIL	0.26	0.69
TERGITOL 15-S-7	SEC. ALCOHOL ETHOXYLATE	PARAFFIN OIL	0.32	1.06
TERGITOL 24-L-50N	PRIM. ALCOHOL ETHOXYLATE	PARAFFIN OIL	0.27	1.20
TERGITOL 24-L-25N	PRIM. ALCOHOL ETHOXYLATE	PARAFFIN OIL	0.49	1.32
APG 550	AKLYLPOLYGLYCOSIDE (DP=1.5)	PARAFFIN OIL	0.52	
NIKKOL C14EO6	HEXAETHYLENEGLYCOL TETRADECYL ETHER	TETRADECANE	1.7	
NIKKOL C14EO6	HEXAETHYLENEGLYCOL TETRADECYL ETHER	HEXADECANE	0.82	2.40
NIKKOL C16EO6	HEXAETHYLENEGLYCOL HEXADECYL ETHER	TETRADECANE	2.2	
NIKKOL C16EO6	HEXAETHYLENEGLYCOL HEXADECYL ETHER	HEXADECANE	1.9	3.1
WITCONATE S 1280	TEA DODECYBENZENE SULFONATE	PARAFFIN OIL	0.38	0.57
ADOGEN 441	TOLLOWTRIMETHYLAMONNIUM BROMIDE	PARAFFIN OIL	0.42	
TTAB	TERTADECYLTRIMETHYLAMONNIUM BROMIDE	PARAFFIN OIL	0.26	
TTAB	TERTADECYLTRIMETHYLAMONNIUM BROMIDE	HEXADECANE	0.32	

phase [10]. An example of this dependence is given in Figure 3. This behavior allows one to control the droplet size by controlling the external phase viscosity which is governed by the emulsifier concentration.

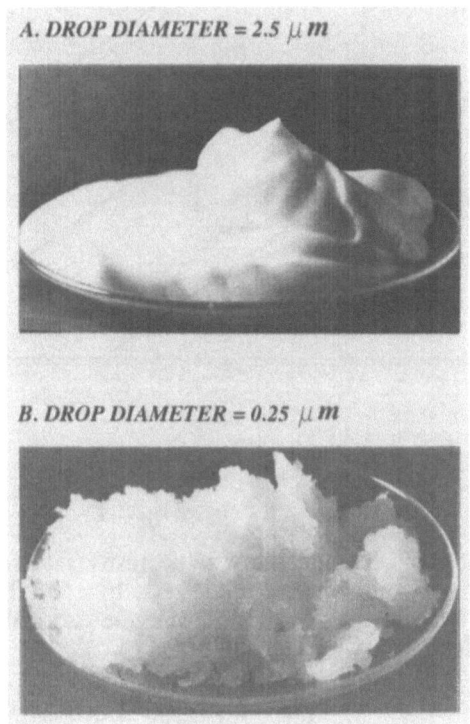

Figure 2. Photographs of Paraffin oil-in-water emulsion concentrates used to prepare model emulsions. Top photograph: 90 Wt% Oil phase and 10 Wt% aqueous phase containing 20 Wt% nonylphenol ethoxylate (10EO). Bottom photograph: 90 Wt% oil and 10 Wt% aqueous phase containing 65 Wt% surfactant.

Emulsions were examined under an optical microscope to gain a rough idea of the droplet diameters. If less than $1\mu M$, droplet sizes were estimated by photon correlation spectroscopy using a Brookhaven BI-90 instrument. A Coulter Counter (TA II) was employed to estimate droplet sizes greater than 1 μM. Both procedures were used for borderline emulsions and a simple average taken. In all case the distributions were unimodal.

Stability was assessed by the sedimentation procedure described previously [10,14]. Stock emulsions were diluted to a standard oil content (typically 25 Wt%) and the desired surfactant and salt concentration, and allowed to equilibrate overnight with stirring. The emulsions were then allowed to sediment for a fixed time (1-24 hours depending on droplet size) and a sample was taken from the bottom and analyzed for total residue.

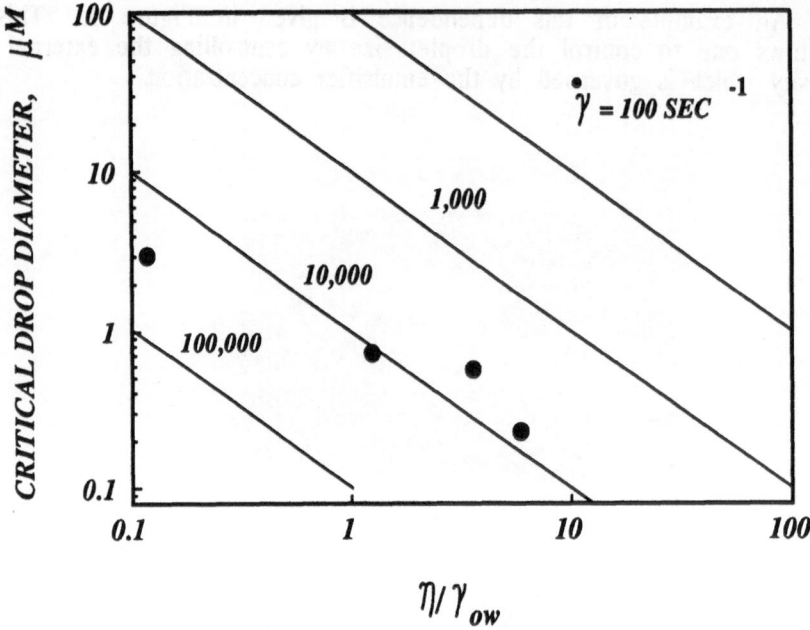

Figure 3. Dependence of droplet size on viscosity/interfacial tension ratio for paraffin oi-in-water emulsions stabilized by nonylphenol ethoxylate (10EO). The solid lines are calculated from the critical droplet size for bursting in laminar flow at various shear rates.

Gravimetric analysis after water evaporation proved adequate for paraffin oil and hexadecane but residues from the tetradecane emulsions were estimated by density (Parr Metler model DMA 46 density meter) with suitable calibration. The differences in creaming rate were striking and clearly visible, so any small errors in the measured residues do not alter the overall result.

3. Results and Discussion

3.1 NONIONIC SURFACTANTS

A variety of commercial nonionic surfactants have been found in earlier studies to induce flocculation of oil-in-water emulsions. These include surfactants based on ethylene glycol as well as glycoside head groups and various types of hydrophobic groups. The process is always accompanied by an increase in viscosity and shear thinning. Typical behavior is shown in Figure 4 for nonylphenol ethoxylate, Tergitol NP-10 ex Union Carbide. A striking feature is the influence of droplet size. As the droplet size increases, the emulsions become increasingly sensitive to flocculation by surfactant.

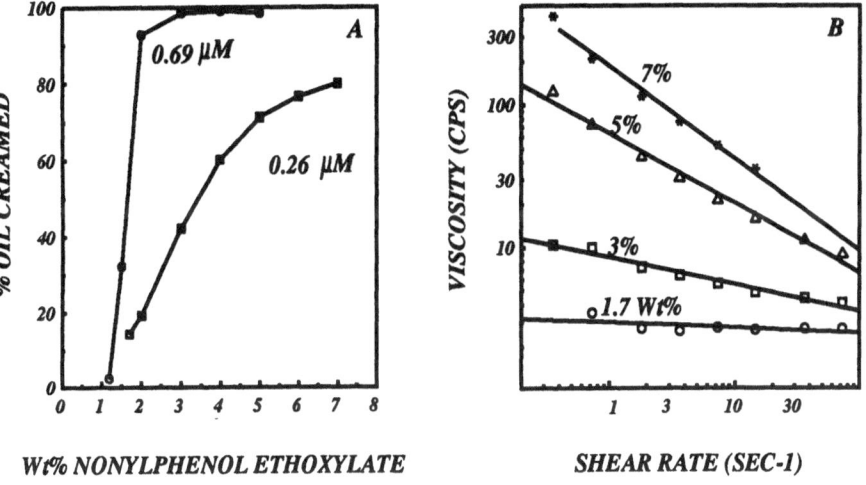

Figure 4. Observed behavior of paraffin oil-in-water emulsions stabilized with Tergitol NP 10 (nonylphenol ethoxylate). (A) Extent of creaming (droplet sizes shown on each curve). (B) Viscosity (Wt% surfactant in each emulsion shown on the curves).

The results for a variety of commercial surfactants are summarized in Table 2. The surfactant concentration in the aqueous phase causing 50% of the oil to cream during the experiment (typically one day) is given in the column labeled $C_{50\%}$. This value was corrected for adsorption from a calculated estimate based on monolayer coverage (see below). It is clear that all the surfactants induce flocculation although the specific concentrations vary with structure. The significance of the last column in Table 2 will be discussed later.

The stabilities of various oil-in-water emulsions made with three pure alcohol ethoxylates ($C_{12}EO_6$, $C_{14}EO_6$, and $C_{16}EO_6$) and two pure alkanes (tetradecane and hexadecane) are shown in Figure 5. The emulsions had droplet diameters in the range of about 0.8 to 2.5 μM. Clearly, these highly pure systems are subject to flocculation by free surfactant and impurities can be ruled out as the fundamental driving force for the process. The $C_{50\%}$ values collected in Table II indicate that the surfactant concentrations producing rapid flocculation ($~1$ Wt%) are in the range of their commercial counterparts although a bit higher than expected considering the fairly large droplet sizes these emulsions. These monodisperse surfactants may exhibit some partitioning in the oils used so that their actual concentration in the aqueous phase may be lower than our estimate which is only corrected for adsorption.

82

Table 2. Concentrations for surfactant induced flocculation of emulsions.

SURFACTANT	DROPLET SIZE (μm)	CMC (Wt%)	ESTIMATED Wt% ADSORBED	Wt% SURFACTANT [d] @ 50% CREAM	$-\dfrac{\Delta G_{OSM}}{kT}$
NP-10 [a]	0.26	0.004	1.9	2.7	1.6
	0.69		0.7	1.4	2.0
15-S-7 [a]	0.32	0.0042	1.3	2.7	1.9
	1.06		0.4	1.2	2.7
24-L-50N [a]	0.27	0.0013	1.4	0.7	0.4
	1.30		0.3	0.7	1.9
APG N550 [a]	0.35	0.0045	1.3	5.0	3.9
C12LAS (TEA salt) [a]	0.38	0.11	0.9	1.9	1.6
	0.57		0.6	1.5	1.8
				Average = 2.0+/-0.9	
C14EO6 [b]	0.83	0.00048	0.25	1.8	3.1
	2.4		0.08	1.6	7.8
C14EO6 [c]	1.7	0.00048	0.17	1.7	5.9
C16EO6 [b]	2.6	0.000076	0.17	1.7	4.4
TTAB [a,b]	0.35	0.121	–	>8.0	>11.8

a) Paraffin oil, 25 Wt% oil
b) Hexadecane, 15 WT% Oil
c) Tetradecane, 15 WT% oil
d) Wt% in aqueous phase, corrected for adsorption

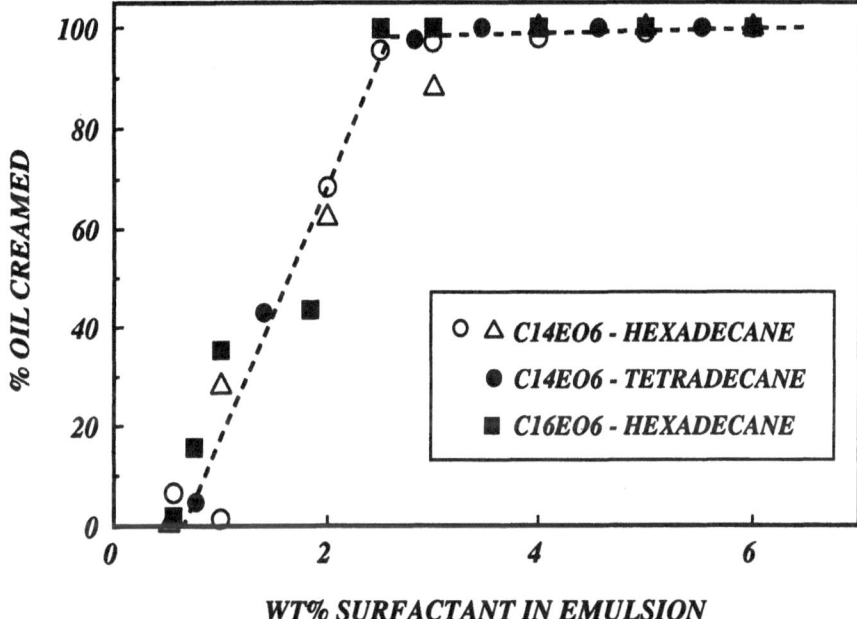

Figure 5. Influence of surfactant concentration on stability of pure alkane-in-water emulsions stabilized with monodisperse alcohol ethoxylates. Oil drop diameters are 0.8 to about 2.5 μM.

3.2. MODELS OF MICELLE EXCLUSION

Even cursory microscopic observations of emulsions indicate that the creaming induced by surfactant is driven by droplet flocculation which arises suddenly as the surfactant concentration is increased. As an example, photographs of hexadeane-in-water emulsions at several concentrations of nonionic surfactant, C14EO6, are given in Figure 6. The sudden appearance of flocs is a characteristic feature of the process and has been observed for a variety of surfactant structures and oil combinations. The process driving surfactant induced flocculation is thought to be the exclusion of micelles from approaching droplets [9-15]. Some aspects of this process are discussed in the following sections.

3.2.1. Exclusion Process. The magnitude of the droplet interaction free energy arising from a micelle exclusion process can be estimated from the theory of Asakura [20] - see also [11,21]. A very simple model assumes that the micelles are spherical, have a single molecular weight, that their internal density is equivalent to that of the surfactant in bulk ($\approx$ 1g/cm^3 for most surfactants), and that they behave with respect to interaction energy as an athermal system [22]. The last assumption allows one to estimate the second virial coefficient from an excluded volume based solely on the geometry of the micelle.

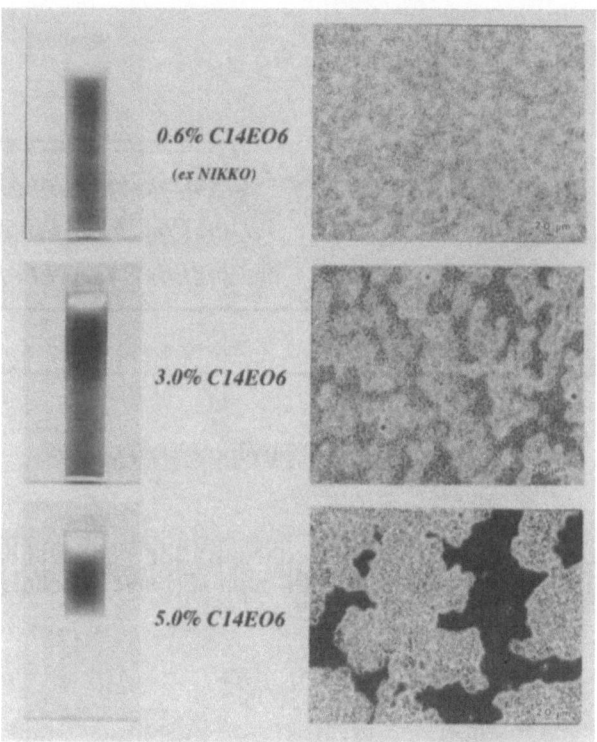

Figure 6. Photographs of hexadecane-in-water emulsions stabilized with Nikkol's hexaethyleneglycol tetradecyl ether. The top row illustrates the creaming of the emulsions after 0.5 hours. The bottom row are photomicrographs. Note the formation of compact flocs in unstable emulsions.

The free energy change attributed to micelle exclusion when two droplets flocculate into a secondary minimum of separation 2δ is given by

$$\Delta G_{osm} = \int_{\infty}^{h} F_{osm} \, dh = \int_{\infty}^{h} \pi_m \, A_{ex} \, dh \qquad (1a)$$

$$= \int_{\infty}^{h} \pi_m \, [(d/2 + r + \delta)^2 - (d/2 + h/2)^2] \, dh \qquad (1B)$$

where π_m is the osmotic pressure arising from the excluded micelles, A_{ex} is the excluded area, and the geometric variables are defined in Figure 7. Equation 1B has an analytical solution if it is assumed that the osmotic pressure does not depend on the interparticle separation [11,21]. Under the assumptions of spherical micelles of known density and molecular weight, Equation 1B can be written as

$$\pi_m = (CRT/M)(1 + 2C/p) \qquad (2)$$

where C is the micelle concentration in the aqueous phase (g/cm^3), M is the micelle molecular weight, and p is the density of a micelle in g/cm^3. In deriving Equation 2, we replace the second virial coefficient by the athermal excluded volume approximation [20] and the volume of an individual micelle by M/pN_a where N_a is the Avagadro number.

Equation 1B can be solved in terms of the easily measured variables of surfactant concentration (C) and drop diameter (d) if the micelle molecular weight, the surfactant density, and the equilibrium separation 2δ are specified. Note that in the current treatment the micelle radius, r, is $(3M/4pN_a)^{1/3}$

A few solutions to Equation 1B are shown in Figure 8A for the case of a droplet separation of 50 Å and a micelle molecular weight of 1.3×10^5 Daltons. The model predicts a substantial attractive potential that can become significant at only a few percent surfactant. There is a pronounced effect of droplet size with droplets less than 0.1 μM being insensitive to surfactant while droplets greater than about 2 μM being very sensitive to micelle depletion.

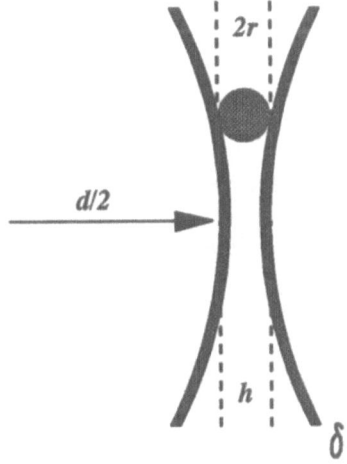

Figure 7. Idealized model of micelle exclusion by approaching emulsion droplets. Not drawn to scale.

The effect of micelle molecular weight on interaction free energy as computed from Equation 1B is shown in Figure 8B.   In this case the drop diameter is fixed at 0.5 μM and the remaining variables have the same values as used in Figure 8A.   The model predicts that as the molecular weight of the surfactant micelles increase, they should have a smaller effect in destabilizing emulsions. This arises because micelle molecular weight has a stronger effect on the osmotic pressure than it does on the extent of the depletion zone.   Thus, the model predicts that depletion effects will be relatively small for surfactants that form association structures with large aggregation numbers. It must be stressed that these

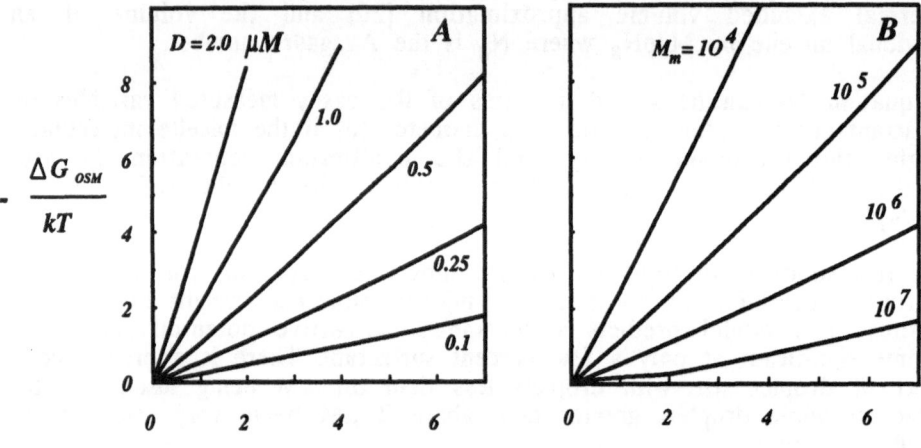

Wt% SURFACTANT IN AQUEOUS PHASE

Figure 8. Interaction free energy due to micelle exclusion calculated from Equation 1. A. Influence of oil drop size. Assumed values are: $2\delta = 50$ Å; micelle molecular weight $= 1.3 \times 10^5$.   B. Influence of micelle molecular weight.   Assumed values are: $2\delta = 50$ Å, oil drop diameter $= 0.5$ μM.

conclusions are a direct consequence of estimating osmotic pressure based on an athermal approximation which can lead to an underestimate of the second virial coefficient and thus a large dependance on molecular weight.

The model can be used to estimate the interaction free energy at rapid flocculation.   For simplicity, this is again assumed to correspond to a surfactant concentration causing 50% of the oil to cream within the given time period.   However, in order to carry out the calculation we require a knowledge of the amount of surfactant that transfers into the oil phase and the amount adsorbed at the oil/water interface. Since all of the commercial surfactants shown in Table 2 have HLB's greater than 12, they have negligible solubility in paraffin oil (confirmed by chromatography). The amount of surfactant adsorbed was not measured directly in these studies and, thus, must be estimated. We assume monolayer coverage and use reasonable values for the molecular area (50-80 $\mathring{A}^2$/molecule [23]).

The resulting interaction free energies calculated in this way are also given in the last column of Table 2. The average value of $\Delta G_{osm}/kT$ is 2.0 +/-0.9. Considering the approximations made, this value is not unreasonable for the interaction energy at the onset of rapid flocculation [24] although it does seem a bit on the low side. The pure nonionics have considerably higher values of interaction energy and this may reflect some partitioning into the oil phase - a possibility ignored in the analysis.

3.2.2. Implications. Figure 9 shows contours of the surfactant concentration in the aqueous phase calculated to produce an attractive potential of 2kT plotted against the oil drop diameter. The asymptotic nature of this relationship is evident. The analysis implies that two extremes of behavior should be encountered in surfactant induced flocculation. We believe this is a key result that provides an underlying explanation for a number of common observations in the emulsion art.

Firstly, almost any concentration of micelles will induce flocculation and creaming of emulsions that have drop diameters greater than about 5 μM except if the surfactant has a large aggregation number or there exists some long range repulsion as may arise with charged surfactants. Thus, emulsions produced by processes that lead to large droplet size are prone to flocculation by free nonionic surfactant unless the surfactant is essentially water insoluble. This probably explains why many oils with limited ability to dissolve surfactants possess a required HLB for oil-in-water emulsions of 8-10 (insoluble) when determined by Griffins original test method [25]. This test involves shaking equal volumes of oil and water in the presence of 5 Wt% emulsifier. Here the predominant instability for high HLB surfactants is creaming as would be expected for surfactant induced flocculation. Such a low shear process typically produces droplets above ~0.5 μM.

The model also provides an alternative explanation for why mixtures of low and high HLB surfactants that form complexes often have better stability than a single surfactant of the same average HLB. Such mixtures have a sizable fraction that is oil soluble and thus, does not contribute to exclusion. The association structures formed in such mixtures tend to be large and this further lowers the osmotic pressure.

At the other extreme are emulsions with droplet sizes less than about 0.2 μM. Figure 9 indicates that such emulsions will be relatively insensitive to micelle concentration. Thus, emulsions made by high shear processes should be able to tolerate considerable levels of surfactant regardless of whether small association structures are present. Such emulsions should not be very sensitive to HLB which is again consistent with practical experience.

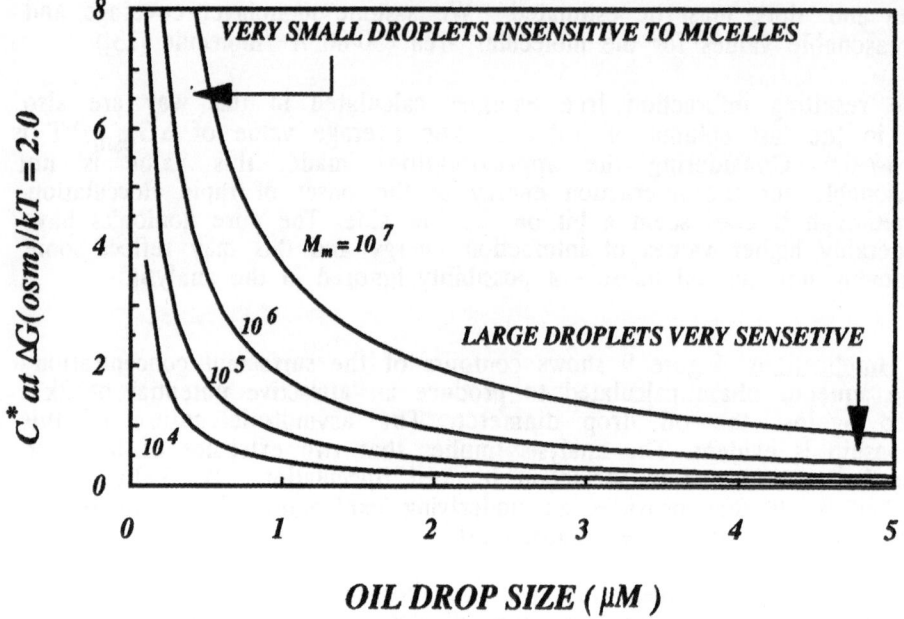

Figure 9. Critical surfactant concentration for rapid flocculation $(C_{50\%})$ versus droplet size and micelle molecular weight $(2\delta=50$ Å).

3.2.3. Contact Angles of Macroscopic Films.

The magnitude of the interaction energy arising from micelle depletion is in principle measurable through its influence on the contact angle of a macroscopic emulsion film with its associated plateau border. Although such contact angle measurements have not been carried out in the present study, it is nevertheless instructive to estimate whether they are of sufficient magnitude for accurate measurement given the current state of the art ($\approx$ 0.1 degree) [26,27].

The situation is depicted in Figure 10. The macroscopic contact angle that the film makes with the extrapolated profile of the oil/water interface is related to the interaction energy by [26]

$$\theta = Cos^{-1} (\Delta G/2\sigma_{0/w} + 1) \qquad (3)$$

where $\sigma_{0/w}$ is the oil/water interfacial tension. The interaction energy is related to the net disjoining pressure by

$$G = -\int_{\infty}^{2\delta} \pi_i \, A_f \, dh \qquad (4)$$

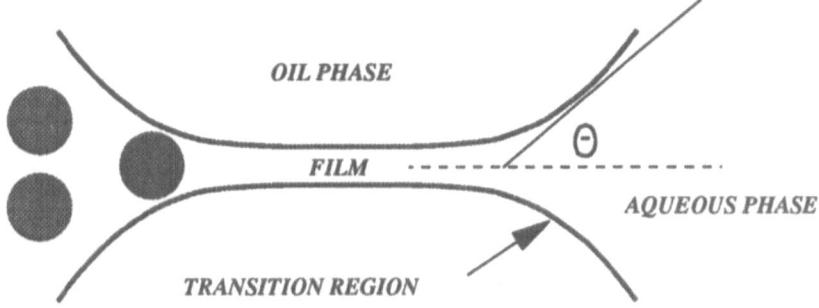

Figure 10. Macroscopic emulsion film in equilibrium with plateau border.

Where A_f is the exclusion area which for a planer macroscopic film is the film area (neglecting end effects). The disjoining pressure can be estimated by the usual approximation [28] of superposition of various interaction forces

$$\pi_i = \pi_{el} + \pi_{vw} + \pi_{osm} + \pi_p + \quad\ldots\ldots\ldots \tag{5}$$

Equation 4 can be solved for flat films if we ignore double layer forces (nonionic surfactants); line tensions (large films); consider only a "hard wall" model for steric interaction ; and use a simple approximation for the dispersion force contribution [28]. The result is

$$\Delta G = -A/48\pi\delta^2 - (2CRT/M)(1 + 2C/p)(3M/4\pi\ N_a) \tag{6}$$

Equation 6 was solved for two values of the interfacial tension (0.5 and 2.4 dyne cm^{-1}), two values for the Hamaka constant (1×10^{-14} and 1×10^{-13} ergs), a film thickness $2\delta = 40$ Å, and a micelle molecular weight of 1.3×10^5 Daltons. The results are given in Figure 11. Since hydrocarbon oil/water interfacial tensions of typical nonionic surfactants are likely to fall within the range selected [29,30], changes in contact angles with surfactant concentration of several degrees are predicted. These changes should be large enough to measure given a sensitive experimental technique.

3.2.4. Limitations. The model used above makes a number of simplifying assumptions which are incorrect. These oversimplifications limit the insight it can provide concerning the connection between dispersion stability and surfactant molecular structure. Unfortunately, the problem quickly gets very complicated as we shall now see. The assumptions of most interest are those relating to: (i) the absence of ionic groups on the surfactant or oil droplet surface; (ii) the spherical shape of the micelles;

90

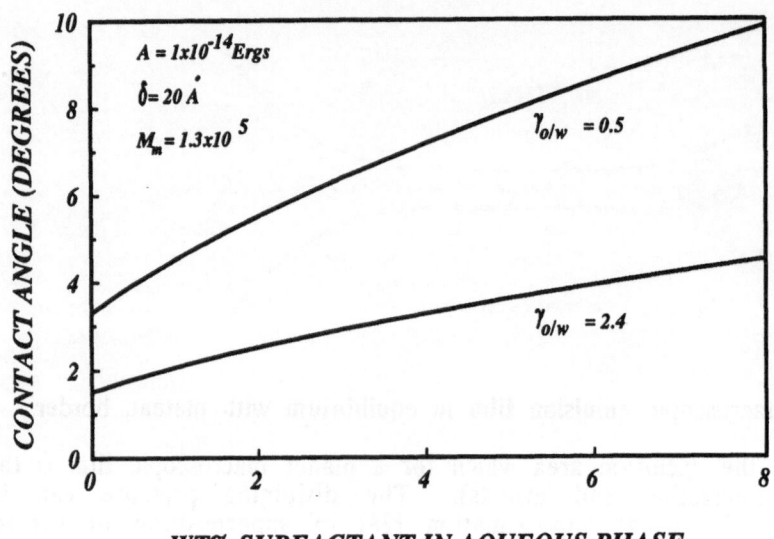

WT% SURFACTANT IN AQUEOUS PHASE

Figure 11. Estimate of the effect of micelle exclusion on the contact angle of macroscopic emulsion films.

and (iii) the magnitude of the osmotic pressure and in particular, the value for the second virial coefficient. Each of these will be examined qualitatively below. A quantitative treatment is the subject of continuing research.

Ionic interactions can have two effects on micelle exclusion. The most obvious effect is the electrical double layer that can arise from adsorption of charged surfactants. The conditions for instability require are that micelles can be excluded, i.e., the droplets approach to distances smaller than a micelle, and that the interaction energy overcomes the thermal energy of the droplets, i.e.,

$$\delta\pi_i/\delta h > 0 \qquad (7A)$$

$$\Delta G_{net} \big|_{at\ \pi_i = 0} < \sim 2kT \qquad (7B)$$

Thus micelle depletion is not expected to occur when a sufficiently strong repulsive force between the droplets extends beyond the dimensions of a micelle. The qualitative effect of electrical double layer is as follows: when the repulsive disjoining force of the electrical double layer is larger than the osmotic pressure attributed to micelles, then the emulsion should be insensitive to surfactant concentration. Conversely, if the double layer is highly compressed ($1/K$ << micelle diameter, where $1/K$ is the double layer thickness), micelle depletion should come into play. In this case the double layer simply provides a short range barrier to coalescence. The intermediate situation when $1/K \approx 2r$ will be difficult to predict because in this case the system is very sensitive to surface charge, solution composition, particle size and other details. One consequence of the electrical double layer is that micelle depletion for charged surfactants is expected to be sensitive to ionic strength - an expectation only partially confirmed by experiments (see below).

A second affect of charge is the Donnan equilibrium that arises from the condition for electrical neutrality, i.e., the surfactant and counterion must be simultaneously excluded. Qualitatively, this increases the osmotic pressure depending upon the degree of counterion binding and the total ionic strength. One might expect that this effect would be most pronounced for surfactants (or their mixtures) that possess low CMC coupled with low micelle counterion binding. The influence of Donnan equilibrium is difficult to estimate for micelles because of the possibility of shifts in CMC with electrolyte concentration.

The second questionable assumption concerns micelle shape. It is well known that surfactant association structures have a range of allowable shapes [31-33] depending of the head group area and the length of the hydrophobic chain. For highly asymmetric structures such as rods, and to a lesser extent disks, the exclusion process becomes much more compli-cated and might be expected to result in repulsion over certain distances and attraction over others. The situation is complicated further when the surfactant can solubilize appreciable oil into micelle since this can not only increase micelle size but alter the shape as well. The role microemulsion droplets in surfactant induced flocculation of water-in-oil emulsions has recently been investigated by Horsup [15].

The last approximation discussed is the estimate of the second virial coefficient. It is based on an analysis of excluded volume for a spherical particle in an **athermal** solvent [22]. Thus, significant interactions of micelle head groups with water as well as with other micelles are ignored. Clearly, such interactions are important for nonionic head groups like polyethylene glycol. For example, it has been found that the osmotic pressure of PEG solutions approach a limit with respect to molecular weight and that the values are much higher than would be expected based only on the Van't Hoff limit [34]. Recently, Van Oss et al [35], showed that this effect was due to a very high second virial coefficient and could be accurately estimated from an analysis of interfacial tension. The net effect would be a much higher osmotic pressure and a weaker dependance on micelle molecular weight.

It would be very interesting to extend this approach to EO containing surfactants and especially to the effect of temperature. Since the excluded volume of nonionic polymers scales as ($1-\theta/T$) [22], temperature may also have a significant effect on surfactant induced flocculation by ethylene oxide based surfactants and might lead to reduced fluocculation at some temperature below the cloud point. This area warrants further study.

3.3. CHARGED SURFACTANTS

The influence of a commercial anionic and cation surfactant on the stability of paraffin oil emulsions are compared in Figure 12. The anionic surfactant is predominantly a triethanolammonium salt of dodecyl benzene sulfonate (LAS) while the cationic is a tallowtrimethylammonium chloride (TTAC). These surfactants have quite different effects on stability. LAS induces flocculation. Similar behavior has been reported recently by Bibette et al [12] for another anionic surfactant, sodium dodecyl sulfate [12]. In contrast, TTAC has no effect on stability even at 10 wt% - a value far in excess of all the surfactants we have studied.

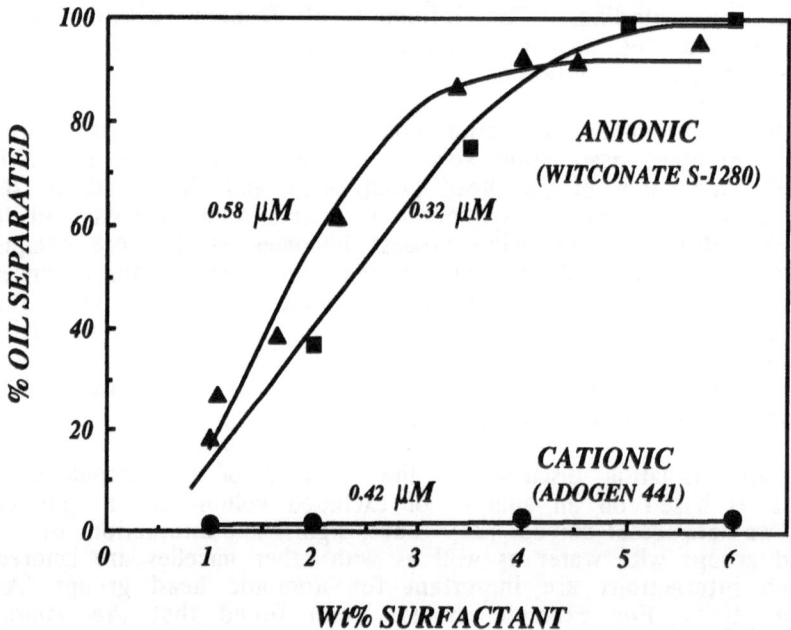

Figure 12. Stabilities of paraffin oil-in-water emulsions in the presence of two charged surfactants.

The results for paraffin oil and hexadecane-in-water emulsions in the presence of a pure cationic, tetradecyltrimethylammonium brobide (TTAB), are collected in Table 3. As with its commercial counterpart, no influence on stability is observed. Microscopic observation revealed no flocculation up to 8 Wt% surfactant.

Table 3. Influence of cationic surfactant concentration on emulsion stability in the Absence of Added Electrolyte.

Wt% Surfactant in Emulsion	Wt% Oil Creamed		
	Adogen 441 Paraffin Oil (0.42 μM)	TTAB Hexadecane (0.30 μM)	TTAB Paraffin Oil (0.26 μM)
2.0	0.8	0.9	0.1
4.0	1.2	3.4	-
6.0	1.9	0.8	1.1
8.0	1.1	0.3	0.7

As discussed above, the depletion process for charged surfactants should in principle, depend on the magnitude of the disjoining pressure contributed by the electrical double layer. This in turn depends on total counterion concentration. This prompted a study of the influence of NaBr with the results shown in Figure 13. Here we have kept the total Br⁻ concentration fixed by assuming a value of 0.27 for the degree of dissociation of Br⁻ counterions from the cationic micelles [29,30]. These results indicate that the cationic surfactant will indeed induce instability when the electrical double layer is suppressed.

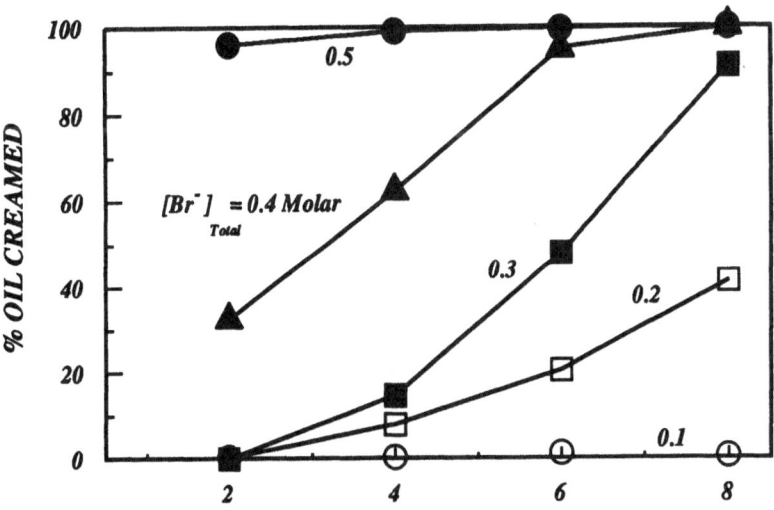

TOTAL SURFACTANT CONCENTRATION (WT%)

Figure 13. Influence of TTAB and total bromide counterions on the stability of paraffin oil-in-water emulsions. The average drop diameter is 0.26μM.

Although it is tempting to ascribe the behavior of the cationics in distilled water to electrical double layer effects, it is unclear why they are not observed for at least two common anionics, LAS and SDS. Also it is disturbing that so much electrolyte is required to switch on the effect for the cationics. The zeta potentials of tetradecane droplets in the presence of various surfactants are recorded in Table 4 but provide little help in reconciling the different behavior of these surfactants. Both charged surfactants produce large negative potentials that are reasonably surfactant independent and both should be sufficient to produce a long range barrier to micelle depletion. Why only the TTAC is stable at all concentrations in <u>distilled water</u> remains to be understood.

Table 4. Zeta potential of emulsion droplets.

Surfactant	Concentration (Wt%)	Zeta Potential (mV)
Dodecyltrimethylammonium bromide	1.0	+66.5
"	2.0	+57.4
"	4.0	+63.3
"	6.0	+55.3
"	8.0	+52.0
Witconate S-1280 (C12LAS, TEA salt)	1.0	-68.5
"	2.0	-71.5
"	4.0	-73.3
"	6.0	-71.7
"	8.0	-72.0
APG 550 (C12-13 Alcohol, DP ˜ 1.8)	1.0	-64.0
Tergitol 24-L-50N (C12-14 Alcohol ˜ 6.4EO)	1.0	-26.0

4. Conclusions

Emulsions stabilized by surfactants are susceptible to flocculation when the surfactant concentration exceeds a critical limit. All the evidence point to micelle depletion and a simple model of this effect predicts many of the experimental features. Possible improvements of the model were pointed out but a rigorous quantitative treatment will not be trivial. The model was used to estimate the potential effect of micelle depletion on the macroscopic contact angles of free emulsion films. The analysis predicts that these angles should depend on surfactant concentration and that the depletion force should increase their value by several degrees.

The behavior of pure oils and monodisperse nonionic surfactants is entirely similar to their commercial and impure analogs. This rules out impurities as the fundamental cause of the instability. Both commercial and pure cationic surfactants do not exhibit micelle depletion in distilled water. However, the addition of electrolyte brings this process into play. Surprisingly, this charge effect was not observed with several anionic surfactants studied previously which induce flocculation even in distilled water. The reason is not known at the present time but would seem important.

The depletion phenomenon has several implications concerning the way in which emulsion stability responds to surfactant structure. There is a significant interplay between surfactant structure, aggregation number, and oil solubility on the one hand and droplet size and thus method of emulsion manufacture on the other.

Acknowledgments

The electrophoresis measurements were carried out by Dr. Wasudevan in Dr. P. Somasudaran's laboratory at Columbia University. Several useful discussions were held with Dr. P. K. Ananthapadmanabhan of Unilever. The author also thanks Dr. Henry Kalinoski and Mr. Alan Silver for their analyses of the Nikkol and hexadecane samples; Mr. Dennis Palatini for image analysis of emulsions; Mr. Walter Samuel for his expert photographic assistance; to Mr K. Kruza for Coulter Counter results; and to Unilever for permission to publish this work.

References

1. Becher, P., Editor, "Encyclopedia of Emulsion Technology", Vol.1 and 2, Marcel Dekker, New York, 1983.
2. Vaughan, C. D. and Rice, D. A., J. Dispersion Sci. Technol., 11, 83 (1990); Schechter, R. S., Colloids Surfaces, 29, 53 (1988).
3. Ruckenstein, E., Langmuir, 4, 1318 (1988).
4. Cavallo, J. L. and Cheng, D., Chem. Eng. Prog. 54 (June 1990).
5. Friberg, S., Langmuir, 2, 121 (1986); Adv. Liq. Cryst., 2, 173 (1976); J. Coll. Interface Sci., 55, 614 (1976).
6. Friberg, S., 8th International Symposium on Surfactants in Solution, Gainesville, Fl. 1989.
7. Friberg, S., J. Colloid Interface Sci., 56, 19 (1976).
8. Graciaa, A., Barakat, Y.,Schechter, R., Wade, W., and Yiv, Y., J. Colloid Interface Sci., 89, 217 (1982).
9. Fairhurst, D., Aronson, M., Gum, M., and Goddard, E. D., Colloids Surfaces, 7,153 (1983).
10. Aronson, M., Langmuir, 5, 494 (1989).
11. Ma, C., Colloids Surfaces, 28, 495 (1987).
12. Bibette, J., Roux, D., Nallet, F., Phys. Rev. Lett., 65, 2470 (1990).
13. M. Packman and Th. Tadros, 8th International Symposium on Surfactants in Solution, Gainesville FL (1990).
14. Aronson, M., International Symposium on Surfactants in Solution, Ottawa, Canada (1988).

15. Fletcher, P. and Horsup, D., 8th International Symposium on Surfactants in Solution, Gainesville FL (1990).
16. Kalinowski, H., Private Communications (1990). See also: Kalinowski, H. T. and Hargiss, L. O., J. Chromatog., 474, 69 (1989); Kalinowski, H. T. and Jensen, A., J. Amer. Oil Chem. Soc., 66, 1171 (1989).
17. Aronson, M. and Petko, M., U.S. Patent 4,606,913 (1986).
18. Taylor, G. I., Proc. R. Soc. A 138, 41 (1932); 146, 501 (1934).
19. Walstra, P. In "Encyclopedia of Emulsion Technology Vol 1"; Becher, P., Ed.; Marcel Dekker: New York, 1983; P86.
20. Asakura, S. and Oosawa, F., J. Chem. Phys., 22, 22 (1954); J. Polym. Sci., 33, 133 (1958).
21. Sperry, P., J. Colloid Interface Sci., 87, 375 (1982).
22. Heimenz, P., "Polymer Chemistry", Chapt. 8, Marcel Dekker, New York, 1984.
23. Rosen, M. J., Surfactants and Interfacial Phenomena", John Wiley, New York, 1989.
24. Hogg, R. and Yang, K., J. Colloid Interface Sci., 56, 573 (1976).
25. Griffin, W., J. Soc. Cosmet. Chem., 1, 311 (1949).
26. Princen, H. M., J. Phys. Chem., 72, 3342 (1968).
27. Kolarov, A., Scheludko, A., and Exerowa, D., Trans. Faraday Soc., 64, 2864 (1968). For a more accurate extrapolation method see: Zorin, Z., Platikanov,D., and Kolarov,T., Colloids Surfaces, 22, 147 (1987).
28. Sonntag, H. and Strenge, K., "Coagulation and Stability of Disperse Systems", Halsted Press, New York, 1970.
29. Aronson, M., Gum, M., and Goddard, E. D., J. Amer. Oil Chem. Soc., 60, 1333 (1983).
30. Dillan, K. W., Goddard, E. D., and McKenzie, D. A., J. Amer. Oil Chem. Soc., 56, 59 (1979); Ibid, 57, 230 (1980).
31. Tanford, C., Proc. Nat. Acad. Sci., 71, 1811 (1974).
32. Isrealachvili, J., Mitchell, D., and Ninham, B., J. Chem Soc. Faraday II, 72 1525 (1976).
33. Mitchell, D., Tiddy, G., Waring, L., Bostock, T., and McDonald, M., J. Chem. Soc., Faraday I, 79, 975 (1983).
34. Gawrisch, K., Ph.D. Thesis, Karl Marx University, Leipzig, 1986.
35. Van Oss, J., Arnold, K., Good, R. J., Gawrisch, K., and Ohki, S., J. Macromol. Sci.-Chem., A27, 563 (1990).
36. Palepu, R., Hall, D.,and Wyn-Jones, E., 8th International Symposium on Surfactants in Solution, Gainesville FL (1990).
37. Sepulveda, L., J. Phys. Chem., 89, 5322 (1985).

THE RESOLUTION OF EMULSIONS, INCLUDING CRUDE OIL EMULSIONS,
IN RELATION TO HLB BEHAVIOUR

R. AVEYARD, B. P. BINKS, P. D. I. FLETCHER AND X. YE
School of Chemistry, University of Hull, Hull, England HU6 7RX.

J. R. LU
Physical Chemistry Laboratory, University of Oxford,
South Parks Road, Oxford, England OX1 3QZ.

ABSTRACT. The stability of emulsions depends on, amongst other factors, the equilibrium phase behaviour in the oil+water+stabilising surfactant system. In high hydrophile-lipophile balance (HLB) systems, where an aqueous micellar solution or an oil-in-water (O/W) microemulsion coexists in equilibrium with excess oil phase, homogenisation produces an O/W (macro)emulsion. Low HLB systems on the other hand, in which a water-in-oil (W/O) microemulsion (or reverse micellar solution) coexists at equilibrium with an excess aqueous phase, produce W/O emulsions on agitation. An effective way to break emulsions is to change the HLB by some means. Here we discuss the resolution of crude oil emulsions by a commercial chemical demulsifier in these terms. We also show how the stability of emulsions (both O/W and W/O) formed from a pure alkane varies with the HLB of the system, which is changed by altering the concentration of inorganic electrolyte (NaCl).

1.   Introduction

1.1. FACTORS AFFECTING EMULSION STABILITY

Emulsions are not thermodynamically stable and there are various sources of instability leading to ultimate phase separation [1]. Creaming or sedimentation of the drops occurs (depending on density differences) and is enhanced by flocculation. Drops continually collide as a result of Brownian motion and of gravitational, shear or electrical fields. For coalescence to occur, the thinning film between two approaching drops must reach a critical rupture thickness. Film drainage depends on a number of factors including interfacial tension and tension gradients, as well as the rheological properties of bulk and surface phases. The critical rupture thickness can be understood in terms of surface forces (as embodied in the well-known DLVO theory of colloid stability) and in addition may well be influenced by the presence of surfactant aggregates (micelles or microemulsion droplets), forming percolation pathways between drops as proposed recently by Hazlett and Schechter [2].
     Emulsion stability is also known to be intimately associated with the equilibrium phase behaviour in surfactant+oil+water systems [3-8]. Emulsions which are stabilised by surfactants which tend to aggregate in the aqueous phase are of the oil-in-water (O/W) type. Conversely, water-in-oil (W/O) emulsions are given when surfactant aggregation occurs in the oil phase. The locus of aggregation can be discussed in terms of the system hydrophile-lipophile balance (HLB) and the preferred curvature of surfactant monolayers at oil/water interfaces (see below). A further important variable in emulsion stability is the surfactant

97

J. Sjöblom (ed.), Emulsions – A Fundamental and Practical Approach, 97–110.

concentration. The stabilising monolayer coating the drops must be close-packed to be effective. This is achieved at concentrations close to the critical aggregation concentration (cmc) of the surfactant. At concentrations in excess of the cmc in the emulsion system, aggregates (micelles or microemulsion droplets) exist in the continuous phase (see above). The aggregates can enhance flocculation rates by a depletion mechanism [9,10] and, as already described, can facilitate coalescence by a percolation mechanism.

In this paper we describe how the demulsification of brine-in-crude oil emulsions by a commercial chemical demulsifier can be understood in terms of the system HLB. Our approach is based on previous work where we studied the demulsification of crude oil emulsions using pure or reasonably well-characterised commercial surfactants [11]. We will also present data on the effects of NaCl on creaming and sedimentationin in emulsions formed from equal volumes of brine and nonane. As background we present below a brief account of the relevant phase behaviour in systems containing roughly equal volumes of hydrocarbon and water together with surfactant above its cmc.

1.2. RELEVANT PHASE BEHAVIOUR IN OIL+WATER+SURFACTANT SYSTEMS

For the purposes of example we consider systems containing equal volumes of a normal alkane and water; to this is added surfactant. At surfactant concentrations below that required for aggregation to occur (i.e. the cmc), surfactant monomer distributes between the oil and aqueous phase. The widely-studied twin tail anionic surfactant Aerosol OT (AOT), sodium diethylhexyl sulphosuccinate, distributes strongly in favour of the aqueous phase, even though AOT is very soluble in alkanes but only slightly soluble in water [12]. On the other hand the nonionic surfactant $C_{12}H_{25}(OCH_2CH_2)_5OH$ ($C_{12}E_5$) below the cmc distributes strongly in favour of the alkane [13]. At and beyond the cmc, aggregates form in either the oil or the aqueous phase, or in a third surfactant-rich phase, depending on conditions. Variables which determine the phase in which aggregation occurs are termed HLB variables and include *inter alia* aqueous phase salt concentration (me) and temperature (T). For example AOT in the heptane + aqueous system at 298K forms O/W microemulsion droplets when me (for NaCl) is less than about 0.05M (giving rise to a so-called Winsor I system, although the detailed behaviour is a little more complex than described here [14]). As me is increased in this regime the O/W microemulsion droplets swell and the (positive) curvature of the stabilising surfactant layers around the droplets falls. For me in excess of about 0.05M, the W/O droplets form in the oil phase (a Winsor II system), and their negative curvature increases (size decreases) as me is increased further. In a fairly narrow range of me around 0.05M, a third surfactant-rich phase forms which contains both oil and water and which is in equilibrium with excess oil and aqueous phases containing unaggregated surfactant (Winsor III system). Inorganic salts have the same effect on nonionic surfactant systems, but much higher salt concentrations are needed to effect the Winsor progression. A similar sequence is also obtained by increasing T in alkane + water systems containing a nonionic surfactant e.g.($C_{12}E_5$). The effect of T on the AOT system is similar but reversed, i.e. low T favours Winsor II systems for ionic surfactants.

The HLB of the system (as opposed to the empirical HLB number of the surfactant) is related to the locus of aggregate formation. High HLB systems are those in which O/W microemulsion droplets (or swollen aqueous micelles) form, i.e. Winsor I systems. Low HLB systems are Winsor II systems where a W/O microemulsion coexists with excess aqueous phase devoid of aggregates. The transition from Winsor I to Winsor II systems is termed microemulsion inversion. A ready indication of the occurrence of inversion is the variation of interfacial tension, γ, between the oil and aqueous phases. As is well-known, γ falls with surfactant concentration up to the cmc and thereafter attains a more or less constant value, γ_c. This post-cmc tension passes through a distinct minimum in the Winsor III region.

Emulsions can be prepared by homogenisation of the equilibrium Winsor systems. It is found that the emulsions formed from Winsor I systems are of the O/W type, the same as the equilibrium microemulsion, which constitutes the the continuous phase of the emulsion. Winsor II systems yield W/O emulsions, the continuous phases of which are W/O microemulsions. The intriguing equivalence between microemulsion and emulsion type is as yet largely unexplained. Microemulsion behaviour can be explained in terms of preferred curvature of surfactant layers, which is modulated in some way by the HLB variables.

On a molecular scale however, emulsion drop surfaces are usually effectively planar and it is less easy to see how curvature arguments apply to preferred emulsion type.

One way of breaking an emulsion is to alter the system HLB through the inversion condition, and we describe the resolution of brine-in-crude oil emulsions by a commercial chemical demulsifier in these terms. But the stability of emulsions of the preferred type can also be substantially changed by HLB variables without crossing the inversion condition. In this connection we consider the effect of salt concentration on creaming (in O/W emulsions) and sedimentation (in W/O emulsions) in alkane + water systems stabilised by AOT. The work on the pure systems is a preliminary to the longer term objective of understanding the influence of the curvature properties of surfactant monolayers on emulsion stability and preferred emulsion type.

2. Experimental

2.1. MATERIALS

The crude oil was a water and additive-free Forties sample supplied by the BP Research Centre, Sunbury-on-Thames, England. The demulsifier was a commonly employed commercial preparation, which contained about 30 wt.% added solvents and 70 wt.% active material. The active components of the demulsifier were believed to be mainly ethoxylated phenol-formaldehyde resins. For the demulsification experiments a solution of about 10 wt. % in toluene was used.

Sodium chloride was BDH (England) AnalaR grade, and AOT (diethylhexyl sodium sulphosuccinate) was 99% pure from Sigma; both were used as received. n-Nonane was a puriss sample (>99% pure) from Aldrich (England), which was passed over chromatographic alumina prior to use to remove surface active impurities. Water was distilled once, passed through an Elgastat deionising column and finally through a Milli-Q reagent water system. The surface tension of the water was always 71.9±0.1 mN/m at 298K.

2.2. METHODS

Both the crude oil and the alkane emulsions were prepared using an Ultra Turrax T25 homogeniser with shaft 18G operating at 8 000 rpm. The standard mixing times were 5 minutes for the crude oil emulsions and 2 minutes for the nonane emulsions.

The crude oil emulsions contained 1 volume of aqueous phase to 5 volumes of oil (i.e.17% aqueous phase by volume). They were prepared and kept at 313K to prevent the formation of wax crystals, which are capable of stabilising the emulsions [15]. The emulsions were aged for 3 hours before demulsifier was added in order that the indigenous surface-active agents could adsorb and the typical rigid interfacial film form around the drops. Demulsifier in toluene was added and distributed through the emulsion by gentle shaking for 30 seconds. It was ascertained that the toluene solvent alone had no discernible effect on emulsion stability. The volume of water resolved (coalesced) was determined visually as a function of time from the height of the clear aqueous phase in the stoppered, graduated tubes. It was shown that crude oil emulsion stability in the absence of demulsifier was not affected by the concentration of NaCl in the dispersed aqueous phase.

The nonane-containing emulsions were prepared by first leaving the alkane+aqueous NaCl+AOT (stabiliser) system in a thermostat at 298K overnight. The aqueous to oil phase volume ratio was 1:1 and emulsion stabilities were determined at 298K. In these emulsions it is possible to determine (again visually) the separation of both clear oil and aqueous phases, giving information about creaming or sedimentation (and hence flocculation) as well as droplet coalescence.

Emulsion type was determined conductimetrically using a simple conductivity bridge and a dip cell. Oil-water interfacial tensions were measured by the spinning drop technique using a Kruss SITE 04 instrument; air-solution tensions were obtained by the du Nouy ring method on a Kruss K10 tensiometer.

3. Results and Discussion

3.1. RESOLUTION OF CRUDE OIL EMULSIONS BY A COMMERCIAL DEMULSIFIER

3.1.1. Effect of salt and demulsifier concentration on resolution rates. The effects of both salt (NaCl) and demulsifier concentration on the rate of formation of a clear water phase (reflecting water drop coalescence) have been determined. Sample results are shown in Figures 1a and 1b. It has been convenient to arbitrarily express demulsification efficiency (which varies greatly) in terms of the logarithm of the initial slopes of plots such as those shown in Figure 1; Figure 2 depicts results expressed in this way. As can be seen, for pure water (as well as for low salt concentrations), demulsification initially increases very rapidly with increasing demulsifier concentration (ppm by weight of total emulsion), and then levels off at a roughly constant value (Type A behaviour). For 1M (and 2M) salt however, the rate passes through a distinct maximum, falling rapidly at higher concentrations of demulsifier (Type B behaviour). The increasing stability beyond the maximum exemplifies the well-known phenomenon of "overdosing".

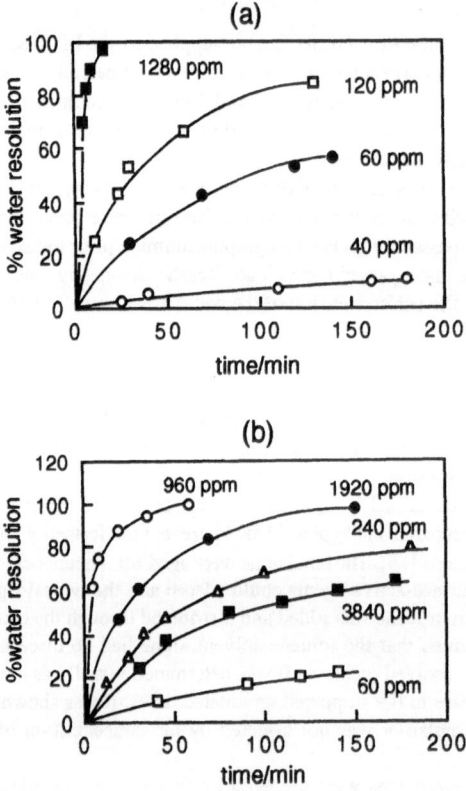

Figure 1. Water resolution curves for water-in-crude oil emulsions at 313K in the presence of varying amounts of commercial demulsifier (see text). (a) is for zero salt concentration, (b) is for 1M aqueous NaCl solution.

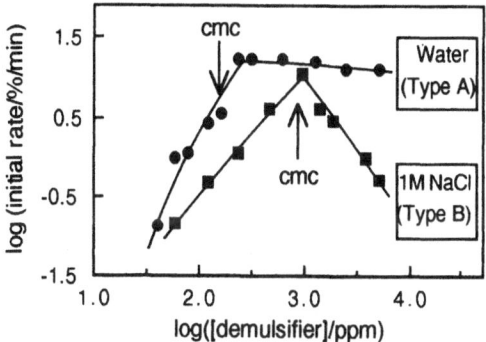

Figure 2. Logarithm of the initial slopes of the water resolution curves (Figure 1) as a function of the logarithm of the demulsifier concentration in the water-in-crude oil emulsions at 313K.

The question we have sought to answer is if the observed behaviour can be explained in terms of simple ideas concerning the HLB (as outlined in Section 1) in crude oil emulsions. For such an interpretation to be valid we need to show in the first instance that the demulsifier, which appears to displace the viscous stabilising layers around the emulsion drops, behaves as a simple nonionic surfactant in crude oil+water systems. If it does, it can be expected that it will exhibit a critical aggregation concentration in the concentration range in which it is an effective demulsifier.

3.1.2. Surfactant behaviour of the commercial demulsifier. To demonstrate the existence of a cmc we have used a tensiometric method. Samples containing demulsifier, with the same volume ratio as the emulsions, were prepared at 313K. The equilibrium clear aqueous phases were separated and their surface tensions determined. Plots are shown in Figure 3 of the surface tension against the logarithm of the demulsifier concentration expressed in ppm by weight (excluding the toluene solvent) of the total system. Remarkably, the behaviour mirrors that of a simple pure surfactant in a pure hydrocarbon+water system. The cmc is given by the break point and rises with salt concentration, as expected for a nonionic surfactant which partitions strongly in favour of the oil phase [16].

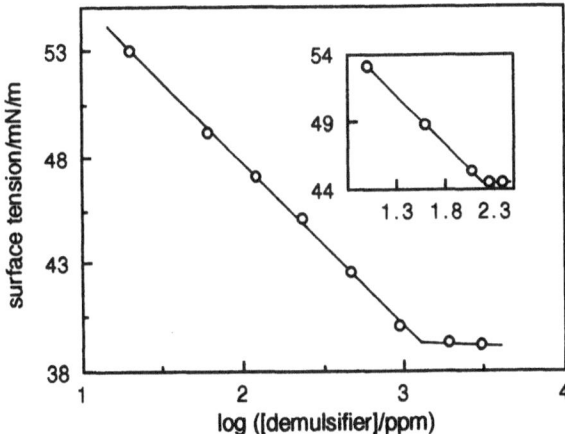

Figure 3. Equilibrium aqueous phase surface tensions vs log [demulsifier] in the original emulsion system (see text). Main graph is for 1M NaCl and inset for zero salt. Breakpoints occur at the cmc.

The cmc values obtained are indicated on the curves in Figure 2. For water (and low salt concentrations) the aggregation concentrations occur close to, but a little lower than the (optimal) demulsifier concentration where the plateau commences. For high salt concentrations the cmc corresponds closely to the maximum in resolution efficiency. This correspondence between the demulsifier concentration at optimal efficiency (C_{opt}) and the cmc of the demulsifier in crude oil systems is also observed when simple surfactants are used as demulsifiers, as seen from Figure 4.

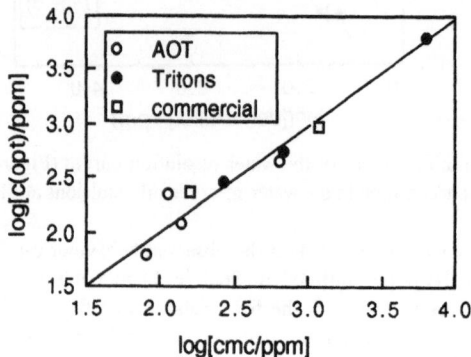

Figure 4.   Correspondence between optimal concentration for demulsification and demulsifier cmc in the emulsion system; line has unit slope. The Triton surfactants are octyl phenyl polyethoxylates with 5 (high cmc), 9-10 and 16 (low cmc) ethoxy groups. Data for AOT and Tritons are from ref [11].

It appears that the demulsifier first displaces the indigenous stabilising layers (the composition of which has been recently discussed by Sjoblom et al.[17]) around the drops. Then, as the concentration increases, the adsorbed demulsifier layer becomes more concentrated until, around the cmc in the system, the layer becomes close-packed. In this case, if the crude oil systems behave in an analogous fashion to pure oil+water+surfactant systems, simple HLB ideas should be applicable. The implications of the data in Figure 2 for concentrations in excess of the cmc are that at low or zero salt concentration, since the water-in crude oil emulsions remain unstable, the preferred emulsion type is oil-in-water (high HLB system). Conversely, for higher salt concentrations, the emulsions become progressively more stable above the cmc of the demulsifier, implying the preferred emulsion type is now water-in-crude oil (low HLB system). It is interesting that the maximum initial rate in Figure 2 (log rate about 1.3) is close to that estimated for diffusion controlled coalescence. This implies surface displacement is fast and the rate determining step is drop coalescence, which proceeds with virtually no energy barrier [11].

3.1.3. *Oil-water interfacial tensions and emulsion conductances.* If the above arguments are reasonable, by analogy with behaviour in pure systems we expect, for crude oil systems with demulsifier present above the cmc, that (a) the oil-water interfacial tensions should pass through a minimum as the salt concentration is varied between zero and 1M, and (b) the preferred emulsion type should invert from O/W to W/O as the salt concentration is increased in the same range. The tension does indeed pass through a minimum at a salt concentration of about 0.6M, as seen in Figure 5. In the same Figure we also show the conductance of the crude oil emulsions formed with different salt concentrations. The emulsions are strongly conducting at low salt concentrations, indicating they are of the O/W type (i.e. water continuous). Inversion to W/O type occurs (i.e. conductance falls drastically) in the region of salt concentration between 0.5 and 0.8M.

In summary, systems of crude oil, aqueous NaCl and the commercial demulsifier behave in all relevant respects like corresponding systems containing pure hydrocarbon and a well-defined low molar mass nonionic surfactant. In high HLB systems (low salt concentration) the efficiency of resolution of water-in crude oil emulsions reaches a plateau close to the cmc of the demulsifier, since the preferred emulsion type is O/W. For higher salt concentrations (in excess of between 0.6 and 0.8M), where the system HLB is low, the W/O emulsions become restabilised above the cmc since W/O is the preferred emulsion type. It is remarkable that such complex systems can be successfully described in such simple terms.

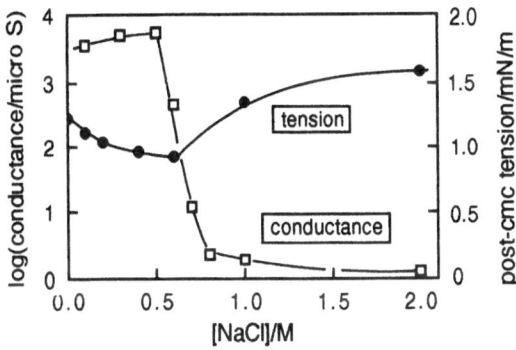

Figure 5. Crude oil-aqueous phase interfacial tensions (above demulsifier cmc) and emulsion conductances, as a function of salt concentration, demonstrating the occurrence of emulsion inversion by salt.

3.2. EQUILIBRIUM PHASE BEHAVIOUR AND EMULSION STABILITY IN SYSTEMS WITH PURE HYDROCARBON AND AQUEOUS NaCl

3.2.1. *Equilibrium phase behaviour.* The demulsification behaviour of crude oil systems is, as described, consistent with the rapid displacement of the indigenous stabilising layers around the emulsion drops by the added surfactant (demulsifier). Thereafter the emulsion stability is determined by the demulsifier and, crucially, the prevailing conditions. We now broaden the discussion to consider equilibrium phase behaviour and concomitant changes in stability of emulsions of the preferred type containing pure hydrocarbon (nonane) and AOT. The system HLB has been varied by changing the aqueous phase salt concentration, me.

It is well-known that emulsion stability is profoundly influenced by proximity of the system to the phase inversion condition [2-8]. We have therefore determined the effect of NaCl concentration on the equilibrium phase behaviour through the Winsor I to III to II progression. We show in Figure 6 the phase

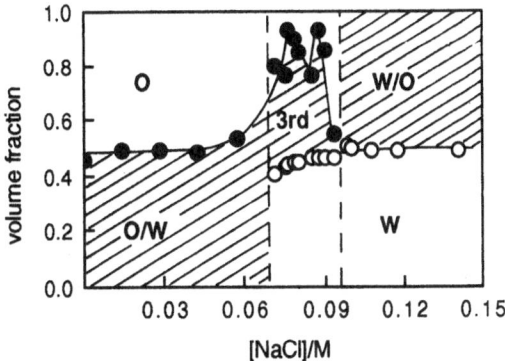

Figure 6. Volume fractions of oil (O), water (W), O/W microemulsions (O/W), W/O microemulsions (W/O), and third phases (3rd) in the aqueous NaCl + AOT system at 298K. Initial volume ratio of 41 mM AOT in nonane to aqueous NaCl was 1:1.

volume fractions of systems prepared by hand-shaking equal volumes of 41 mM AOT in nonane and aqueous NaCl and then leaving the systems for 4 weeks to separate at 298K. The Winsor I/III and III/II boundaries occur at 0.069 and 0.096M NaCl respectively. It is clear that the middle, surfactant-rich, phases (which are fluid and isotropic) contain much more oil than water. This is in marked contrast to similar systems containing dodecane, where the third phases contain mainly water [18].

The surfactant content has been determined in all coexisting phases, using the Hyamine titration method [19], together with the water content of the W/O microemulsions in the Winsor II region (measured by the Karl Fischer technique). In Figure 7 we show the molar ratios R of water to AOT in the Winsor II region and of oil to AOT in the Winsor I region (data from ref. [20]). In Winsor I and II systems the ratio R is directly proportional to the microemulsion droplet core radius [12]. It is clear therefore that the O/W microemulsion droplet size increases with salt concentration up to the I/III boundary and the size of the W/O microemulsion droplets falls with salt concentration beyond the III/II boundary.

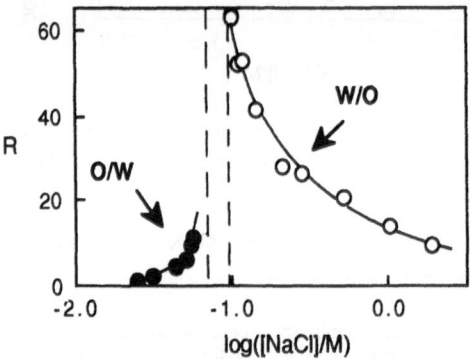

Figure 7. Molar ratio R of dispersed phase to surfactant in O/W (Winsor I region) and W/O (Winsor II region) microemulsions at 298K. In these systems (which are the same as those represented in Figure 6), R is directly proportional to the core radius of the microemulsion droplets [12].

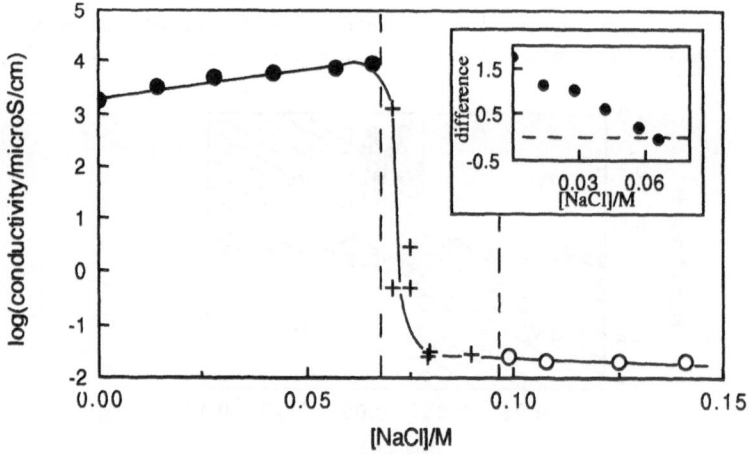

Figure 8. Conductivities of O/W (filled circles) and W/O (open circles) microemulsions and third phases (crosses). Systems same as in Figure 6. Inset shows differences in conductivities (in milliS/cm) between microemulsions and parent salt solutions in the Winsor I region.

The electrical conductivities of the O/W and W/O microemulsions and of the third phases have been determined, and are shown in Figure 8. The O/W microemulsions are highly conducting as expected. It is interesting however that the differences in conductivities of the microemulsions and the corresponding pure NaCl solutions (inset in Figure 8) tend to zero around the Winsor I/III boundary. A possible explanation for this is that the NaCl progressively suppresses the ionisation of the AOT stabilising the microemulsion droplets, this ionisation becoming zero around the phase inversion condition. This effect has been discussed previously in connection with changes in oil/water interfacial tensions with salt concentration through phase inversion [12]. The conductivities of the third phase microemulsions are low (except very close to the Winsor I/III boundary), which is consistent with the high oil content. Conductivity remains very low in the Winsor II region (W/O microemulsions).

3.2.2. Effects of salt on the stability of emulsions formed from equilibrium Winsor systems. In these systems (as opposed to the crude oil emulsions) it is possible in the Winsor I and Winsor II regions to follow the appearance of separated layers of both the continuous phase (reflecting either creaming or sedimentation) and the disperse phase (resulting from the later stages of coalescence). In the 3-phase regime the stabilities of the triphasic and the 3 biphasic emulsions can be studied.

The effects of HLB variables on emulsion stability in the region of the Winsor I/III and III/II boundaries appear to be system dependent. In previous work the systems studied have often contained commercial or impure surfactant [4-6], impure oil [4,5] and added cosurfactant [2,4-6]. In a careful study by Hazlett and Schechter [2], using toluene, aqueous NaCl and sodium dodecyl sulphate, the HLB was varied, to give a Winsor progression through the 3-phase region, by addition of butanol (cosurfactant). The position of the Winsor boundaries with respect to butanol concentration were precisely known. It was found that stability to coalescence was low in the triphasic region but rose sharply at both the Winsor III/I and III/II boundaries. Stability in the Winsor II region at higher butanol concentrations was then found to fall off.

Baldauf et al. [6] studied emulsions containing commercial grade sodium dodecylbenzene sulphonate, aqueous NaCl, nonane and a cosurfactant (secondary butanol or isopropanol), and induced the Winsor progression by changing the salt concentration. They followed both creaming/sedimentation and coalescence rates, and reported that in 2-phase regions the rate of creaming or sedimentation decreased (stability increased) as the Winsor I/III and II/III boundaries were approached whereas the coalescence rates followed the opposite trend. Again, coalescence in the 3-phase region was very rapid and occurred before creaming could be detected. In contrast to these findings, Anton and Salager [5] found (for systems containing ionic or nonionic surfactant, kerosene, cosurfactant and aqueous NaCl of varying concentration) that stabilities with respect to both creaming/sedimentation and coalescence fell as the 3-phase boundaries were approached from the 2-phase regions. Again, low stabilities were encountered in the Winsor III region.

In our present work we have chosen to study a system containing as few components as possible, all of them pure. We can subsequently add other components (e.g. cosurfactant) and hence separate out effects directly due to the added materials. Further, we have studied emulsions in the Winsor I and Winsor II regions where coalescence is slow and followed creaming or sedimentation which are the major sources of instability. We have also determined the stability of the various possible emulsions in the Winsor III region.

Winsor I and Winsor II systems

In the Winsor I region, no separation of oil phases was observed within 6 days, except for 0.06M NaCl where a small amount of oil phase separation became evident after 5 days. However, much more rapid creaming was apparent which was found to be very dependent on salt concentration (Figure 9). For low salt concentrations, in the range 0.02 to 0.04M, there is a long induction period (between 20 and 60 hours) for aqueous phase separation. The creaming rate also passes through a minimum in the same range of salt concentration. This can be appreciated from Figure 10 where the % resolution of the continuous aqueous (i.e. O/W microemulsion) phase after 50 hours is shown as a function of [NaCl].

It is known that the oil-O/W microemulsion interfacial tension falls to very low values as the salt

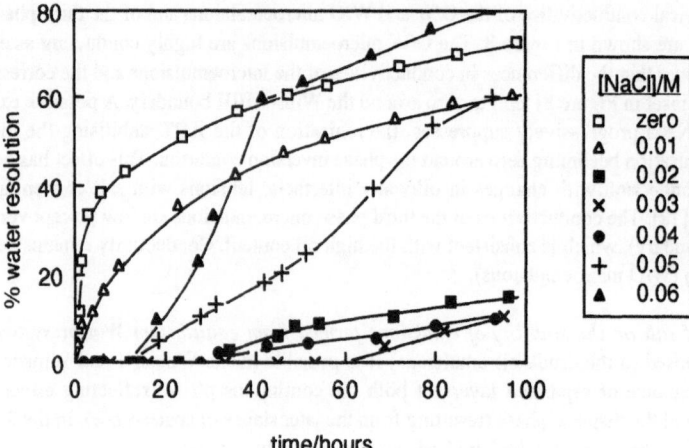

Figure 9. Water phase resolution as a function of time for various salt concentrations for O/W emulsions formed in the Winsor I region (see text).

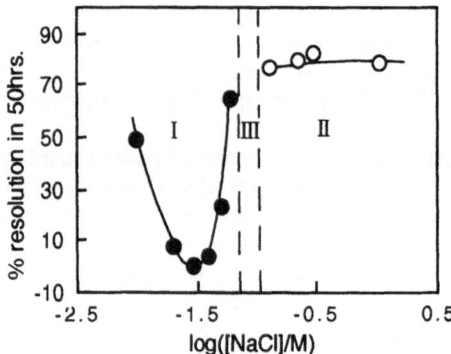

Figure 10. Percentage resolution of continuous phases in 50 hours for emulsions formed from equilibrium Winsor I systems (filled circles) and Winsor II systems (open circles).

concentration increases in this range [18] and we can suppose that the emulsion drop size falls concomitantly [21]. This is likely to contribute to the initial increase in stability (up to 0.03M salt). For the present systems the tension in the absence of salt is about 2mN/m and has fallen to 0.06mN/m at 0.03M salt. The fall in tension from 0.03 to 0.06M salt however is only about 0.05mN/m. Thus, the effect of tension on drop size, and hence creaming rate, is likely to be most pronounced in the region of lower salt concentration.

Salt can be expected to progressively reduce the zeta potential for the O/W emulsions by compressing the electrical double layer [22], which will result in increased flocculation and hence creaming. Further, we have argued earlier that the degree of dissociation of the adsorbed AOT may be falling with increasing salt concentration, again favouring flocculation. However, it should be noted that the surface concentration of AOT increases substantially with salt concentration [23], although this may not result in an increased surface charge density if the degree of dissociation is falling.

In conclusion, the maximum in stability with respect to salt concentration in the Winsor I regime may arise from competing effects due to drop size (increasing stability) and falling electrical repulsion between drops (decreasing the stability).

Stability with respect to sedimentation in the Winsor II region does not vary significantly with [NaCl] in the range 0.13 to 1.05M, about 80% of the continuous W/O microemulsion phase separating in 50 hours (Figure 10). The change in W/O microemulsion-aqueous phase interfacial tension in this range of salt concentration is only about 0.5mN/m, much smaller than over the Winsor I range studied, so the drop sizes are not expected to change much on this account. In addition one expects that the electrostatic effects will be much less pronounced where the continuous phase (of low relative permittivity) contains an insignificant concentration of free ions.

In summary, our findings with respect to creaming/sedimentation rates are at variance with previous results (for different systems) [5,6]. Indeed the results in refs. 5 and 6 themselves show opposite trends. We would remark however that our results are not complicated by the presence of cosurfactant. We plan to determine cosurfactant effects on stability by comparing the present results with those to be obtained for appropriate 5-component systems.

Winsor III systems

In general, emulsion stability in the Winsor III region is low, in agreement with previous reported work [2,4-6]. Homogenisation of the 3-phase systems produces white, non-foaming emulsions. Almost immediately an "upper phase" with a blue appearance resolves, the lower layer remaining as a white emulsion (resolution curves in Figure 11). As seen in the Figure, emulsion stability decreases continuously with increasing [salt]. At longer times (>3 hours), when the boundary between the upper and lower phase is

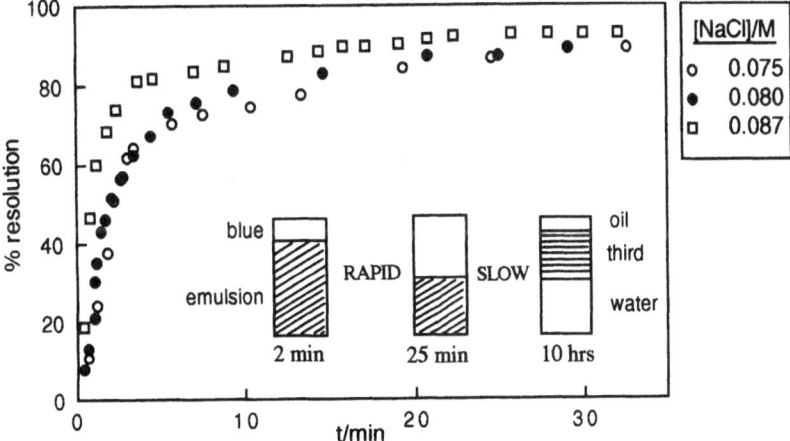

Figure 11. Resolution of the "upper phase" (see text) with time for 3 different salt concentrations in the Winsor III region. The % resolution is expressed relative to the total volume of the third plus oil phases in the equilibrium Winsor systems, to which the upper phase ultimately resolves. The inset shows the appearance of systems over a longer time scale (see text).

effectively stationary, another sequence of visual changes occurs (inset to Figure 11). After about 3 hours the lower phase begins to lose its white intensity throughout the whole phase, taking on a greyish appearance. The phase clears after 8 hours, becoming the equilibrium excess aqueous phase. At longer times (about 10 hours) the upper blue phase resolves into an equilibrium oil phase (top) and the (blue) surfactant-rich phase. The slow resolution of these top two phases must in part be due to the similar densities. It will be recalled that the surfactant-rich phase contains large amounts of nonane (Figure 6).

Homogenisation of the surfactant-rich and excess oil phase pairs, in the same volume ratios as in the equilibrium Winsor systems, produces greyish unstable emulsions which become bluish in a matter of minutes. Then, as in the triphasic systems, a clear oil layer appears after about 10 hours. Emulsions formed from the excess aqueous phases and third phases behave in line with the resolution of the triphasic systems. That is, in the first few minutes a bluish upper phase appears, growing in volume over a period of an hour. The lower phase emulsion becomes completely clear in about 8 hours. Emulsions formed from the excess oil and aqueous phases are again unstable. Initially they are white, and the clear oil phase rapidly resolves. The lower layer then becomes greyish and finally clears over a period of 3 hours.

3.2.3. *Emulsion conductivities*. In Figure 12 we show conductivities of emulsions formed from all the Winsor systems. It is interesting that the conductivities in the Winsor III region remain high (suggesting the emulsions are water-continuous) until very close to the Winsor III/II boundary, where the conductivity attains the same value as that observed for the W/O emulsions in the Winsor II regime. This behaviour contrasts with that observed for the microemulsions (Figure 8), which over much of the Winsor III region have very low conductivities.

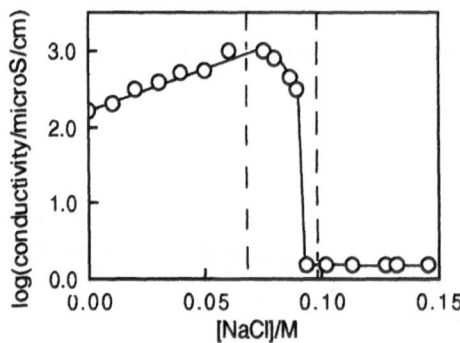

Figure 12. Conductivities of emulsions prepared from Winsor systems at 298K.

4. Summary and remarks on origins of emulsion inversion

The resolution behaviour of brine-in-crude oil emulsions by a commercial demulsifier can be explained, at least for the present systems, in terms of simple HLB concepts. The demulsifier has been shown to display the properties of a simple nonionic surfactant. The crude oil, once the indigenous stabilising layer has been displaced, in relevant respects appears to be like a pure hydrocarbon.

For systems containing pure alkane (nonane), aqueous NaCl and AOT, we have investigated the equilibrium phase behaviour with respect to salt concentration (through the Winsor progression), together with some aspects of emulsion stability for the same systems. For the emulsions formed from Winsor I and II systems we have restricted our study to creaming or sedimentation processes. In the Winsor I system we find a distinct maximum in stability with respect to creaming, with instability rising towards the Winsor I/III boundary. In the Winsor II region, the sedimentation is relatively rapid and virtually independent of salt concentration. These findings contrast with previously reported behaviour of systems containing alcohol cosurfactants. The surfactant-rich phases in the Winsor III region have for the most part low conductivities and high oil content. The triphasic emulsions however remain highly conducting over most of the Winsor III regime.

A number of avenues have been opened up for future research. From a knowledge of equilibrium droplet sizes we can make a quantitative investigation of size effects in depletion flocculation occurring in W/O emulsions. We also plan to identify more clearly the effects which cosurfactants have on emulsion stability. As an extension of the emulsion work we are investigating lifetimes of single drops (both oil and water) at planar O/W interfaces through the phase inversion condition. In this context we will also determine film drainage rates, critical rupture thicknesses and surface forces using the technique described by Francis et al.[24].

Finally we remark on the origins of the phase inversion phenomenon in emulsions whereby equilibrium Winsor I systems yield O/W emulsions and Winsor II systems give W/O emulsions. It has often been suggested that the inversion is associated with the Gibbs elasticity of the interdroplet films. In emulsions where the surfactant is largely in the dispersed phase (rather than in the films), the surfactant can substantially reduce the dilational elasticity, contributing to the ease of droplet coalescence [1]. It has been assumed that this, taken together with Bancroft's rule to the effect that the preferred emulsion type is that where the surfactant resides mainly in the continuous phase, explains the origin of emulsion inversion. We have pointed out in §1.2 however that emulsion type is associated not with the overall distribution ratio of surfactant between oil and water, but with the phase in which surfactant aggregation occurs. That phase, which is the continuous phase of the preferred emulsion type, may contain negligible amounts of surfactant if the surfactant concentration in the system is just above the cmc. On the other hand, the droplet phase in the preferred emulsion type may contain a high concentration of unaggregated surfactant. This is the case for some nonionic surfactants in (O/W emulsion) systems at temperatures below the phase inversion temperatures, as discussed in §1.2.

The correspondence between macro- and microemulsion inversion strongly suggests the possibility that monolayer curvature properties are implicated in determining preferred emulsion type. These properties are now being quantified from direct studies of monolayer behaviour [18] and from the investigation of microemulsion kinetic processes [25]. Future research should therefore be directed towards linking equilibrium phase behaviour, and associated surfactant monolayer properties, with preferred emulsion type.

Acknowledgements

We express our sincere thanks to BP Research (Sunbury, England) for considerable support including grants for J.R.L. and B.P.B.

References

1. Carroll, B. J. (1976) Stability of Emulsions and Mechanisms of Emulsion Breakdown, in Surface and Colloid Science, E. Matijevic (ed.), Wiley, New York, pp 1-67.
2. Hazlett, R. D. and Schechter, R. S. (1988) Colloids and Surfaces, 29, 53-69.
3. Shinoda, K. and Friberg, S. (1986) Emulsions and Solubilisation, Wiley, New York.
4. Vinatieri, J. E. (1980) Soc. Petroleum Eng. J. 20, 402-406.
5. Anton, R. E. and Salager, J-L. (1986) J. Colloid Interface Science, 111, 54-59.
6. Baldauf, L. M. , Schechter, R. S., Wade, W. H. and Graciaa, A. (1982) J. Colloid Interface Science, 85, 187-197.
7. Selle, M. H. , Sjoblom, J. and Skurtveit, R. (1991) J. Colloid Interface Science, 144, 36-44.
8. Friberg, S.E. and Solans, C. (1986) Langmuir, 2, 121-126.
9. Ma, C. (1987) Colloids and Surfaces, 28, 1-7.
10. Aronson, M. P. (1989) Langmuir, 5, 494-501.
11. Aveyard, R., Binks, B. P., Fletcher, P. D. I. and Lu, J. R. (1990) J. Colloid Interface Science, 139, 128-138.
12. Aveyard, R., Binks, B. P., Clark, S. and Mead, J. (1986) JCS Faraday Trans. I, 82, 125-142.
13. Aveyard, R., Binks, B. P., Clark, S. and Fletcher, P. D. I. (1990) JCS Faraday Trans., 86, 3111-3115.
14. Ghosh, O. and Miller, C. A. (1987) J. Phys. Chem, 91, 4528-4535.
15. Thompson, D. G., Taylor, A. S. and Graham, D. E. (1985) Colloids and Surfaces, 15, 175-189.
16. Horsup, D. I. (1991) Ph. D. Thesis, University of Hull.

110

17. Sjoblom, J., Urdahl, O., Hoiland, H., Christy, A. A. and Johansen, E. J. (1990) Progr. Colloid Polym. Sci., 82, 131-139.
18. Binks, B. P., Kellay, H. and Meunier, J. (1991), Europhys. Lett., in press.
19. Reid, V. W., Longman, G. F. and Heinerth, E. (1967) Tenside, 4, 292-304.
20. Fletcher, P. D. I. (1987) JCS Faraday Trans. I, 83, 1493-1506.
21 Shinoda, K. and Saito, H. (1969) J. Colloid Interface Science, 30, 258-263.
22. Van den Tempel, M. (1953) Rec. Trav. Chim., 72, 419-432.
23. Aveyard, R., Binks, B. P. and Mead, J. (1986) JCS Faraday Trans. I, 82, 1755-1770.
24. Francis, G. W. , Fisher, L. R. , Gamble, R. A. and Gingell, D. (1987) J. Cell Science, 87, 519-523.
25. Clark, S., Fletcher, P. D. I. and Ye, X. (1990) Langmuir, 6, 1301-1309.

INTERFACIAL RHEOLOGY OF SURFACTANT SOLUTIONS

J.C. EARNSHAW AND A.C. McLAUGHLIN
Department of Pure and Applied Physics
The Queen's University of Belfast
Belfast BT7 1NN
Northern Ireland

ABSTRACT. Surface light scattering has been developed as a probe of the viscoelastic response of liquid surfaces. Results from experiments on dilute solutions of a homologous series of primary alcohols indicate that the effective value of the surface dilational viscosity is negative for these systems. Such viscosities indicate that the dilational modes of the film on the surface are destabilized by competition of diffusive exchange between bulk and surface with adsorption/desorption kinetics.

1. Introduction

The factors affecting the stability of foams are not yet completely understood, but they must certainly include interfacial rheology. The relationship between the two is, as yet, somewhat obscure. This is partly due to the complexity of the phenomena which must be considered within the sphere of interfacial viscoelasticity. In recent years some progress has been made in establishing a fundamental basis for this subject [1, 2], but quantitative calculations *ab initio* are still not possible. It has been shown that a fluid interface may exhibit up to five separate, independent viscosities, probably best regarded as interfacial excess quantities. It is not possible to define these viscoelastic moduli uniquely, but a useful approach follows Goodrich's ideas, originally devised for molecular surface films [3]. Of the five possible moduli for a surface or interface we will here only consider two: details will be given below.

Dynamic processes involved in solutions of surfactant molecules can affect the interfacial rheology, and hence may influence foam and emulsion stability. In particular it has been shown that the stability of the surface modes may be affected by such processes. The destabilising influence depends upon the processes involved, and the time scales associated with them. Recent theoretical studies [4, 5] have suggested that relatively high frequency surface modes are more susceptible to such destabilisation. Thus in experimental investigations of these matters observation of high frequency surface deformations may help in the identification of the processes involved.

Light scattering from thermally excited capillary waves is well established as an experimental probe of the high frequency response of liquid surfaces and interfaces. For a liquid surface, as considered below, the light scattering can yield information on both the tension and the dilational modulus of the surface, due to the coupling between capillary and dilational modes.

The remainder of this paper comprises a summary of the basic theory involved, followed by a brief résumé of the experimental background. The results to be considered derive from a study of aqueous solutions of the higher alcohols, and present novel and instructive aspects. The discussion elucidates their origin and significance, drawing together the various strands sketched above.

J. Sjöblom (ed.), Emulsions – A Fundamental and Practical Approach, 111–121.
© 1992 *Kluwer Academic Publishers.*

2. Theoretical Background

Various theoretical treatments of interfacial waves [6, 7, 8] have been shown to be equivalent. We require only the briefest summary, and adapt Kramer's treatment [8]. While we will be concerned with liquid surfaces, it is convenient to write the equations in the form for a fluid interface. The dispersion equation governing interface waves derives from the boundary conditions imposed at the interface. These can be written as:

$$[\eta(q-m) - \eta'(q-m')] E + \left\{ \frac{\varepsilon q^2}{\omega} + i\left[\eta(q+m) + \eta'(q+m')\right] \right\} D = 0, \tag{1}$$

$$\left\{ \frac{\gamma q^2}{\omega} + \frac{g(\rho - \rho')}{\omega} - \frac{\omega(\rho + \rho')}{q} + i\left[\eta(q+m) + \eta'(q+m')\right] \right\} E$$
$$+ [\eta(q-m) - \eta'(q-m')] D = 0. \tag{2}$$

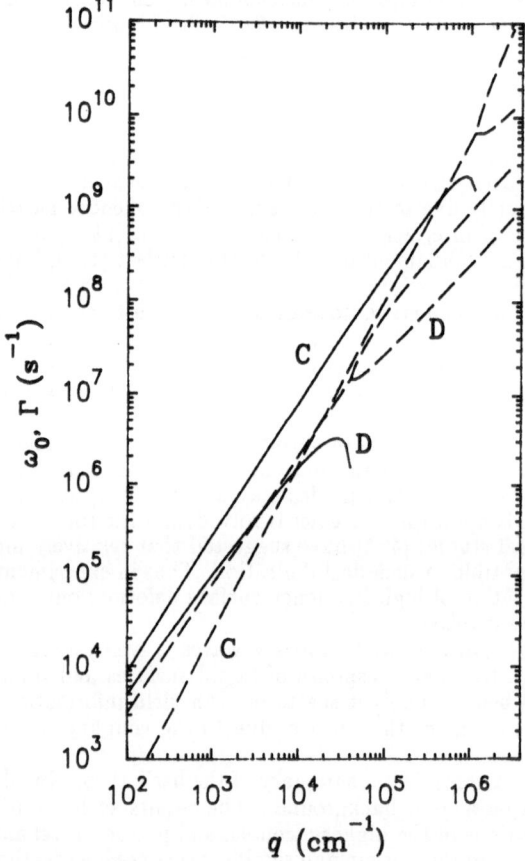

Figure 1: The dispersion of capillary (C) and dilational (D) waves for a surface having $\gamma_0 = 65.0$ mN/m and $\varepsilon_0 = 5$ mN/m. ———, ω_0; — — —, Γ.

The primed/unprimed quantities refer to the upper/lower fluid; $m = \sqrt{q^2 + i\omega\rho/\eta}$, m' being similar. The physical properties of the surface enter the equations through the tension γ and the surface dilational modulus ε ($= d\gamma/d\ln A$, A being the surface area per surfactant molecule); other symbols have their usual meanings.

The dispersion equation of the surface waves is obtained by setting the determinant of the coefficients of the above homogeneous set of equations equal to zero. This equation relates the temporal evolution of waves (represented by the complex frequency $\omega = \omega_0 + i\Gamma$) to the surface wavenumber q. It is now well known that this equation has two sets of physically significant roots. At low q these roots are complex conjugates, representing damped oscillations. One set of roots represents the established capillary waves, the exact dispersion behaviour being approximated by

$$\omega = \sqrt{\gamma q^3/\rho} + i2\eta q^2/\rho. \tag{3}$$

The other set of roots describes dilational surface waves, for which

$$\omega = \frac{1}{2}(\sqrt{3} + i)(\varepsilon^2 q^4/\eta\rho)^{1/3} \tag{4}$$

provides an approximation to the dispersion behaviour [8].

Figure 1 shows the behaviour of both types of wave; at relatively low q the approximate dispersion behaviours are clearly visible. Above a critical value of q (different for each case) both modes become overdamped, the dispersion equation having two imaginary roots.

2.1 SURFACE VISCOSITIES

Surface or interfacial dissipative effects can be incorporated into the formalism by expanding the tension and dilational modulus as linear response functions [1]:

$$\gamma = \gamma_0 + i\omega\gamma', \tag{5}$$
$$\varepsilon = \varepsilon_0 + i\omega\varepsilon'. \tag{6}$$

The tension and dilational elastic modulus are now written as γ_0 and ε_0 respectively, whilst the primed quantities represent interfacial viscosities. Neither is the conventional surface viscosity which governs shear in the interfacial plane: γ' is the surface viscosity affecting shear transverse to the interface, while ε' governs dilation within the interface. There is as yet no concensus on the nature of the surface viscosities; they are perhaps best regarded as surface excess quantities.

The primary effect of each surface viscosity is to increase the dissipation, and hence the temporal damping of that mode which it directly affects. Due to the coupling of the two surface modes, they do however affect the propagation of the 'other' mode to some extent.

2.2 MODE COUPLING

Equations 1 and 2 resemble those governing two classical coupled oscillators, D and E relating to the coordinates of the oscillators. These quantities are in fact closely connected to the velocities of an interfacial element of the fluid: $D = v_x/iq$ and $E = v_z/q$. From equations 1 and 2 it is clear that, unless the density and viscosity of the two fluids adjoining the interface are equal, the capillary (commonly regarded as transverse) waves couple to the dilational (predominately longitudinal) waves. As we will here be concerned with air/water interfaces we will necessarily have to consider the effects of this coupling. The first point of note is that the capillary wave involves horizontal as well as transverse motion (and vice versa for the dilational mode): the vertical and horizontal motions cannot be treated in isolation. The other established consequence [6] is the occurrence of a resonance between the two modes when their frequencies are equal (Figure 2).

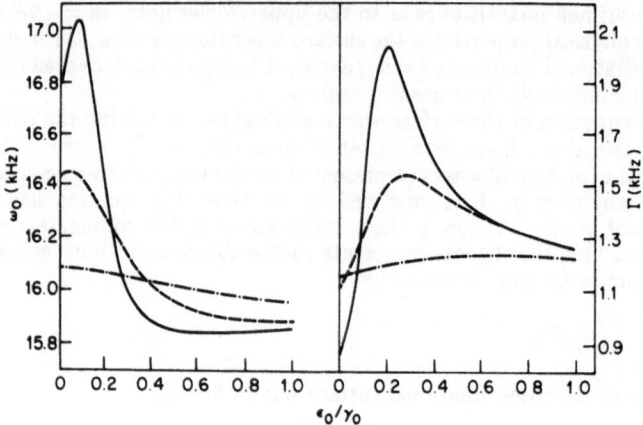

Figure 2: Changes in ω_0 and Γ for $q = 540$ cm^{-1} induced by changes in ε_0 relative to γ_0. The series of lines show how the resonant interaction is modified by increasing ε' (0, 1 and 5×10^{-4} mN s/m).

It has recently been shown [9] that for certain values of the surface viscosities the coupling between the two surface modes can cause the appearance of mixed modes, not identifiable as capillary or dilational in nature. These may arise when the damping values of the surface modes approach each other, enhancing the effect of the coupling terms of equations 1 and 2. Now it happens that the damping of the dilational modes generally exceeds that of the capillary waves, so convergence of the Γ values can only be achieved by increase of γ' (since ε' will increase Γ for the dilational mode, causing divergence). Under the appropriate combination of surface properties mode mixing occurs, and the modes swap their character. This situation is illustrated in Figure 3: that mode which is predominately capillary in nature at low q becomes dilational at high q and vice versa.

2.3 SURFACTANT SOLUTIONS

The above formalism is appropriate to an insoluble molecular film at a fluid surface or interface, whose physical properties can be encapsulated by the four surface parameters (γ_0, γ', ε_0 and ε'). When the surfactant molecules constituting the film are soluble, other processes come into play. Foremost amongst these is diffusive interchange of surfactant molecules between the surface and solution in a bulk fluid [6]. This leads to only relatively minor conceptual changes. In particular the dilational modulus ε must be treated as frequency dependent. The particular form derived is [6]

$$\varepsilon_0 = \frac{-\partial \gamma_0}{\partial \ln \Gamma_s} \frac{1 + \Omega}{1 + 2\Omega + 2\Omega^2} \tag{7}$$

$$\varepsilon' = \frac{-\partial \gamma_0}{\partial \ln \Gamma_s} \frac{\Omega}{\omega(1 + 2\Omega + 2\Omega^2)}. \tag{8}$$

In these equations Γ_s is the interfacial adsorption of the surfactant, and

$$\Omega = \frac{\partial c}{\partial \Gamma_s} \sqrt{\frac{D}{2\omega}}, \tag{9}$$

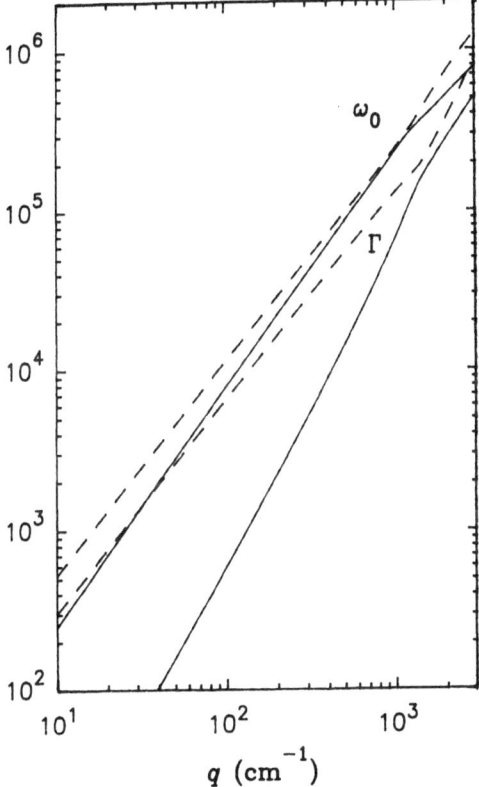

Figure 3: The dispersion of both surface modes for a surface having $\gamma_0 = 65.0$ mN/m and $\varepsilon_0 = 15$ mN/m as modified by a non-zero transverse shear viscosity: $\gamma' = 4 \times 10^{-5}$ mN s/m. The solid/dashed lines are identifiable as capillary/dilational wave-like at low q; changing to the opposite character at high q.

where c is the bulk concentration of the surfactant in solution, relates to the time constant for the diffusive interchange. The above functional form, incorporated into the standard surface wave dispersion equation, has found some degree of support in various studies (e.g. [10]). However, in some other experiments such agreement has not been forthcoming (e.g. [11]).

Various workers have elaborated this simple picture by introducing further processes, such as desorption to a vapour phase, convection and adsorption/desorption kinetics. The primary conclusion has been that under some circumstances the competition between diffusive interchange and these further processes can lead to destabilisation of the surface modes [4, 5]. In particular it appears that competition between diffusion and adsorption/desorption kinetics can destabilise the dilational waves at rather high frequencies, ultimately leading to their instability and nonlinear growth. The full theoretical treatment of this competition unfortunately contains several parameters whose values are by no means well established, and hence is not at present suited to rigorous computational analysis. However the point of present concern is clear: if the dilational waves are destabilized, their associated Γ values must be decreased below those expected in the absence of the pro-

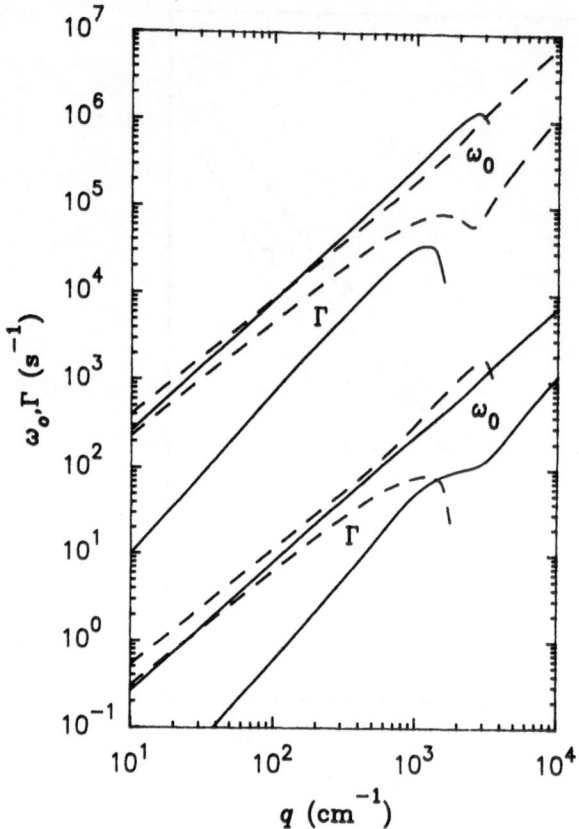

Figure 4: Surface wave dispersion for $\gamma_0 = 70$ mN/m and $\varepsilon' = -2 \times 10^{-5}$ mN s/m (solid lines represent capillary and dashed line dilational waves). The lower set of lines (scaled down by 10^3 for clarity) are for $\varepsilon_0 = 15$ mN/m. The upper set of lines is for $\varepsilon_0 = 10$ mN/m: mode mixing evidently occurs for $q \sim 10^3$ cm^{-1}. All other fluid properties as for water at 29.0°C.

cesses leading to the destabilisation. Therefore the dilational wave damping will approach that of the capillary waves, leading to an increase in the effect of the mode coupling, and potentially to the appearance of mode mixing.

As a full computational investigation of the consequences of the novel processes postulated for soluble surfactants is not possible, we have pursued an alternative. This seeks to approximate the effects of destabilising influences upon the dilational waves by making the dilational viscosity ε' negative. It will be recalled that positive values of ε' caused the dilational wave dissipation (and hence Γ) to be increased: it follows that negative ε' (while physically quite unrealistic) will mimic the destabilisation of that mode. Indeed allowing $\varepsilon' < 0$ does lead to a reduction of Γ for the dilational mode.

The effects of negative ε' upon the surface wave propagation are illustrated in Figure 4 for two cases. In both cases the Γ values of the surface modes are brought closer together. For $\varepsilon_0 = 15$ mN/m the convergence is insufficient to induce mode mixing. The dilational wave damping progressively deviates from the expected dependence upon q, becoming negative (mode instability) at $q \sim 1800$ cm^{-1}. The capillary wave damping departs from the expected q^2 variation over the range of q at which the destabilisation of the dilational waves becomes noticeable, but thereafter revert to this capillary wave like behaviour. Basically Γ for the capillary waves is raised at low q due to the resonance with the dilational modes, but at high q recovers its 'uncoupled' value as by then the dilational wave is completely destabilized, and is no longer resonant with the capillary wave. When ε_0 is rather lower the two Γ values are initially closer, the two modes being nearer the resonant state, and in this case the coupling suffices to induce mode mixing at about $q \sim 1000$ cm^{-1}. This is accompanied by a cross-over in mode characteristics, that mode which is initially dilational at low q becoming capillary in nature and vice versa.

3. Experimental Methods

The dispersion of high frequency ($10^3 < \omega_0 < 10^6$ s^{-1}) thermally excited capillary waves can be observed by quasi-elastic light scattering. In principle dilational waves will also scatter light, but the intensity is much lower than that scattered by the capillary waves except perhaps in exceptional circumstances. In our experiments [12] light from an Ar$^+$ laser incident upon the liquid surface is scattered about the specular reflection by capillary waves. The scattered light, shifted in frequency by the ripple, is mixed at the photodetector with a beam of light at the original laser frequency. The detector output is modulated by beating between scattered and reference light; its frequency spectrum is just the power spectrum of the thermally excited waves, and is determined in our experiments by photon correlation. The observed correlation functions can be analysed to yield unbiased estimates of ω_0 and Γ.

These quantities depend upon, but do not suffice to determine, the surface properties. The exact spectrum of thermally excited capillary waves on a monolayer covered surface [7], which is a function of these four properties, can be used to estimate their values from the observed correlation functions [13]. Now this functional form cannot be correct for soluble surfactants, because the various processes cited above are not incorporated into it. However the surface properties estimated in this way are useful in understanding the observed dispersion behaviour for surfactant solutions.

4. Results and Discussion

4.1 RESULTS

The variations of ω_0 and Γ have been measured (e.g. [12]) in many studies of capillary waves on the free surfaces of various liquids. In all cases they agree well with the predictions of the dispersion equation derived from Equations 1 and 2, with only the tension out of the four surface properties being non-zero. As an example the dispersion behaviour measured in our experiments for heptane is compared in Figure 5 with the exact theoretical predictions based on the known properties of that fluid. The agreement of the data with theory indicates the essentially correct functioning of our apparatus.

Similarly there have been several studies of *insoluble* molecular monolayers (e.g. [7, 14, 15, 16]). In several cases the complex frequency of the capillary waves has been interpreted in terms of the four surface properties [7, 14]: this is obviously only possible using *a priori* assumptions, which are often unjustified, and possibly unjustifiable. When the photon correlation functions have been analysed in terms of the surface properties directly, more consistent and reliable results have been obtained [15, 16]. Two features of such studies which are noteworthy in the present context are that, in at least some areas of the monolayer phase diagrams, the transverse shear viscosity γ' was found to non-zero, and that the dilational viscosity ε' was positive (occasional negative values were always zero within errors, and occurred in regions where $\langle\varepsilon'\rangle$ was negligible).

Our studies of soluble surfactants have concentrated upon dilute solutions of some members of the homologous series of normal alcohols. Here rather different results were found. Under all conditions for all of these systems the fitted values of γ' were zero. Further, in many cases the dilational viscosity was found to be negative. The latter finding, which we understand in the sense outlined above as indicating some destabilisation of the dilational waves, is particularly significant.

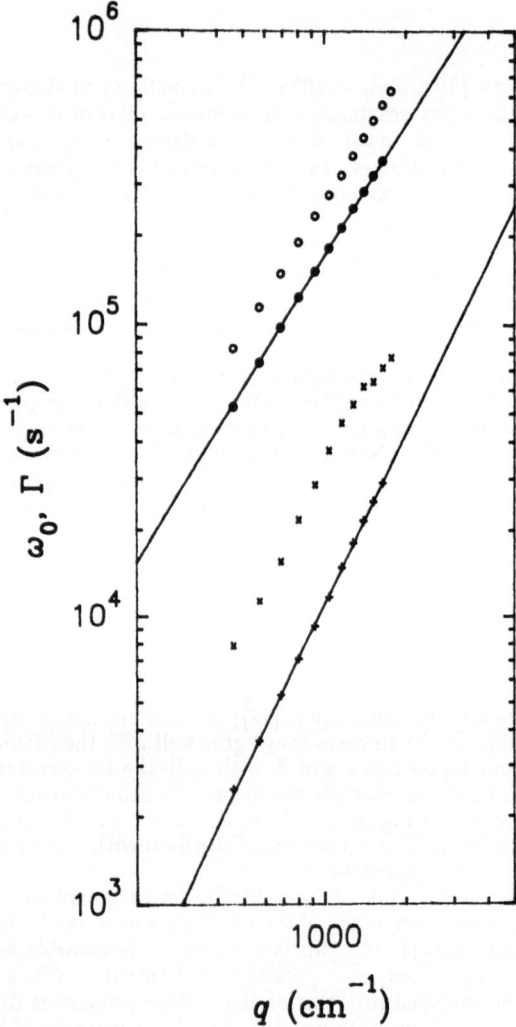

Figure 5: Capillary wave dispersion observed for two fluids. (a) Heptane at 17.5°C: • and +; the lines are predictions from the dispersion equation, with no fitting involved. (b) 4.53×10^{-4}M aqueous solution of n-heptanol at 29.0°C: o and ×.

The measured dispersion behaviour over a range of surface wavenumbers ($400 < q < 2000$ cm^{-1}) for one particular solution is shown in Figure 5. Whilst the frequencies accord reasonably well with the expected $q^{3/2}$ variation, the Γ values depart markedly from the expected q^2 behaviour as q rises above 1200 cm^{-1}. Such deviations from the 'classical' behaviour have never been reported before for any system, to our knowledge.

When this data was analysed to extract the four surface properties, the dilational viscosity was, indeed, found to be negative. The tension (70.3 ± 0.3 mN/m) agreed with Wilhelmy plate values, and the dilational modulus ε_0 was 11.2 ± 0.3 mN/m. Given the probable inappropriateness of the capillary wave spectrum used, due to possible presence of destabilising influences upon the dilational modes, the exact values of the 'effective' dilational properties may not be very significant. We concentrate upon the apparent negative nature of ε'.

4.2 DISCUSSION

We associate the apparent anomalous dispersion behaviour of capillary waves on these surfactant solutions with the associated negative values of the dilational viscosity. Useful comparisons can be made between the observed behaviour for the 4.53×10^{-4}M n-heptanol solution (Figure 5) and the theoretical predictions of Figure 4. The data clearly resemble the latter variations for $\varepsilon_0 = 15$ mN/m. The results do not appear to exhibit mode mixing, but do reflect the unusual consequences of the enhanced coupling between the capillary and dilational modes.

Quantitative comparisons between these two graphs are somewhat difficult, and so the experimental data are replotted in Figure 6, normalised by the first order approximations for the capillary wave ω_0 and Γ (equation 3). The theoretical variations referred to are also replotted, similarly normalised. The qualitative agreement of the data with the theory is very good. No attempt has been made to 'fit' the theoretical behaviour to the data; better agreement can be achieved by varying the values of ε_0 and ε'. However it seems that it is not possible to find a single set of constant surface properties which yield a fit with the data over all q. This does not appear unreasonable, as the degree of destabilisation of the dilational waves will vary with frequency, so that the 'effective' surface properties will not be constant over the range of q studied.

Unfortunately the extended theory incorporating the various effects which can compete with the diffusional exchange between surface and bulk solution, contains several parameters whose values are as yet rather ill-defined. It is, therefore, not yet in a form for convenient computation. Similarly there is not yet available a mathematical form for the spectrum of thermally excited capillary waves on the surface of a solution in which all these processes are occurring. However the accord of the data with the approximate theoretical approach sketched above is impressive. In particular it is only possible to reproduce the abrupt change in behaviour of the capillary wave damping seen in Figure 6 if we assume $\varepsilon' < 0$.

The data can thus only be understood if we postulate that the stability of the dilational waves is reduced. The capillary wave frequency at which this destabilisation occurs exceeds 10^5 s^{-1}, in good accord with theoretical predictions of the range within which the main destabilising influence is the competition between the diffusive exchange and adsorption/desorption kinetics.

5. Conclusions

In summary we have shown computationally that effects which may occur in surfactant solutions, leading to destabilisation of the dilational surface waves, may have profound consequences upon the observable dispersion properties of the capillary waves. These effects can be particularly significant near the resonance between the two surface modes, where the coupling between the modes has its greatest effects. Experiments on dilute solutions of long-chain alcohols have indeed revealed anomalies in the capillary wave behaviour. These are of the exact form expected for the effects of destabilisation of the dilational

120

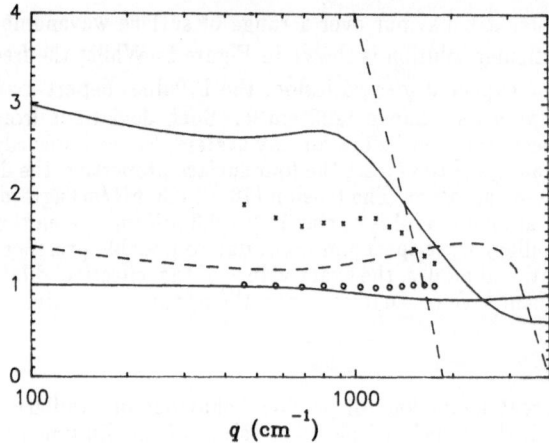

q (cm^{-1})

Figure 6: The heptanol data of Figure 5 compared to the theoretical dispersion of Figure 4 for $\varepsilon_0 = 15$ mN/m. The capillary wave behaviour is shown be the solid lines, the dilational waves by the dashed lines.

waves. The anomalous behaviour occurs at wave frequencies similar to those predicted for a specific destabilisation mechanism: competition between the diffusive exchange and adsorption/desorption kinetics. It thus appears as if the system is undergoing diffusional interchange between the bulk solution and a fluid layer immediately beneath the surface, at which point there apparently occurs some barrier opposing immediate adsorption at the surface itself. More extensive experimental studies are under way, to further elucidate this picture.

Various other workers have observed unusual effects in the propagation of capillary waves on the surfaces of surfactant solutions (e.g. [11]). In particular it has often been found difficult to reconcile the experimental data with Lucassen's theory based on simple diffusional exchange between surface and bulk solution [6]. The interpretation of these effects has not always been transparent, and rather *ad hoc* explanations have been advanced. These difficulties may in part have arisen because the data were only analysed in terms of the capillary wave dispersion, and no attempt was made to infer the effective surface properties in the direct way which yielded $\varepsilon' < 0$ here. It may also be remarked that the present study has involved larger surface wavenumbers ($400 < q < 2000$ cm^{-1}) than those previously studied for such systems. As we have seen, the destabilisation of the dilational waves appears most marked at high frequencies, not observed in the previous work. Reconsideration of these earlier studies in the light of the present work may suggest a common origin for the various discrepancies.

A more general conclusion which can be drawn from the present work is that the coupling between the surface modes cannot be neglected. This implies that theoretical studies of the stability of the modes must be reconsidered, as it has been common to treat them separately. This cannot be correct. There will be further implications for considerations of the stability of foams and interfaces. The novel processes in which surfactants may participate must markedly afffect the entire question of interfacial stability.

References

[1] Goodrich, F.C. (1981) 'The theory of capillary excess viscosities', Proc. R. Soc., A374,

341–370.

[2] Baus, M. (1982) 'Elastic moduli of a liquid-vapour interface', J. Chem. Phys., 76, 2003–2009.

[3] Goodrich, F.C. (1962) 'On the damping of water waves by monomolecular films', J. Phys. Chem., 66, 1858–1863.

[4] Chu, X.-L. and Velarde, M.G. (1986) 'Sustained transverse and longitudinal waves at the open surface of a liquid', Physicochem. Hydrodynamics, 10, 727–737.

[5] Hennenberg, M., Chu, X.-L., Sanfeld, A. and Velarde, M.G. (1991) 'Transverse and longitudinal waves at the air liquid interface in the presence of an adsorption barrier', preprint.

[6] Lucassen-Reynders, E.H. and Lucassen, J. (1969) 'Properties of capillary waves', Adv. Colloid Interface Sci., 2, 347–395.

[7] Langevin, D. (1981) 'Light-scattering study of monolayer viscoelasticity', J. Colloid Interface Sci., 80, 412–425.

[8] Kramer, L. (1971) 'Theory of light scattering from fluctuations of membranes and monolayers', J. Chem. Phys., 55, 2097–2105.

[9] Earnshaw, J.C. and McLaughlin, A.C. (1991) 'Waves at liquid surfaces: coupled oscillators and mode mixing', Proc. R. Soc., in the press.

[10] Lucassen, J. and van den Tempel, M. (1972) 'Dynamic measurements of dilational properties of a liquid interface', Chem. Eng. Sci., 27, 1283–1291.

[11] Thominet, V., Stenvot, C. and Langevin, D. (1988) 'Light scattering study of the viscoelasticity of soluble monolayers', J. Colloid Interface Sci., 126, 54–62.

[12] Earnshaw, J.C. and McGivern, R.C. (1987) 'Photon correlation spectroscopy of thermal fluctuations of liquid surfaces', J. Phys. E: Sci. Instrum., 20, 82–92.

[13] Earnshaw, J.C., McGivern, R.C., McLaughlin, A.C. and Winch, P.J. (1990) 'Light scattering studies of surface viscoelasticity: direct data analysis', Langmuir, 6, 649–660.

[14] Hård, S. and Neuman, R.D. (1987) 'Viscoelasticity of monomolecular films: a laser light scattering study', J. Colloid Interface Sci., 120, 15–29.

[15] Earnshaw, J.C., McGivern, R.C., and Winch, P.J. (1988) 'Viscoelastic relaxation of insoluble monomolecular films', J. Phys. Paris, 49, 1271–1293.

[16] Earnshaw, J.C. and Winch, P.J. (1990) 'Viscoelasticity of monolayers of n-pentadecanoic acid: a light scattering study', J. Phys: Condensed Matter, 2, 8499–8516.

STUDIES OF INTERACTIONS BETWEEN SURFACES IMMERSED IN CRUDE OILS

H.K. CHRISTENSON
Department of Applied Mathematics
Research School of Physical Sciences
Australian National University
Canberra, A.C.T. 2601
AUSTRALIA

ABSTRACT. The surface force apparatus is an instrument that allows accurate measurements of the force as a function of separation between two surfaces to be carried out. Both equilibrium forces and dynamic interactions due to viscous effects can be studied and the refractive index of liquids in thin films can be determined. This paper describes some results of relevance to the study of crude oil systems, in particular measurements with mica surfaces immersed in crude oils from Australian reservoirs. The force between two solid surfaces across a crude oil is quite different from the oscillatory solvation forces found in pure nonpolar liquids or the short-range repulsive forces encountered in mixtures of low-molecular weight hydrocarbons. In crude oils the interaction is dominated by repulsive forces of longer range, due to the adsorption of high-molecular weight components such as asphaltenes to the surfaces. There are time-dependent effects related to slow adsorption and subsequent rearrangement of adsorbed species. In contrast to the case with pure hydrocarbon liquids, where water acts to make the interaction everywhere attractive, the addition of water has negligible influence on the long-range forces in crude oils because the surfaces are rendered hydrophobic by asphaltene adsorption. Only if the surfaces are forced together under very high loads does adsorption of water in some cases lead to adhesion between the surfaces. The viscosity of crude oils in films down to thicknesses of the order of tens of nanometres is equal to the bulk value and the only measurable deviation is an outward shift in the plane of shear of a few nanometres caused by adsorption.

1. Introduction

The interaction between two surfaces across a liquid medium governs the stability of colloidal systems such as dispersions, emulsions and foams. Until recently, the forces responsible for the stability of such systems could in most cases only be calculated by use of some appropriate theoretical model or inferred by rather indirect means. The last fifteen years have seen the emergence of a new area of surface science concerned with direct measurements of the force between surfaces. It is now possible to directly measure the forces that are ultimately responsible for the stability of colloidal systems by using surfaces of macroscopic dimensions. At the same time a wealth of other information relevant to the behaviour of such systems can be obtained. This includes the viscosity of liquids in thin films, the extent of solute adsorption to the surfaces from the liquid as well as the refractive index of the intervening medium. Most of the investigations have been carried out on systems of pure liquids or solutions, but in this paper I will summarise some results of experiments with naturally occurring crude oils.

J. Sjöblom (ed.), Emulsions – A Fundamental and Practical Approach, 123–134.
© 1992 *Kluwer Academic Publishers.*

2.   Principles of Direct Force Measurements

A number of devices designed to measure the force as a function of separation between two surfaces in liquid media has been described in the literature (1-6). The most widely used of these are designed and constructed at the Australian National University in Canberra and are known as the "surface force apparatus" (2,6). The principle of these instruments is very simple. The separation between two surfaces is measured to within 0.1-0.2 nm with multiple-beam interferometry and the force between them is detected by monitoring the deflection of a double-cantilever spring on which one of the surfaces is mounted (see Figure 1).

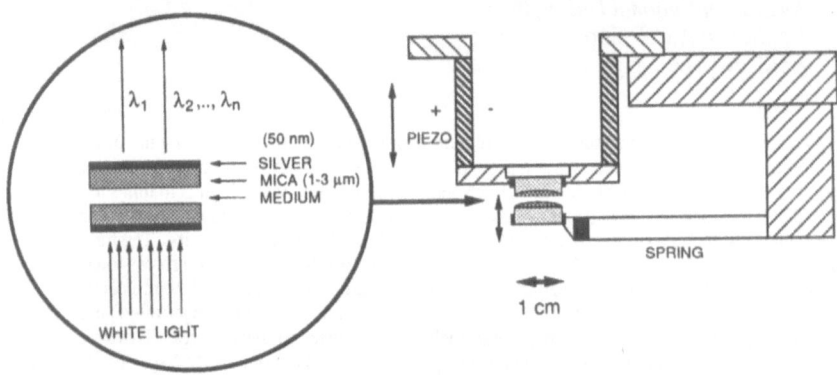

Figure 1   Principle of measurements with the surface force apparatus. The mica sheets are silvered on the back side and glued to supporting silica discs. The surface separation is measured by observing the wavelengths of transmitted fringes when white light is passed through the surfaces. The separation is controlled by applying a voltage to a piezoelectric cylinder on which the upper surface is mounted. The force is detected by measuring the deflection of the spring on which the lower surface is mounted.

The use of multiple-beam interferometry limits the surfaces to transparent substrates, but in practice most substances except metals are sufficiently transparent if the material is thin enough. Most work to date has been carried out with molecularly smooth mica surfaces (7), often with adsorbed or deposited surfactants, lipids or polymers (8,9), but more recently measurements have been carried out with silica (10) and alumina surfaces (11). In all cases the surfaces are silvered on the back side to approximately 95-98% reflectivity. When white light is passed through the system of back silvered mica surfaces (with or without intervening medium) only discrete wavelengths are transmitted. These "fringes of equal chromatic order" can be resolved in a grating spectrometer and their wavelengths are sensitive to the separation between the opposing mica surfaces and the refractive index of the medium in between.

The force between the surfaces can be measured to 10^{-7} N or better, since the accuracy of the separation measurement allows the detection of a very small (≤ 1 nm) deflection of the force-measuring spring (stiffness typically ≈ 100 Nm^{-1}). The surfaces are mounted in a crossed-cylinder configuration with mean radius of curvature R. This has the advantage of easy alignment and convenient comparison of the measured force F(D) with the free energy of interaction E(D) between parallel flat surfaces via the Derjaguin approximation (12),

$$F(D) = 2\pi RE(D).$$ (1)

The force is measured by first calibrating the movement of the surfaces at large separations, where no force acts. One surface is moved, for example by changing the voltage across a piezoelectric tube on which it is mounted (see Figure 1). In this manner the relative position of the surfaces can be changed with almost arbitrary accuracy and its displacement can be measured to yield a calibration. At smaller separations, where a force operates between the surfaces, the actual surface separation will deviate from the one deduced from the calibration due to deflection of the force-measuring spring. This allows the force to be calculated from the spring constant k and the difference between extrapolated and actual change in surface separation ΔD according to Hookes' law,

$$F = k\Delta D.$$ (2)

If the measurements are done at equilibrium, i.e. sufficiently slowly, this force is just the equilbrium surface force due to, for example van der Waals forces, double-layer forces or solvation interactions. If the measurements are carried out rapidly, it includes a contribution from the hydrodynamic forces due to viscous drag by the liquid medium on the moving surfaces. A comparison of the static and dynamic measurements can thus be employed to measure the viscosity of the liquid between the surfaces (13-15).

Since one surface is mounted on a spring, the system is subject to a mechanical instability whenever the gradient of the force law exceeds the spring constant,

$$\partial F/\partial D \geq k,$$ (3)

and the surfaces will jump to the next stable position. In practice, this creates problems when measuring strongly attractive forces, when "jumps" of the surfaces to the next stable position occur.

The optical interference technique allows the refractive index of the medium between the surfaces to be determined independently of the surface separation. It also gives information on the shape of the two surfaces and how it changes during the measurements, thus allowing surface deformations induced by strong forces to be studied. A further feature is the possibilty of studying frictional forces at carefully controlled surface separations by shearing the two surfaces against each other. In this manner some fundamental aspects of lubrication have recently been investigated (16,17).

3. Forces between surfaces across simple hydrocarbons

The force measured between two molecularly smooth mica surfaces immersed in n-decane is shown in Figure 2. The interaction is a decaying oscillatory function of separation with a period close to the mean thickness of the alkyl chain (0.4 - 0.5 nm). Beyond about 3 nm the force appears to merge into a weak attraction, possibly consistent with the attractive force expected from the theory of van der Waals forces. The forces in other n-alkanes from hexane to hexadecane are virtually identical to the force measured in decane; in all cases about 4-5 oscillations and a weakly attractive regime in the range 2.5 - 5 nm are found (18). Such oscillatory solvation forces have been measured in many different systems and are found across all simple nonpolar liquids confined between mica surfaces (7,18,19). They arise from density variations as the molecules are forced to accommodate themselves between two solid surfaces at separations of the order of only a few molecular diameters. The range of the oscillations (number of measurable periods) varies from about 10 with liquids consisting of near-spherical molecules (such as cyclohexane, tetrachloromethane and even benzene, where the period of the oscillations is close to the mean molecular diameter of the liquid) to only one or two in the case of highly branched or asymmetric molecules. The fact that the period is equal to the mean thickness of the alkyl chains shows that the n-alkane molecules are oriented largely parallel to the mica surfaces. In no case does the measured force at small separations bear any resemblance to the monotonically attractive van der Waals force that is still often assumed to dominate the interaction of surfaces in a nonpolar medium. There are, however, indications that surface roughness even on a molecular scale may at least in some cases lead to a smearing of the solvation force and a resultant interaction that is net attractive (20).

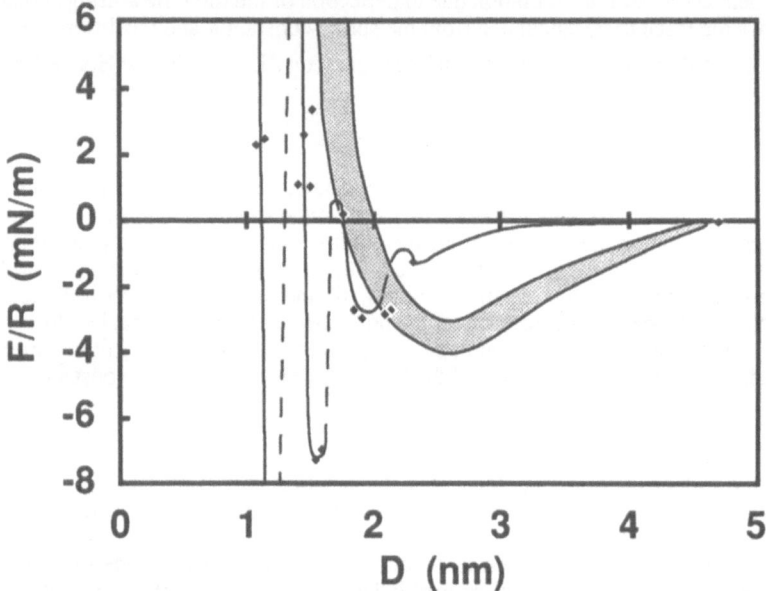

Figure 2.   Measured force (normalised by the mean radius of curvature R of the surfaces) as a function of separation between mica surfaces immersed in pure n-decane (solid line) and in white oil, a mixture of mainly branched C_{20}-C_{22} alkanes (shaded area - results of several different experiments).

Similar oscillatory solvation forces are also found in mixtures of simple nonpolar liquids, although the range of the solvation interaction is in general shorter (21).

In both pure liquids and mixtures the presence of trace amounts of water has a drastic influence on the force between the surfaces (22). Adsorption of water to the hydrophilic mica surfaces leads to a smearing of the density variations near the walls and a consequent reduction in the range and magnitude of the solvation forces. At higher water concentrations capillary condensation of water around the surfaces takes place at small separations and this gives rise to a very large attractive force due to the negative Laplace pressure in the capillary-condensed water. Even at water concentrations as low as 50% of saturation the adhesion between the surfaces is dominated by capillary condensation of water. Similar effects are found with other sparingly soluble and hydrophilic solutes such as methanol.

The force in a mixture of many different, mainly branched hydrocarbons (white oil) shows an attractive regime followed by a short-range repulsion, presumably the remnant of the oscillatory solvation forces (23). This is shown in Figure 2 as the shaded area. Note that the very large forces needed to overcome the repulsive force barriers often make it impossible to measure the forces at separations of less than a few molecular diameters, both in pure liquids and in mixtures. In all systems the force is expected to become attractive close to contact as the last layer of liquid molecules is forced out from between the surfaces. In white oil, too the introduction of trace (ppm) amounts of water causes the force between the surfaces to become attractive at all separations.

4.   Forces between surfaces immersed in crude oils

The force measured between two mica surfaces immersed in a crude oil (24) is qualitatively very different from that found in simple hydrocarbons or mixtures thereof. Figures 3-5 show the results of force runs at different times after immersion in three different Australian crude oils. These oils are all light crudes (see Table I) and the force curves all show similar characteristics. The Harriet oil sample was taken directly from the well head and stored under pressure until immediately before the experiment. The other oils were stored at atmospheric pressure before use. The samples were centrifuged at ≥ 4000 rpm for ≥ 1 hr to remove particulates such as clay particles and large asphaltene aggregates, the presence of which was readily apparent from visual inspection of some samples. Apart from this, no additional treatment or purification was used. The measurements were performed with small droplets (≤ 0.05 cm^3) of crude oil held between the two mica surfaces by surface tension. This was necessary to allow passage of the light beam for the multiple-beam interferometry through the samples. Unfortunately, this leads to some evaporation of lighter fractions with time as the light beam heats the droplet more than the surrounding nitrogen atmosphere in the chamber, even though this is presaturated with vapour from the crude oil.

TABLE I. Physical properties and origin of crude oil samples (at 22° C, Fortescue 26° C)

sample	origin	density	ref. index	viscosity (bulk,cP)	viscosity (SFA,cP)
Harriet	W.A.	0.828	1.472	3.38	4.0
Saladin	W.A.	0.723	1.4455	1.36	1.4
Barrow	W.A.	0.861	1.4782	4.13	4.1
Fortescue	Bass Strait	0.816	1.4667	2.81	3.1

The general features of the force curves may be summarised as follows: There is a weak, rather long-range repulsion and a then a very steeply repulsive force barrier closer in. The range of the repulsion varies between the different oils (from ≈ 20 nm down to only a few nm:s) as does the location of the force barrier (from ≈ 20 nm down to only 3 nm). Unlike the force curves measured in simple liquids or their mixtures mentioned above, there is both hysteresis on approach and separation (see Figures 3 and 4), as well as time-dependence of the measured interaction. When force measurements are carried out in rapid succession, the location of the steep repulsion always shifts towards smaller separations on subsequent approaches (Figure 4). The force measured on decompression, i.e. on separation after the surfaces have been squeeezed together, is always smaller than on compression (approach), as shown in Figures 3 and 4. In some cases a small (≤ 1 mN/m) adhesion between the surfaces is measured on separation (not shown). Time-dependence and hysteresis are never observed in pure liquids or mixtures of simple liquids. They are, however, commonly encountered in measurements of the force between surfaces in solutions of macromolecules such as long-chain polymers and proteins and are related to either slow adsorption-desorption equilibria or molecular entanglement (bridging) of species adsorbed to opposing surfaces (8,9).

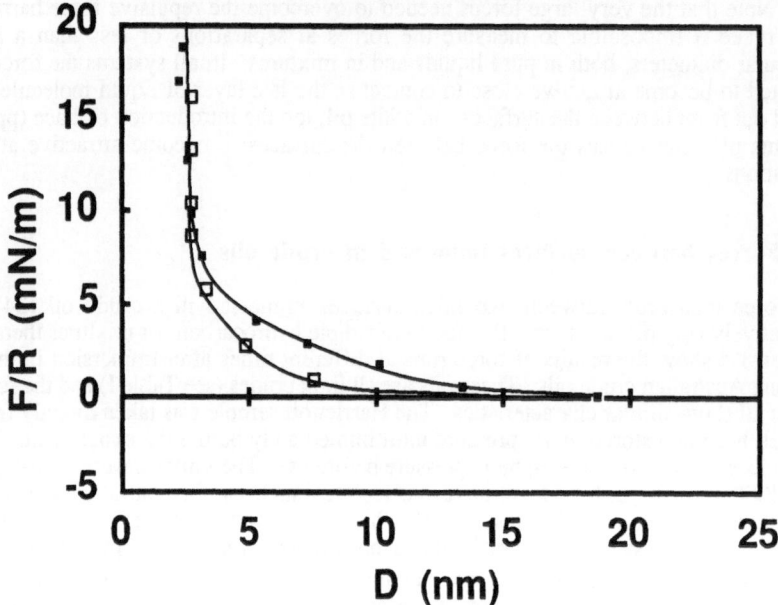

Figure 3.   Measured force as a function of separation between mica surfaces across a crude oil from Western Australia (Harriet). Filled symbols are measurements made on compression (approach), open symbols on separation.

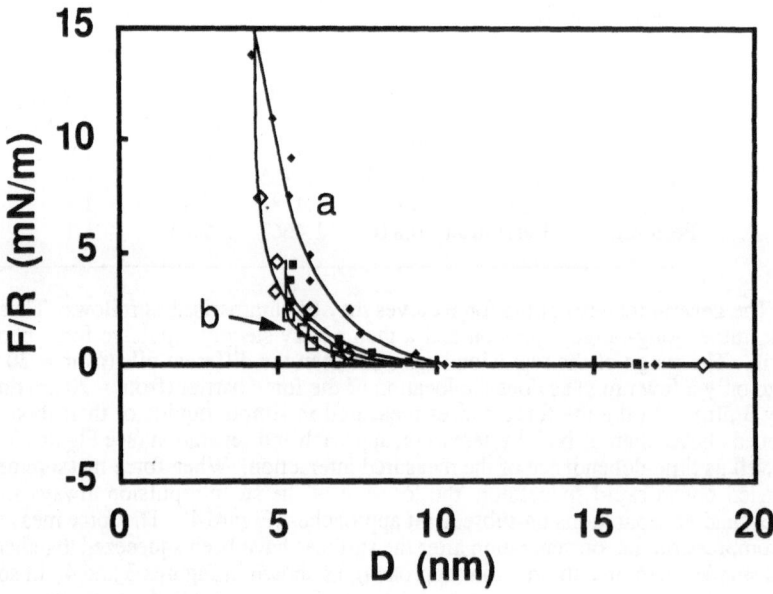

Figure 4   Force as a function of separation in a crude oil sample from Bass Strait (Fortescue). **a** is the first measurement 1/2 hr after immersion of the surfaces (showing approach and separation as filled and open symbols, respectively) and **b** is a second approach and separation (filled and open squares) 3 hrs after immersion.

With time after immersion the location of the steep force barrier shifts outwards to larger surface separations (see Figure 5). As this happens the range of the repulsion outside the steep barrier often remains roughly constant and the measured hysteresis is more or less unaffected (not shown). The surfaces cannot be forced into contact and even under applied loads that are an order of magnitude larger than those shown in the figures the surfaces are only observed to move in by less than a nanometre. The force barriers shown are effectively "hard walls" that prevent any contact between the surfaces.

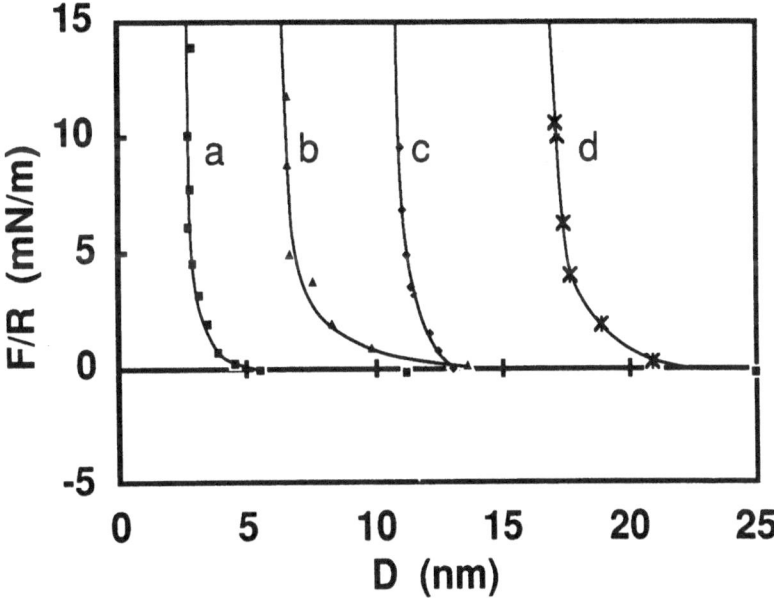

Figure 5   Force as a function of separation (showing approach only) in oil from Saladin field in Western Australia. **a** was measured 2 hrs after immersion, **b** after 24 hrs, **c** after 29 hrs at a different contact position of the surfaces and **d** at this second position 24 hrs later.

Evidently, some components in the crude oils are adsorbing very strongly to the mica surfaces. Separate tests have shown that these compounds are not removed by washing with water or hexane, and that they are very hydrophobic, with an advancing contact angle of water close to 90° (the receding angle, however, is close to zero) (25). The refractive index of the medium between the surfaces at the steep barrier is considerably higher than the bulk refractive index of the oils, up to 1.6-1.7 in some cases, and only at separations beyond about 100 nm does the measured refractive index become indistinguishable from the bulk value (see Figure 6). The high refractive index indicates that these are mainly asphaltenes with polycondensed aromatic rings, possibly with compounds containing various heteroatoms as well (26).

The measurements discussed above and shown in the figures were carried out on dry oils (the samples are kept dry by equilibration through vapour with phosphorous pentoxide in the measuring chamber). On addition of water to the samples (again by equilibration through vapour) the long-range part of the force curve hardly changes. Only on compressing the surfaces considerably, by loads of 0.5 N/m or more, does the force change significantly and become adhesive. The adhesion measured on separation

130

may become substantially larger - up to 60-80 mN/m. The position of the hard wall, however, does not change. In spite of the largely hydrophobic nature of the surface water is clearly able to adsorb to certain sites and increase the adhesion substantially if the surfaces are forced together. The presence of hydrophilic sites is consistent with the small receding contact angle of water.

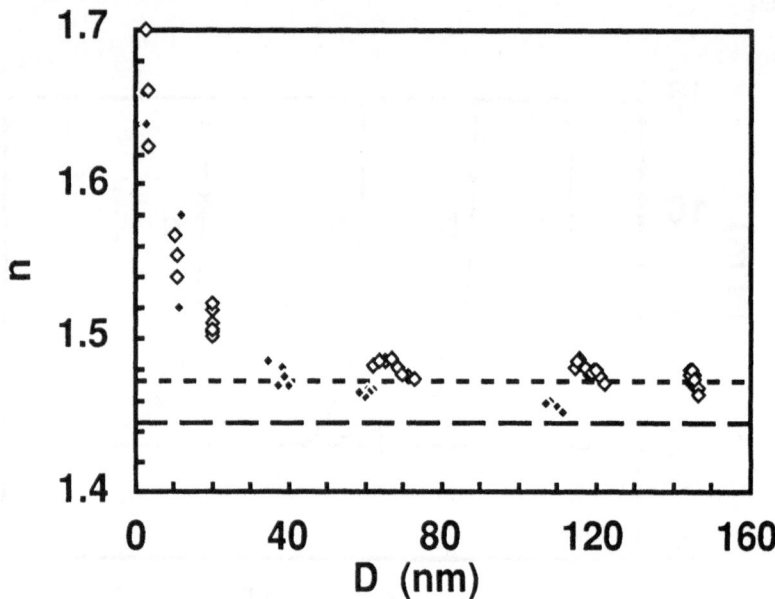

Figure 6   Measured refractive index of films of crude oil (open points - Harriet oil, filled points - Saladin oil) between mica surfaces as a function of the surface separation. The bulk refractive indices are shown as the lower dashed line (Saladin) and the upper dashed line (Harriet).

5.   Viscous forces in crude oils

The hydrodynamic force on the surfaces may be measured by studying the time evolution of the surface separation if one of the surfaces is subjected to some known displacement. If a sinusoidal voltage is applied to the piezoelectric tube on which the upper surface is mounted the resulting motion of this surface couples to the motion of the lower surface via the viscous medium separating the two. The lower surface will also oscillate sinusoidally about its mean (equilibrium) position, but with a reduced amplitude (for finite viscosities) and a phase lag (phase angle). The viscosity of the liquid may be calculated either from the phase angle (15,24) or the measured amplitude (14,24) of the sinusoidally varying surface separation.

Experimentally, the phase angle can easily be measured by replacing the spring with a piezoelectric bimorph and using a lock-in amplifier to read the phase difference directly (15). Alternatively, the amplitude of vibration of the two surfaces can be measured with a video recorder. The phase angle Φ is related to the viscosity of the medium η by the equation

$$\tan\Phi = kD/(12\pi^2 R_h^2 \nu\eta) \tag{4}$$

where D is the surface separation, ν is the frequency of the driving oscillations, and R_h is the hydrodynamic radius of the surfaces defined by

$$R_h^2 = 2(R_1 R_2)^{3/2}/(R_1 + R_2), \tag{5}$$

where R_1 and R_2 are the principal radii of curvature of the surfaces (i.e., $R=(R_1 R_2)^{1/2}$). A plot of $\tan\Phi$ vs. D thus yields a straight line from whose slope the viscosity can be calculated. An example of such a plot is shown in Figure 7.

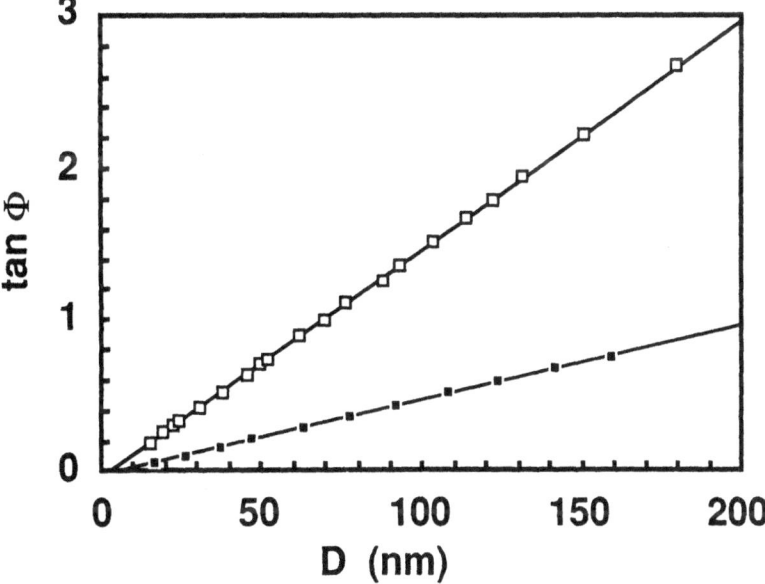

Figure 7   Tangent of the measured phase angle Φ (phase lag) between upper and lower surfaces immersed in Saladin oil with the upper surface subjected to a sinusoidal driving frequency ν of 1 Hz (open points) and 3 Hz (filled points). The results give a viscosity of 1.4 cP and the location of the plane of shear at 2-3 nm out from either surface.

The results of viscosity measurements show that, as for simple liquids (13-15), there is no measurable change in viscosity down to separations of the order of 10 nm, provided that the shift in the plane of shear (slipping plane) caused by adsorption is taken into account. Numerical results for four crude oil samples are given in Table I. The slightly higher than bulk viscosities measured in some cases are due to evaporation of lighter fractions. (Evaporation of approximately 20% of bulk samples of these oils usually causes the viscosity as measured by an Ostwald viscometer to double).

Care must be taken when attempting to interpret the results of such viscosity measurements. Firstly, the viscous drag on the surfaces has contributions from the entire system, so that a large part of it is from those areas were the separation greatly exceeds the measured, or minimum surface separation (the point of closest approach of the two crossed cylinders) which is the quantity plotted in Figure 7. It is hence not correct to claim that the viscosity is being measured in films of this precise thickness (27). Furthermore, there are problems with defining the phase angle correctly. The phase angle in contact is not usually, as might be expected, zero. This is partly a feature of the electronic circuitry and is related to the decay time of the signal from the piezoelectric bimorph. Also, the bending mode of the bimorph changes with the applied force (i.e., in contact, or when a steep repulsion is encountered, it is different) (28). By simply subtracting the phase angle in contact the results usually appear to become self-consistent (as those of Figure 7). These problems notwithstanding, these and other experiments have demonstrated for the first time that liquid flow in very thin films is well described by bulk viscosity values.

6. Discussion

The fact that accurate and reproducible force measurements at the nanometre level can be caried out in untreated crude oils is certainly quite remarkable. It is fair to say that before the measurements were carried out none of us really believed that they would yield any useful information!

The force curves presented in Figures 3-5 show that the adsorption of surface active fractions from the crude oils to the mica surfaces results in an overall repulsive interaction that is of considerable magnitude. There is no doubt that this repulsion is sufficient to confer considerable stability on a disperse system in which the particles or droplets are coated with similar compounds from the crude oils (29,30). For comparison, a force barrier of 5-15 mN/m is measured in aqueous systems where there is very efficient double-layer stabilisation - here the barriers are for all practical purposes infinitely high. Obviously, some features of liquid emulsion droplets are not reproduced in our model system, such as interfacial mobility and deformability of the droplets. Similarly, our surfaces are smooth, unlike the rough surfaces of dispersed particles. Nevertheless, it is clear that surface active components in crude oils are able to practically eliminate interparticle adhesion found in pure nonpolar liquids as well as in most mixtures of pure liquids. In particular, the lack of any great effect of dissolved water, except under unrealistically high loads, is a very important difference. The roughness of most surfaces together with the presence of considerable amounts (in activity terms) of dissolved water in nonpolar liquids probably means that solvation forces are not of any practical importance in determining interparticle forces in dispersions in nonpolar liquids. The viscosity measurements show the importance of the adsorbed layers for the flow of liquid between droplets or particles immersed in crude oils but largely confirm the validity of bulk values of the viscosity and their use in calculations of coagulation rates.

There are potentially a great many additional applications of the surface force apparatus in the study of crude oils. The behaviour of heavier crude oils remains to be investigated. There is the possibility of isolating and studying individual fractions of the oil in a surface force apparatus. Surfactant adsorption to surfaces exposed to crude oils is capable of providing very useful data for emulsion studies as well as oil recovery (31). It is the interaction between surfactants and asphaltene-coated surfaces, not bare mineral surfaces, that is of importance in enhanced oil recovery by micellar flooding.

7. References

1. Derjaguin, B.V., Rabinovich, Ya.I. and Churaev, N.V. (1978) 'Direct measurement of molecular forces', Nature 272, 313-318.
2. Israelachvili, J.N. and Adams, G.E. (1978) 'Measurement of forces between two mica surfaces in aqueous electrolyte solutions in the range 0-100 nm', J. Chem. Soc. Faraday Trans. 1 74, 975-1001.

3.   Peschel, G., Belouschek, P., Müller, M.M., Müller, M.R. and König, R. (1982) 'The interaction of solid surfaces in aqueous systems', Colloid & Polymer Sci. 260, 444-451.

4.   Knapschinsky, L., Katz, W., Ehmke, B. and Sonntag, H. (1982) 'Interaction forces between crossed quartz filaments in presence of adsorbed poly(vinyl alcohol)', Colloid & Polymer Sci. 260, 1153-1156

5.   Tonck, A., Georges, J.M. and Loubet, J.L. (1988) 'Measurements of intermolecular forces and the rheology of dodecane between alumina surfaces', J. Colloid Interface Sci. 126, 150-163.

6.   Parker, J.L., Christenson, H.K. and Ninham, B.W. (1989) 'Device for measuring the force and separation between two surfaces down to molecular separations', Rev. Sci. Instrum. 60, 3135-3139.

7.   Christenson, H.K. (1988) 'Non-DLVO forces between surfaces - solvation, hydration and capillary effects', J. Dispersion Sci. Technol. 9, 171-206.

8.   Klein, J. (1988) 'Surface forces with adsorbed and grafted polymers' in Nagasawa, M., (ed.) Molecular conformation and dynamics of macromolecules in condensed systems, Studies in polymer science 2, Elsevier, Amsterdam, pp 333-352.

9.   Patel, S.S. and Tirrell, M. (1989) 'Measurement of forces between surfaces in polymer fluids', Annu. Rev. Phys. Chem. 40, 597-635.

10.   Horn, R.G., Smith, D.T. and Haller, W. (1989) 'Surface forces and viscosity of water measured between silica sheets', Chem. Phys. Lett. 162, 404-408.

11.   Horn, R.G., Clarke, D.R. and Clarkson, M.T. (1988) 'Direct measurement of surface forces between sapphire crystals in aqueous solutions', J. Mater. Res. 3, 413-416.

12.   Derjaguin, B.V. (1934) 'Friction and adhesion. IV. The theory of adhesion of small particles', Kolloid Z. 69, 155-164.

13.   Chan, D.Y.C. and Horn, R.G. (1985) 'The drainage of thin liquid films between solid surfaces', J. Chem. Phys. 83, 5311-5324 .

14.   Israelachvili, J.N. (1986) 'Measurement of the viscosity of liquids in very thin films', J. Colloid Interface Sci. 110, 263-271.

15.   Israelachvili, J.N., Kott, S.J. and Fetters, L.J. (1989) 'Measurements of dynamic interactions in thin films of polymer melts: the transition from simple to complex behaviour', J. Polymer Sci.-Phys. 27, 489-502.

16.   Homola, A.M., Israelachvili, J.N., Gee, M.L., and McGuiggan, P.M. (1989) 'Measurements of and the relation between the adhesion and friction of two surfaces separated by molecularly thin liquid films', J. Tribology 111, 675-682.

17.   Van Alsten, J. and Granick, S. (1988) 'Molecular tribometry of ultrathin liquid films', Phys. Rev. Lett. 61, 2570-2573.

18.   Christenson, H.K., Gruen, D.W.R., Horn, R.G. and Israelachvili, J.N. (1987) 'Structuring in liquid alkanes between solid surfaces: force measurements and mean-field theory', J. Chem. Phys. 87, 1834-1841.

19.   Christenson, H.K. (1983) 'Experimental measurements of solvation forces in nonpolar liquids', J. Chem. Phys. 78, 6906-6911.

20.   Christenson, H.K. (1986) 'Interactions between hydrocarbon surfaces in a nonpolar liquid - effect of surface properties on solvation forces', J. Phys. Chem. 90, 4-6.

21.   Christenson, H.K. (1985) 'Force between surfaces in a binary mixture of nonpolar liquids', Chem. Phys. Lett. 118, 455-458.

22.   Christenson, H.K. (1985) 'Capillary condensation in systems of immiscible liquids', J. Colloid Interface Sci. 104, 234-249.

23.   Israelachvili, J.N., Kott, S.J., Gee, M.L. and Witten, T.A. (1989) 'Forces between mica surfaces across hydrocarbon liquids: effects of branching and polydispersity', Macromolecules 22, 4247-4253.

24.   Fang, J. and Christenson, H.K. (1990) 'Viscosity and adsorption studies of Australian crude oils in thin films', J. Dispersion Sci. Technol. 11, 97-114.

134

25. Christenson, H.K. and Israelachvili, J.N. (1987) 'Direct measurements of interactions and viscosity of crude oils in thin films between model clay surfaces', J. Colloid Interface Sci. 119, 194-202.
26. Speight, J.G. and Moschopedis, S.E. (1981) 'On the molecular nature of petroleum asphaltenes' in J.W. Bunger and N.C. Li (eds.), Chemistry of Asphaltenes, Adv. in Chem. 195, American Chemical Society, Washington, D.C. pp 1-15.
27. Van Alsten, J., Granick, S. and Israelachvili, J.N. (1988) 'Concerning the measurement of fluid viscosity between curved surfaces', J.Colloid Interface Sci. 125, 739-740.
28. Parker, J.L. (1990) to be published.
29. Menon, V.B. and Wasan, D.T. (1986) 'Particle-fluid interactions with applications to solid-stabilized emulsions', Colloids Surf. 19, 89-122.
30. Sjöblom, J., Urdahl, O., Høiland, H., Christy, A.A. and Johansen, E.J. (1990) 'Water-in-crude oil emulsions. Formation, characterization, and destabilization', Progr. Colloid Polym. Sci. 82, 131-139.
31. Dawe, R.A. and Egbogah, E.O. (1978) 'The recovery of oil from petroleum reservoirs', Contemp. Phys. 19, 355-376.

INTERFACIAL ASPECTS OF WATER-IN-CRUDE OIL EMULSION STABILITY

Andrew J McMahon

BP Research
Sunbury Research Centre
Chertsey Road
Sunbury-on-Thames
Middlesex, TW16 7LN
United Kingdom

ABSTRACT

In certain crude oils the wax crystal size has a major influence
on the stability of water-in-crude oil emulsions. Interfacial
viscosity and other data suggest that the crystals form a barrier
at the water/oil interface which retards the coalescence of
colliding water droplets. In order to associate with the
interface in this way the normally hydrophobic wax must acquire
some affinity for the water phase, possibly via adsorption of
polar asphaltenes and resins from the crude oil. Studies with
octacosane (n-$C_{28}H_{58}$), a model crude oil wax, show that a limited
wax/asphaltene/resin interaction does exist. However, the
adsorbed layer does not confer hydrophilicity to the surface of
either octacosane or a real crude oil wax. Therefore, the effect
of wax on emulsion stability does not appear to be through action
at the interface. Instead, the wax may act in the bulk oil phase
by inhibiting film thinning between approaching droplets or by
scavenging demulsifier. It is the asphaltenes and resins which
were found to affect stability via interfacial action. They can
adsorb in either dissolved or solid form and thereby inhibit
water separation.

1.   INTRODUCTION

A crude oil reservoir consists of gas, oil and water within a
porous matrix, retained by a geological trap such as an
impermeable anticline. A variety of circumstances, including
well bore position and reservoir permeability, can lead to
co-production of water along with the crude oil. Shearing
forces in the well and across control values then cause intense
mixing of the fluids and production of water-in-crude oil
emulsions.

J. Sjöblom (ed.), Emulsions – A Fundamental and Practical Approach, 135–156.
© 1992 Kluwer Academic Publishers.

This water is unwanted. It occupies space in the processing equipment and increases the overall viscosity of the oil phase. Furthermore, because it is highly saline (in some cases even approaching saturation) it invariably leads to scaling and corrosion problems.

Consequently, an essential function of the wellhead processing facilities is to rapidly break the produced emulsions and then dispose of the separated water. Specialised demulsifier chemicals are used for this task. Given the high production temperatures (> 50°C) and long vessel residence times of typical fields in the Middle East, demulsifier concentrations of <30 ppm (on total fluids) are normally effective. Residence times are shorter on North Sea offshore installations in order to minimise equipment weight and space. Nevertheless, the production temperatures are still normally sufficient to allow good demulsification with <30 ppm demulsifier.

There is, however, an increasing trend in the North Sea to use long sea-bed pipelines to connect widely spaced wells, or a collection of marginal fields, to a central processing platform. Cooling can occur in these lines leading to a reduction of the temperature in the emulsion breaking vessel. A critical region exists around 40°C below which the precipitation of crude oil waxes can enhance the stability of water-in-crude oil emulsions.

The important stabilising role of wax crystals has been elegantly demonstrated in previous work at BP [1, 2]. This paper further examines the influence of wax. It reviews previous work and considers whether the wax effect might arise from a wax/asphaltene/resin interaction conferring hydrophilicity to the normally hydrophobic wax thereby allowing it to adsorb at the oil/water interface.

2.   EXPERIMENTAL

2.1   Materials

(a)   Crude Oil - Dry, stabilised and additive free samples of several North Sea crude oils were used.

(b)   Water - Pure water was obtained from a Milli-Q cartridge filtration system. Analyses of oilfield produced waters, ie "formation water", enabled synthetic formation waters to be prepared using laboratory salts.

(c)   Demulsifiers - Commercial demulsifier formulations were obtained from U.K. suppliers.

(d)   Wax - Wax was extracted from crude oil by dilution in a a fourfold excess of dichloromethane, cooling to -40°C and isolating the precipitate.

In some experiments a pure alkane, n-octacosane ($C_{28}H_{58}$, 99% w/w, Sigma Chemicals), was used as a model of crude oil wax. It has a carbon number around the mid point of the C_{15} to C_{40} range found for oilfield wax deposits [3]. It must be remembered, however, that n-alkane is just one of the numerous components of an actual deposit [4].

(e)   Asphaltenes/Resins - Asphaltenes and resins were extracted by adding crude oil to a forty fold excess of pentane, stirring for 24 hours, and filtering the precipitate. A nitrogen purge was maintained throughout extraction and storage to minimise any oxidation.

Pentane was used in this procedure in order to produce as much precipitated material as possible with a solvent which was convenient to handle. For example, butane would have provided more material but would have required pressurised equipment. Heptane would have given less material and is a solvent often recommended for asphaltene extraction. Indeed, "asphaltenes" are sometimes <u>defined</u> as the heptane insoluble and benzene soluble portion of crude oil (eg Institute of Petroleum method no <u>143</u> [5]).

The use of pentane in this work was intended to isolate the materials formally defined as asphaltenes by the standard method (ie heptane insoluble fraction) and also the material which would be described as the most polar of the "resin" fraction. Therefore, the pentane insoluble extract is a mixture of "asphaltenes" and the most polar "resins" (ca one third of the total resins present). It will be referred to as asphaltene/resin material in this paper.

2.2   <u>Methods</u>

(a)   Emulsion Stability - Water-in-oil emulsions were prepared using an Ultraturrax blender (T25 drive, S25 N-18G shaft) at up to 21000 rpm. Stability was monitored by observing the quantity of separated water over time or by measuring the level of residual water in sub-samples from the oil phase, using Karl Fischer analysis.

TABLE 1

ANALYSIS OF WILHELMY PLATE CONTACT ANGLE DATA FOR OCTACOSANE PLATE AT 25°C

Additives	Movement of Oil Phase on Plate	Interfacial Tension (N m^{-1})	ΔW (kg)	$\cos \theta$	θ (°)
no additives	RECEDING	30.2 x 10^{-3}	1.21 x 10^{-4}	1.00	0($\pm$11)
" "	ADVANCING	30.2 x 10^{-3}	1.19 x 10^{-4}	0.99	8($\pm$8)
with asphaltene /resin	RECEDING	24.8 x 10^{-3}	1.01 x 10^{-4}	1.02	~0($\pm$11)
" "	ADVANCING	24.8 x 10^{-3}	0.93 x 10^{-4}	0.94	20($\pm$3)

Notes (a) 7.5% w/w octacosane-in-xylene/pure water system used.

(b) Contact angle of oil on plate calculated using equation (1) ie

$$\cos \theta = \frac{g\Delta W}{p\gamma}$$

(c) p equals 3.92 x 10^{-2} m

(d) g equals 9.81 m s^{-2}

(e) interfacial tension obtained using separate PTFE plate; note the reduced tension with asphaltene present

(f) ΔW from Figure 8

(g) measurement errors on p, γ and ΔW are 1% maximum which compound to 1.7% for $\cos \theta$ which gives the errors on θ shown.

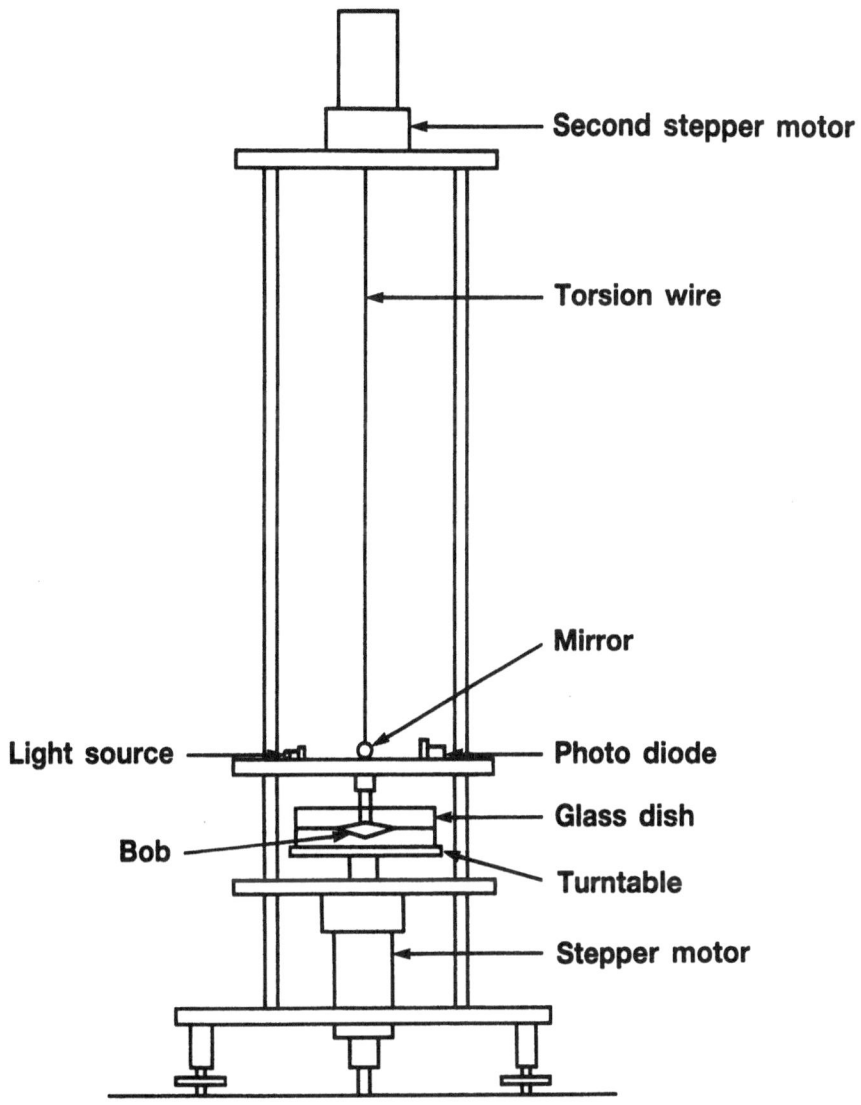

Figure 1. Schematic drawing of interfacial rheometer.

140

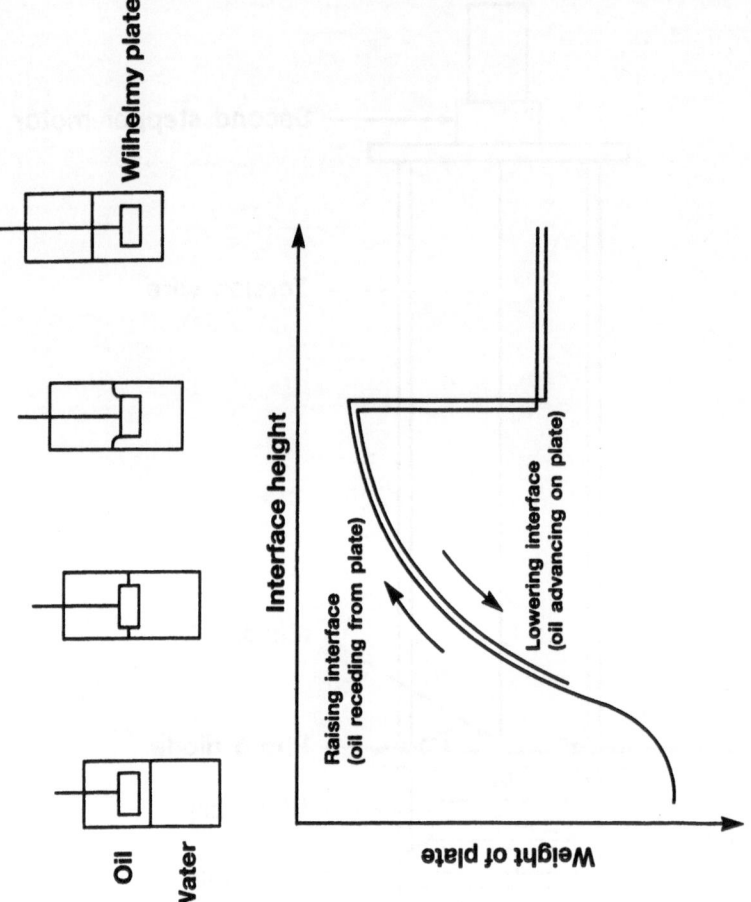

Figure 2. Schematic diagram of Wilhelmy plate contact angle experiment.

(b)   Interfacial Viscosity - A custom built biconical bob
       rheometer (Figure 1) was used. This is described in
       detail elsewhere [6]. Briefly, the razor edge of the bob
       was positioned at the crude oil/water interface of
       interest and its angular deflection measured as the outer
       dish was rotated at speeds up to 0.8 rpm. Control
       experiments allowed these deflection values to be
       corrected for the effect of bulk crude oil viscosity and
       hence the contribution due to the interface alone could
       be obtained. The bulk water phase was found to have a
       negligible effect on the deflection value. The dish was
       thermostated by an external water jacket.

(c)   Asphaltene/Resin Adsorption onto Wax - Xylene which was
       slightly super-saturated with octacosane at 25°C (ie 7.5%
       w/w octacosane-in-xylene) was used in the adsorption
       experiments. An aliquot of octacosane powder (specific
       surface area 3.14 m^2/g) could be added to this liquid
       without any of the solid dissolving. This ensured that
       the surface area available for adsorption was well
       defined. The saturation concentration of octacosane was
       noticeably temperature dependent, therefore, all
       experiments were carried out in a thermostat bath at 25 ±
       0.1°C.

       Adsorption was assessed by measuring the depletion of
       asphaltene/resin in solution one hour after adding a
       large surface area of wax powder. This period is
       sufficient to observe any adsorption effects relevant to
       crude oil emulsion stability since in most offshore
       installations the produced fluids travel from reservoir
       to separator exit in less than one hour.
       Asphaltene/resin concentrations were determined using
       UV/visible spectrophotometry.

(d)   Wetting Properties of Wax Surface - The hydrophilicity
       (or hydrophobicity) of wax surfaces was assessed by
       measuring the contact angle between an oil/water
       interface and a wax plate. The plate was suspended from
       a microbalance (Cahn AD2Z) and initially immersed in the
       upper (oil) phase of a vessel containing oil and water.
       The vessel was then slowly lifted so that the oil/water
       interface rose up to and then above the plate until
       detachment. The interface was then lowered to return to
       the starting conditions. The microbalance operated in a
       manner that prevented plate movement during these
       changes.

       A typical profile of weight against interface height is
       shown in Figure 2. Typically the interface was raised in
       1 mm increments and then left for up to 60s after each
       step for the weight to stabilise. Abrupt changes in
       weight occurred after interface detachment or re-
       attachment to the plate.

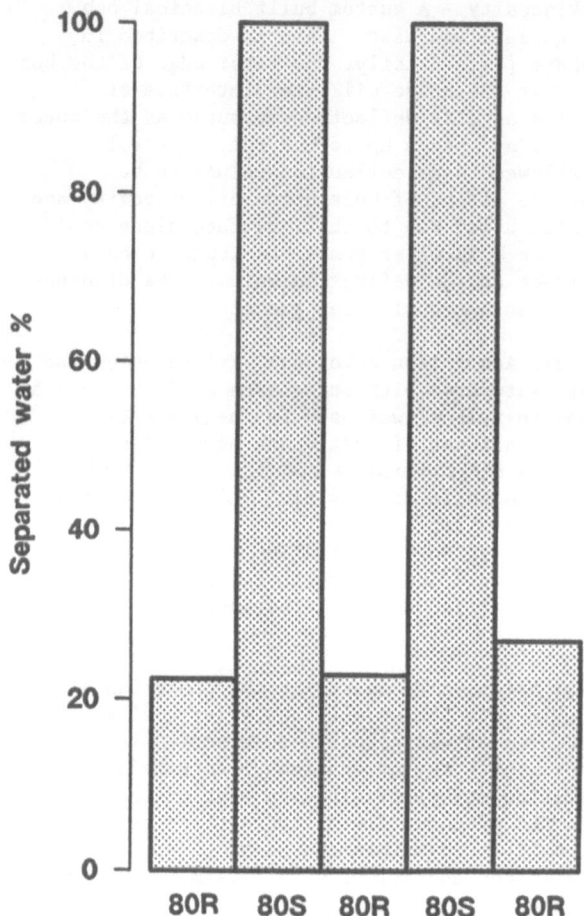

Figure 3. The influence of consecutive crude oil thermal pretreatments on the water separation after 24 hours from a 20% water-in-North Sea crude oil emulsion (80R rapid cooling at 10°C/min from 80 to 20°C, 80S slow cooling at 0.1°C/min from 80 to 20°C).

143

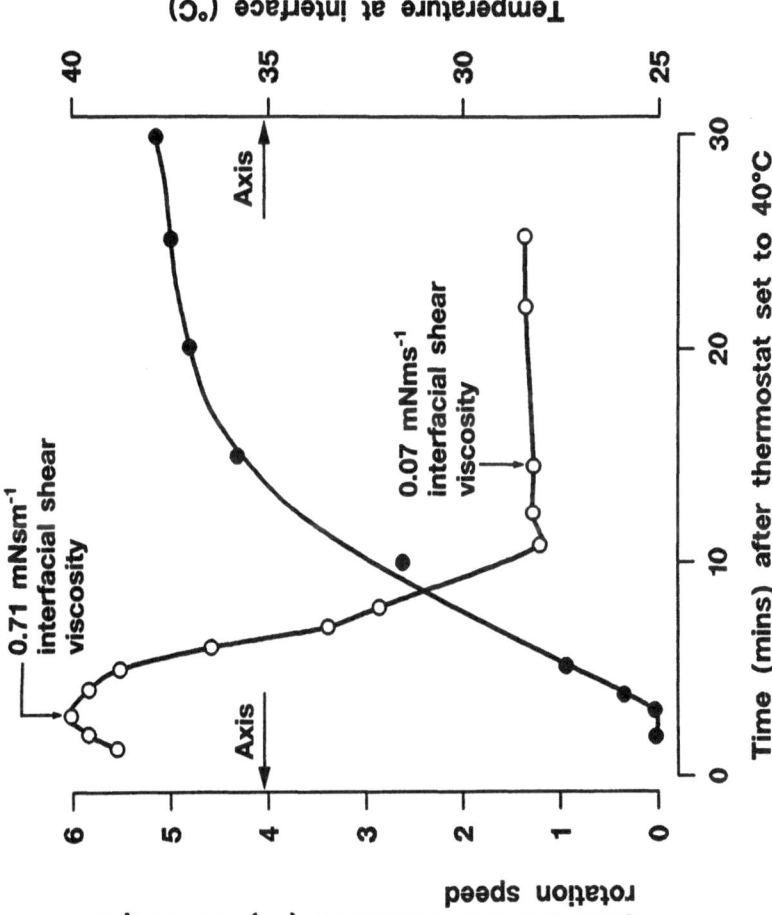

Figure 4. Interfacial rheology data for crude oil/formation water interface; effect of temperature jump from 25 to ca 40°C.

The magnitude of these changes is described by the following equation,

$$\Delta W = \frac{p\gamma \cos \theta}{g} \qquad (1)$$

where ΔW is weight change (kg)
        p is plate perimeter (m)
        g is gravitational constant (9.81 m/s^2)
        γ is oil/water interfacial tension (N m^{-1})
        θ is contact angle of oil on plate (°)

Therefore, θ can be calculated from ΔW_{wax} if p and γ are known. The perimeter length (p) is measured using a travelling microscope. The interfacial tension (γ) is determined using a PTFE plate. This is possible because the PTFE remains oil wet during the experiment ie θ_{oil} = 0°C (for receding oil phase). Therefore γ can be calculated from ΔW_{PTFE}. All measurements were carried out in a glass cell thermostatted at 25 ± 0.1°C.

The wax plates were made by dipping a PTFE plate into molten wax, just above its freezing point, and then quickly withdrawing. This ensured an unbroken and even coating on all parts of the plate including the edges. The coating remained intact during all experiments confirming that it did not dissolve into the oil phase.

3.   <u>RESULTS AND DISCUSSION</u>

3.1   <u>The Influence of Wax on Crude Oil Emulsion Stability</u>

The existence of viscous interfacial films which stabilise water-in-crude oil emulsions has been recognised for some time [7]. Numerous attempts have been made to isolate and characterise the interfacially active species. As well as the expected polar materials in the crude oil such as asphaltene and resin fractions [8], and napthenic acids and their salts [9], some workers have identified long chain, aliphatic waxes [10,11]. The important role of wax in stabilising emulsions has been demonstrated directly by Graham et al [1,2]. They found that thermal pretreatment of a crude oil could produce either large wax crystals (from slow cooling) or small wax crystals (from fast cooling). Emulsions were significantly more stable with small crystals than with large which is consistent with solid stabilisation of the oil/water interfaces (Figure 3). Indeed, examination of the effect of temperature on the interfacial viscosity between a North Sea crude oil and its formation water (Figure 4) shows that the viscosity declines sharply when the temperature rises sufficiently to melt the wax (ca 30°C).

a

$100\mu m$

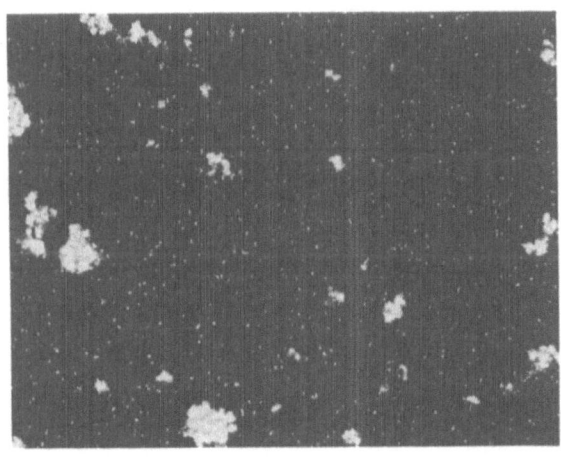

b

FIGURE 5   POLARISED LIGHT MICROGRAPHS OF NORTH SEA CRUDE OIL AT

21°C (a) AFTER FAST COOLING FROM 65°C, (b) AFTER SLOW

COOLING FROM 65°C

146

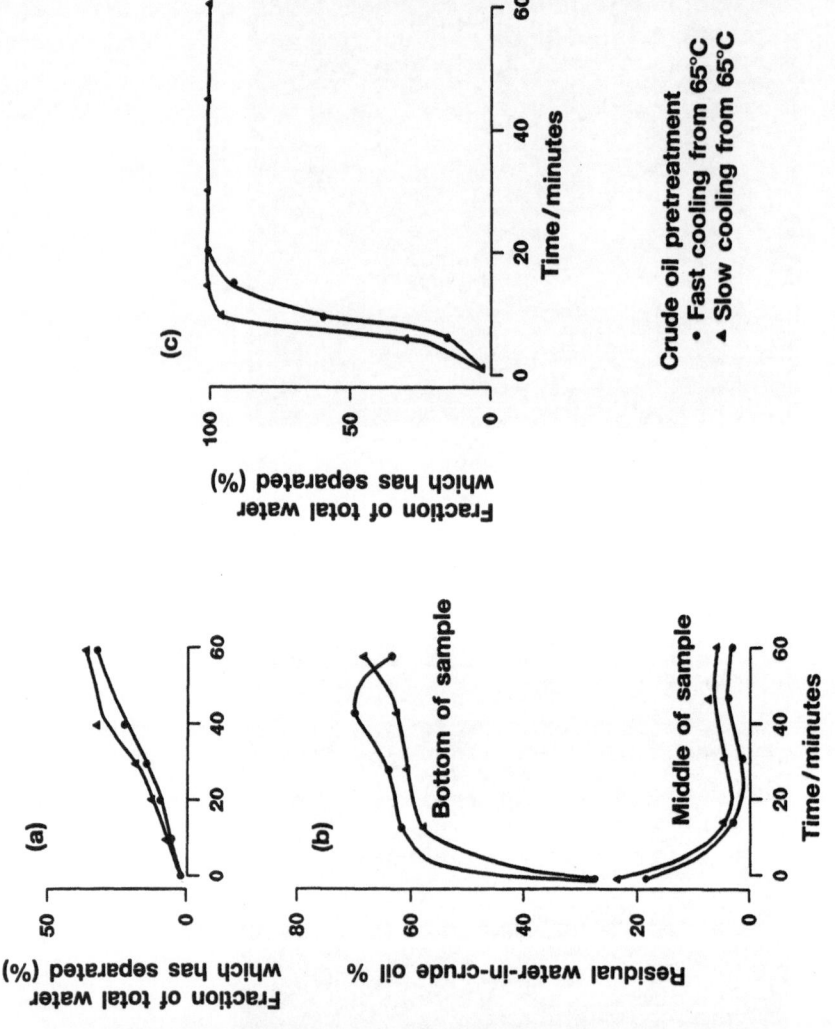

Figure 6. Resolution at 20°C of 20% v/v water-in-crude oil emulsions (a), (b) no additives, (c) with 20 ppm demulsifier.

147

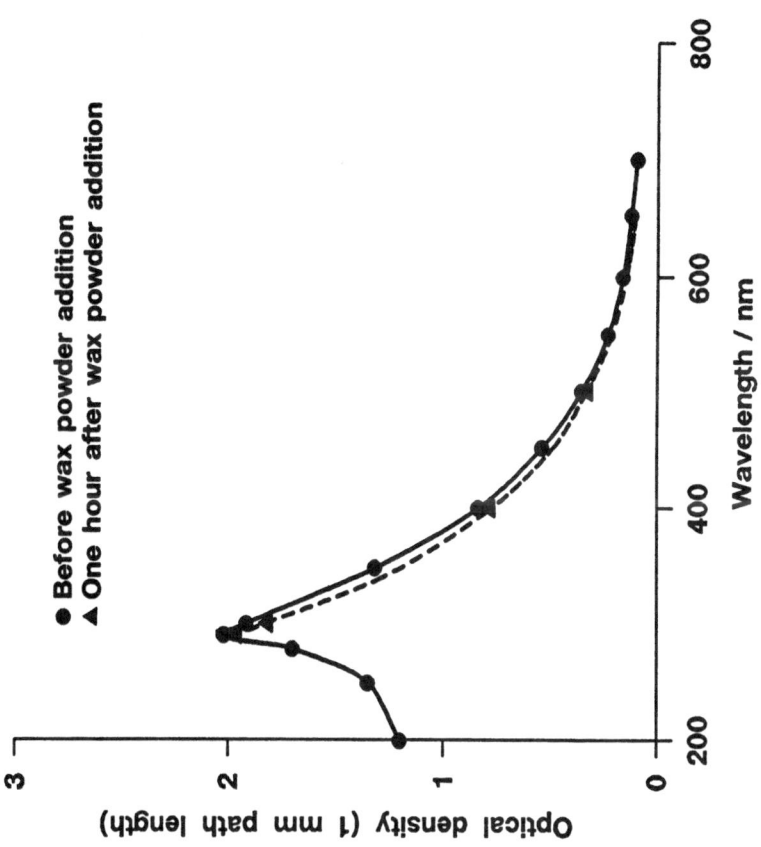

Figure 7. UV/Visible spectra of asphaltene/resin solution before and after addition of n-octacosane powder at 25°C.

Wax adsorption at the interface requires that the normally
hydrophobic solid surface is made partially hydrophilic
possibly via adsorption of an amphipathic species [11]. Since
asphaltenes and resins are the most polar components in crude
oil, and since they have been found together with waxes in
interfacial extracts, it is plausible that they fulfil this
role. Their ability to alter the wetting properties of a solid
in crude oil has been observed previously [12]. However, in
that case the solid was hydrophilic (sodium montmorillonite)
and the asphaltene conferred partial hydrophobicity. This
allowed the particles to adsorb at a water/oil interface and
thereby stabilise emulsions.

The possibility of an wax/asphaltene/resin interaction
producing a partially hydrophilic wax crystal is the subject of
this paper.

3.2   The Prevalence of The Wax Effect

The effect of wax crystal size distribution on water-in-crude
oil emulsion stability at low temperature is not apparent for
all crude oils. For example, the data in Figures 5 and 6 are
for a North Sea crude oil (a different oil from the one in
Figure 3). There is no difference in stability between the
sample containing large crystals (slow cooled) and that
containing small crystals (fast cooled).

The reason why the wax effect occurs in some crude oils but not
others is not yet understood. No doubt it depends on the
interfacial activity of the wax (or wax/asphaltene/resin
combination) relative to that of the other possible stabilising
species in any particular case.

3.3   The Possibility of a Wax/Asphaltene/Resin Interaction

A model crude wax, n-octacosane, was used to examine the
interaction with asphaltene/resin material. Addition of a
large surface area of octacosane powder to asphaltene/resin
solution (up to 1050 ppm w/w) had only a small effect on the
concentration (Figure 7). The small change observed could be
converted to an area per adsorbed molecule (1721 Å^2/molecule)
by assuming a molecular weight of ca 2000 for asphaltene
monomer [13]. This corresponds to a rather diffuse surface
film. However, adsorption might be greater at the 10000-100000
ppm asphaltene/resin typical of crude oils. Unfortunately at
these high concentrations the optical densities of the test
solutions were too high for accurate measurement. Therefore,
an alternative approach was adopted.

3.4   The Effect of Asphaltene/Resin Adsorption on Wax Surface
Properties

Given the observation of a limited, but potentially larger,
wax/asphaltene/resin interaction, the next step was to assess
the effect of the coating on wax crystal wetting properties.

TABLE 2

ANALYSIS OF WILHELMY PLATE CONTACT ANGLE DATA FOR
CRUDE-OIL-WAX PLATE AT 25°C

Movement of Oil Phase on Plate	Interfacial Tension $(N\ m^{-1})$	ΔW (kg)	$\cos\theta$	θ (°)
RECEDING	24.6×10^{-3}	9.32×10^{-5}	1.00	0 (±11)
ADVANCING	24.6×10^{-3}	8.87×10^{-5}	0.96	16(±3)

Notes (a)   crude oil/pure water system used.

(b)   contact angle of oil in plate calculated using equation (1) ie

$$\cos\theta = \frac{g\Delta W}{p\gamma}$$

(c)   p equals 3.70×10^{-2} m

(d)   g equals 9.81 m s^{-2}

(e)   interfacial tension obtained using separate PTFE plate

(f) ΔW from Figure 9

(g)   measurement errors on p, γ and ΔW are 1% maximum which compound to 1.7% for $\cos\theta$ which gives the errors on θ shown.

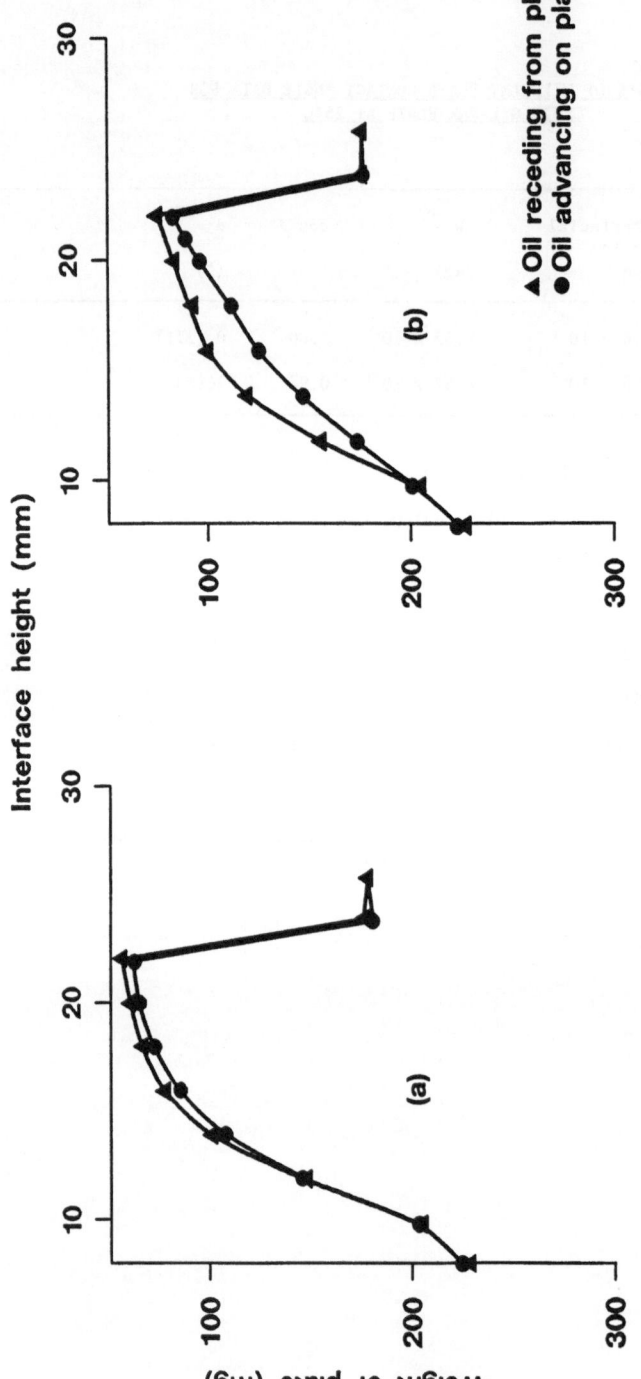

Figure 8. Wilhelmy plate contact angle experiment at 25°C using octacosane plate at 7.5% w/w octacosane-in-xylene/pure water interface, (a) no additives (b) with 1.1% w/w asphaltene/resin in oil phase.

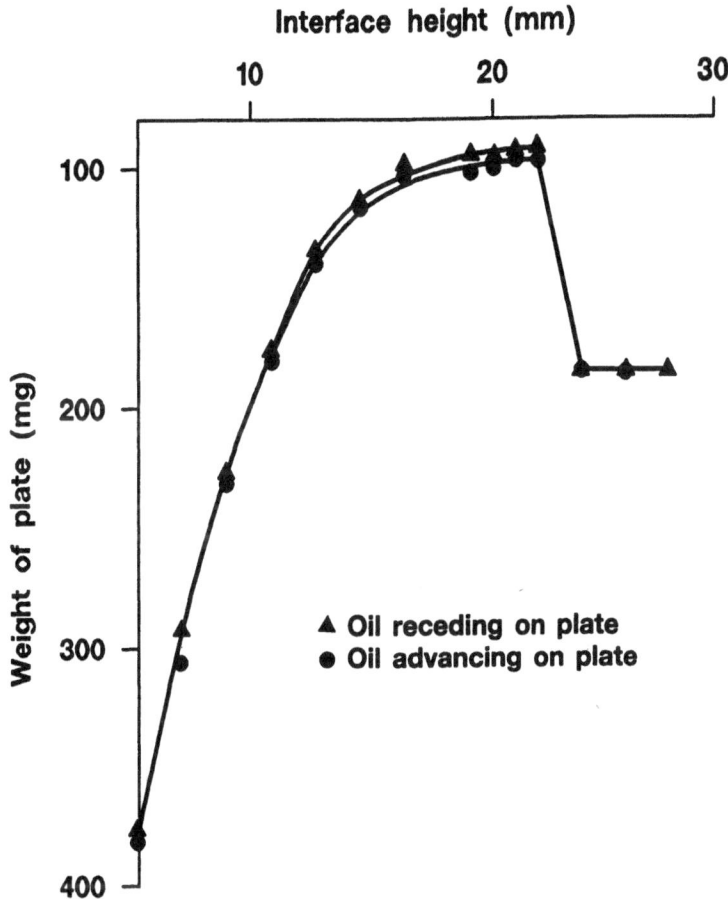

Figure 9. Wilhelmy plate contact angle experiment at 25°C using crude-oil-wax plate at crude oil/pure water interface.

The contact angle experiment used for this work also allowed realistic concentrations of asphaltene/resin (ca 1% w/w) to be examined.

The model wax, n-octacosane, was studied first. An octacosane plate was immersed in the oil phase and the oil/water interface then raised as shown in Figure 2. This was first carried out in a clean oil/pure water system to determine the inherent affinity of a clean octacosane surface for water. Then 1.1% asphaltene/resin was added to the oil and allowed 20 minutes to adsorb onto the wax, in order to gauge whether any extra water affinity developed. The oil phase was xylene, slightly supersaturated in octacosane in order to avoid dissolution of the wax plate, as used in the adsorption experiments.

The weight against height profiles for the clean and asphaltene/resin containing systems are shown in Figure 8. The hysteresis in the data is due to slightly different contact angles when the oil phase first recedes and then advances along the plate (Table 1). The receding angle is the most relevant to wax stabilisation of water-in-crude oil emulsions. This is because in a production system the wax crystals will form in the crude oil as it cools and the oil must recede from the crystal to allow association with the oil/water interface.

The receding angle in the clean system is 0-11°C indicating that the octacosane is almost completely oil wet and has little or no water affinity. Addition of asphaltene/resin does not alter this. The asphaltene/resin may well adsorb on to the octacosane, as shown in Section 3.3, but it does not confer any water affinity to the surface.

Wax extracted from a crude oil was examined next. The crude oil used was one in which the wax crystal size _did_ influence emulsion stability. The wax plate was first bathed in its own crude oil for 20 minutes to allow any asphaltene/resin adsorption to occur. Then a weight/height profile was obtained at the crude oil/pure water interface (Figure 9). The wax gave a receding contact angle of 0-11°C for the crude oil (Table 2) showing that, as with octacosane, little or no water affinity had developed.

What then is the mechanism for the influence of wax crystal size on emulsion stability? Several possibilities are plausible. For example, the crystals may act as buffers between approaching water droplets without actually adsorbing at the interfaces but simply remaining in the thinning film. Alternatively they may act as a surface which scavenges demulsifier thereby reducing the efficiency of this additive. In both these cases emulsion stability would be greater with small than with large wax crystals as is observed.

3.5   Interfacial Activity of Asphaltene/Resin

The ability of asphaltene/resin material to form interfacial films between water and oil is easily demonstrated. Mixing

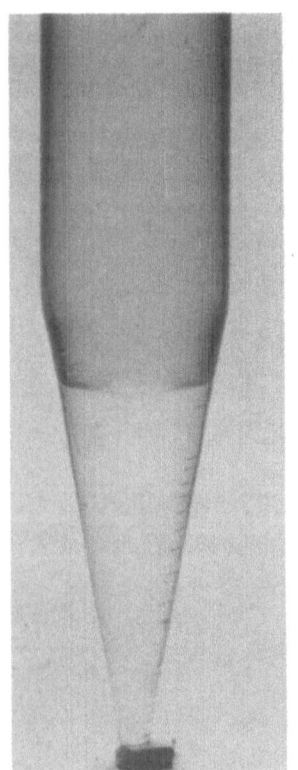 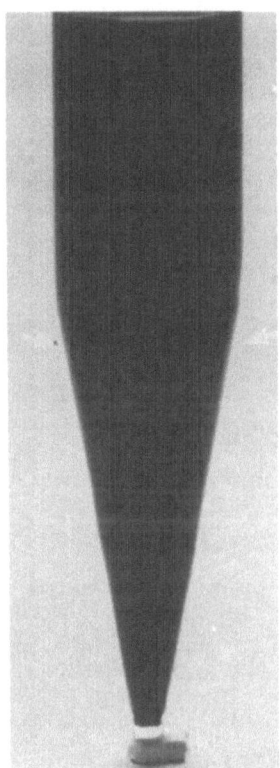

a   b

FIGURE 10   APPEARANCE OF MODEL EMULSIONS AFTER STABILITY TEST

(a) NO ADDITIVES; AFTER ONE HOUR STATIC

(b) 1% w/w ASPHALTENE/RESIN; AFTER ONE HOUR STATIC

(ca 100 ml LIQUID IN EACH TEST)

154

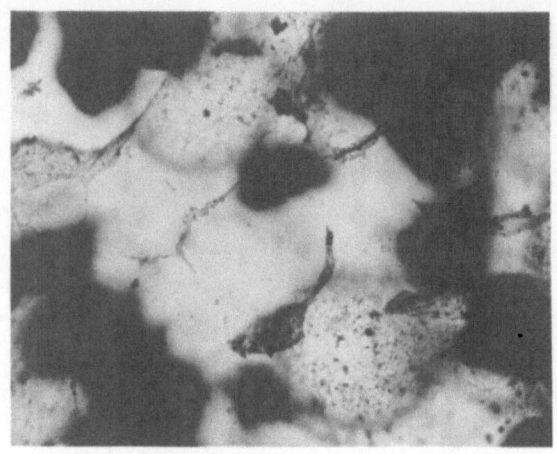

30 μm

FIGURE 11   SAMPLE FROM THE BOTTOM OF A 5% v/v WATER-IN-XYLENE EMULSION

CONTAINING 1% w/w ASPHALTENE/RESIN

8:2 v/v pure water and xylene using high shear produces an
unstable water-in-oil emulsion which completely separates
within a few minutes (Figure 10(a)). With 1% asphaltene/resin
in the oil, the water droplets sediment and form a water-in-oil
mousse which is stable for at least 24 hours because of the
viscous films between the droplets (Figure 10(b)). It is films
like these which contribute to the stability of the water-in-
crude oil emulsions encountered in oilfield operations.

In this example the asphaltene/resin material was completely
soluble in the oil phase and the interfacial film was formed
through adsorption of dissolved components. Asphaltene/resin
material present as a solid can also adsorb at an oil/water
interface and provide solid stabilisation [14]. The micrograph
in Figure 11 shows a water-in-xylene emulsion with
asphaltene/resin in the oil phase. Not all of the
asphaltene/resin has dissolved and the residual solids can be
seen adhering to the water/oil interfaces. Black
asphaltene/resin particles can also be seen in the shreds of
interfacial skin remaining from droplet-droplet coalescence.
This emulsion was considerably more stable than one with only
dissolved material present.

Therefore, solids stabilisation can occur not only with wax
crystals (in some crude oils) but also with asphaltene/resin
particles.

4.   CONCLUSIONS

4.1   Wax crystal size influences water-in-crude oil emulsion
      stability in some crude oils but not in others. The effect is
      not universal.

4.2   A limited interaction between wax, asphaltene and resin does
      exist. However, it does not confer hydrophilicity to the wax
      to allow it to adsorb at an oil/water interface.

4.3   In cases where wax crystal size does influence stability it
      could well be through acting as a buffer between approaching
      water drops, or by scavenging demulsifier, rather than by
      adsorbing at the oil/water interface.

4.4   Asphaltene/resin material can stabilise emulsions both when
      dissolved and also when present as small particles.

5.   ACKNOWLEDGEMENT

      The author would like to acknowledge the assistance of the many
      colleagues at BP Research who helped in this work, and also BP
      for permission to publish this paper.

6.   REFERENCES

      1.   Graham D E, Stockwell A, Thompson D G, In 'Chemicals in
            the Oil Industry', ed Ogden P H, RSC Special Publication
            No 45, 1983, p73.

156

2.   Thompson D G, Taylor A S, Graham D E, Colloids and
        Surfaces, 15 (1985) 175.

3.   Kinghorn R R F, 'An Introduction to Physics and Chemistry
        of Petroleum', Wiley, Chichester, 1983, p72 and 91.

4.   Gilby G W, in 'Chemicals in the Oil Industry', ed Ogden P
        H, RSC Special Publication No 45, 1983, p108.

5.   Methods for Analysis and Testing, Vol 1, Institute of
        Petroleum, Wiley, London, 1990.

6.   Grist D M, Neustadter E L, Whittingham K P, J Can Petrol
        Tech, 20 (1981) 74.

7.   Bartell F E, Niederhauser D O, in "Fundamental Research
        on the Occurence and Recovery of Petroleum 1946-1947",
        American Petroleum Institute, New York, 1949, p57.

8.   Reisberg J, Doscher T M, Producers Monthly, 20 (1956) 43.

9.   Dodd D G, J Phys Chem, 64, (1960) 544.

10.   Hasiba H H, Jessen F W, J Can Petrol Tech, Jan-Mar
        (1968), 1.

11.   Denekas M O, Carlson F J, Moore J W, Dodd C G, Ind Eng
        Chem, 43 (1951) 1165.

12.   Menon V B, Wasan D T, Colloids and Surfaces, 19 (1986)
        89.

13.   Speight J G, "Proceedings of the Symposium on Analytical
        Chemistry of Heavy Oils and Residues (Dallas, 4-
        14/4/89)", American Chemical Society, p321.

14.   Eley D D, Hey M J, Lee M A, Colloids and Surfaces, 24
        (1987) 173.

WATER-IN-CRUDE OIL EMULSIONS FROM THE NORWEGIAN CONTINENTAL SHELF
PART-VI - DIFFUSE REFLECTANCE FOURIER TRANSFORM INFRARED CHARACTERIZATION OF INTERFACIALLY ACTIVE FRACTIONS FROM NORTH SEA CRUDE OIL

Li Mingyuan, Alfred A. Christy and Johan Sjøblom
Department of Chemistry,
University of Bergen,
N-5007 Bergen,
Norway.

ABSTRACT. Interfacially active components in crude oils have been isolated and characterized using diffuse reflectance infrared spectroscopy. Asphaltenes were seperated using pentane precipitation and resins by adsorption onto silicagel from the maltene fraction. Different components of the asphaltenes and resins were selectively seperated by desorption from silicagel using mixtures of benzene and methanol. These fractions were then characterized by infrared spectroscopy and tested with regard to their emulsion forming ability in model systems of decane and water.
    The interfacially active components that stabilize model emulsion are found both in asphaltene and resin fractions. However, asphaltenes contain a higher content of these components as compared to resins and consequently form relatively stable emulsions. The interfacially active components seem to contain high concentration of carbonyl groups, specially open chain carbonyls. Furthermore, it appears that molecular size and aromaticity of the components also play an important role in the stability of the model w/o emulsions.

1. Introduction

Understanding the chemistry involved in the stabilization of water-in-crude oil emulsions is important both for economic and environmental reasons. Water-in-crude oil emulsions are responsible for the enormous increase in the viscosity of the crude oils produced in reservoirs. Transportation of the viscous crude oil through pipe lines is difficult and adds to the cost of the production of crude oil. Environmental pollutions after oil spillages can be reduced if effective clean-up techniques are developed. Past incidents have shown that the water-in-oil emulsions are responsible for the difficulty in effective clean-up of the oilspillages using techniques such as burning, use of sorbants, use of dispersants and pumping [1].

A series of articles dealing with North Sea water-in-crude oil emulsions has been published [2,3]. Topics like formation, characterization, and destabilization aspects together with chemical destabilization and interfacial tensions have been covered [2,3]. In addition monolayer properties of interfacially active fractions have been investigated by means of Langmuir - Blodgett technique [4,5].

Crude oil is a mixture of aliphatic , aromatic hydrocarbons and oxygen, nitrogen

157

J. Sjöblom (ed.), Emulsions – A Fundamental and Practical Approach, 157–172.
© 1992 *Kluwer Academic Publishers.*

and sulphur containing compounds such as resins and asphaltenes. Many scientists have shown that the interfacially active components come from the polar fraction of the crude oil. For examples carboxylic acids [6,7], phenols [8] and waxes [3] have been identified as interfacially active components. Furthermore, resins and asphaltenes have also been shown to be responsible for the stabilization of water-in-crude oil emulsions [3,9].

Resins and asphaltenes are polymeric in nature and there are structural similarities between them [10]. They are differentiated by their solubility in light hydrocarbons such as pentane. This differentiation clearly shows that the resins are smaller in molecular size compared to asphaltenes. Within these fractions there is a spectrum of molecular weight components with differing functionalities. Smaller molecules such as carboxylic acids are components of resins [10]. Unfortunately resins and asphaltenes are not obtained using the same precipitation technique. However, the terms resins and asphaltenes are used regardless of the precipitation technique. The fractions such obtained will have different molecular weight spectra and functional group differences. Hence, a comparison between the reported works becomes difficult.

Infrared spectroscopy has traditionally been used to determine structural features of asphaltenes [10,11]. Layrisse et al [12] have used infrared spectroscopy to study interfacially active components of water-in-crude oil emulsions. Most of the study involves traditional transmission technique which requires solvent medium for measuring the spectrum. New sampling technique such as diffuse reflectance has been successfully used to study the structural features of resins and asphaltenes[10]. This technique provides an easy and elegant way to measure the spectra of resins and asphaltenes.

In the present study we have extracted resins and asphaltenes from crude oils and further seperated the resins and asphaltenes into three different fractions using different mixtures of benzene and methanol. These fractions were characterized by infrared spectroscopy using the diffuse reflectance technique. Furthermore the stability of the water-in-oil emulsions stabilized by these fractions was determined. In the next step characterization of the fraction giving rise to stable and unstable emulsions, respectively, was undertaken. Here an extraction procedure with a subsequent analysis by means of infrared spectroscopy was utilized in order to identify variations in the intensity of the functional groups.

2. Experimental

Three different crude oils from the North Sea were used in these experiments. The oils were denominated as D, G and B. Seperation procedures for the seperation of resins and asphaltenes, and further fractionation into their separate fractions are shown in Fiqure 1.

2.1 SEPARATION OF ASPHALTENES AND FURTHER FRACTIONING

Twenty ml crude oil (15g) was agitated with 600 ml of pentane at room temperature for 10 minutes. The mixture was then left to stand for 24 hours. The precipitated asphaltene fraction was filtered and washed with smaller portions of pentane. The filterate was combined with the supernatant. The asphaltene fraction was then dried,

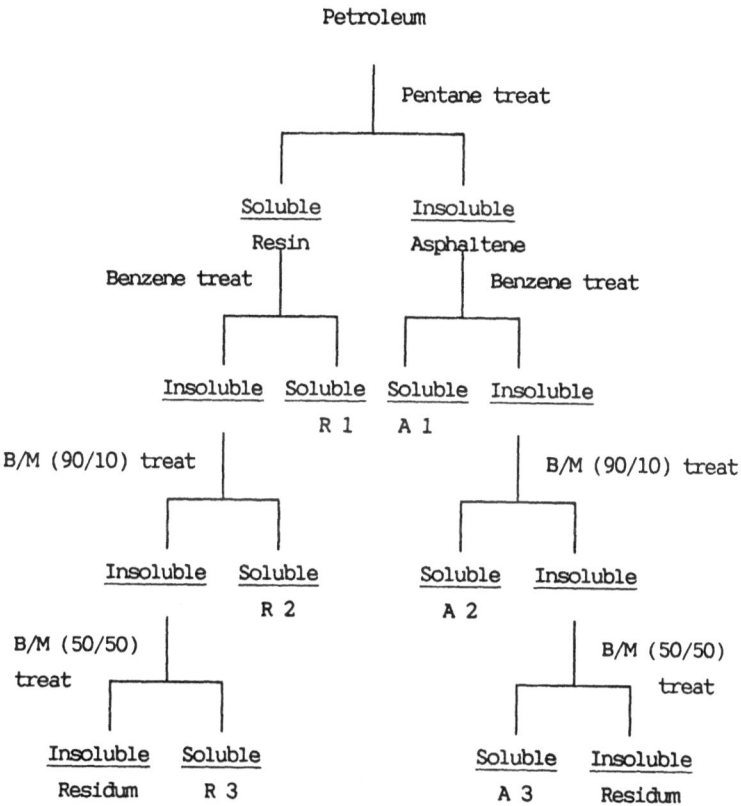

Fig. 1. Separation of polar fractions from crude oils

redissolved in dichloromethane and poured onto 10 g of silica. The mixture was allowed to stand for an hour to allow the solvent evaporation. The silica containing adsorbed asphaltenes was then agitated with benzene (200 ml) at room temperature by a standard propeller rotor at 400 rpm for 10 minutes. The mixture was then allowed to stand for 30 minutes. The extract from the mixture was filtered and asphaltene 1 was obtained by evaporating the solvent.

The remaining asphaltenes adsorbed onto silica were desorbed using similar procedure described above using mixtures of bezene/10% methanol and benzene/50% methanol respectively. The asphaltene fractions so obtained were classified as asphaltenes 2 and asphaltenes 3.

2.2 SEPARATION OF RESINS AND FURTHER FRACTIONING

The filterate obtained from the precipitation of asphaltenes was poured onto silica and the resins were adsorbed by agitating the mixture with a standard propeller rotor for 10 minutes at room temperature. After 30 minutes of standing the solution was filtered using a vacuum pump. The components of the resin fraction were seperated using the same procedure for the separation of asphaltene components. The resin components were classified as resin 1, 2 and 3.

2.3 EMULSION FORMATION AND STABILITY TESTING

Emulsions of the components were prepared by dissolving the components in decane and mixing with equal volume of distilled water. The emulsification process was carried out using a rotor for 1 minute at 1000 rpm. This procedure was repeated with synthetic formation water.
   Stability of the emulsion systems were determined by measuring the water seperated from the systems with time at 21^0 C and 50^0 C.

2.4 SEPARATION OF INTERFACIALLY ACTIVE COMPONENTS IN REST EMULSION

Interfacially active components that remained in the emusion systems were seperated carefully by decanting the oily organic fraction and removed from the top of the water layer using a syringe. The fraction was heated to 120^0 C to remove water.

2.5 SPECTROSCOPIC ANALYSIS

Infrared spectroscopic analysis was carried out using a Perkin Elmer 1720 X Fourier-transform infrared spectrophotometer. All the samples were measured using the diffuse reflectance technique. 10 micro litres of fractions in dichloromethane solution were carefully dripped onto firmly packed potassium bromide powder using a chromatographic syringe. The solvent was evaporated by placing the sample cup in an oven at 50^0 C for two minutes [10]. The spectrum was measured by scanning the deposited material on the potassium bromide powder with previously scanned pure potassium bromide powder as the background. All samples were scanned 10 times using a DTGS detector at a resolution of 4 cm $^{-1}$. Spectra were then obtained both in reflectance format and Kubelka-Munk [10] format for further analysis. A short description of the diffuse reflectance technique, its theoretical background, sampling procedure and applications can be found in the references [8,13-15].

3. Results and discussion.

An outline of the seperation procedures used in this paper for resins and asphaltenes are given in Fig.1. Infrared absorptions of the functional groups of resins and asphaltenes are summarized in Table 1. Diffuse reflectance spectra of resin and asphaltene components in reflectance format are shown in Fig 2. Partial spectra of asphaltene and resin components and of interfacially active components of total stability are shown in Kubelka-Munk format in Fig 3-5. The spectra are identified by a combination of alphabets. The first letter of the names refer to the oil used, the second letter refer to whether the fraction is resin or asphaltenes (R=resin and A=asphaltenes) and the third letter X is given to the fraction of rest emulsion. The number in the names refer to whether the fraction was extracted by benzene or

Table 1. Infrared band assignments of asphaltenes and resins

Functional groups	Absorption bands (1/cm)
NH, OH stretch	
H - bonded	3500 - 3300
CH stretch aromatic	3050 - 3000
CH stretch aliphatic	
methyl asymmtric	2950
methylenic CH stretch	2920 and 2947
methyl symmetric	2865
Carbonyl stretch	
Open chain	1735
aldehyde, ketone and acid	1720 - 1690
amide carbonyl	1700 - 1650
conjugated C=C and	
aromatic C=C	1600
C-CH$_3$ and methylenic	
asymmetric	1465
C-CH$_3$ asymmetric	1377
ether -O- linkage	1050
Aromatic bending	
one hydrogen on the ring	880
two hydrogen - aadjacent	840
three hydrogen - adjacent	820

benzene/10% methanol or benzene/50% methanol (1= fraction extracted by benzene, 2=benzene/10% methanol, and 3= benzene/50% methanol (Example: DRX2 - is the fraction from rest emulsion formed by the interfacially active components of resins extracted by benzene/10% methanol of oil D). Emulsion stability of the resins and asphaltene components are given in Tables 2 - 6.

3.1 GENERAL CONSIDERATIONS

The stability tables of emulsions formed by asphaltenes and resins show that the asphaltenes and resins vary in their emulsion forming abilities. Comparison

Table 2. Volume fraction of water seperated as a function of time from model emulsion stabilized by polar fractions of oil **D** at 21⁰ C.

% (w/w)		Separation of Water (%)					T = 21° C	
		Separation Time			(hour)			
		0.5	1.0	2.0	3.0	4.0	5.0	6.0
DA 1	1.0	50	60	70	70	70	70	70
DA2	1.0	0	20	30	40	50	60	60
	0.5	40	50	60	65	70	70	70
	0.25	65	70	85	85	85	85	85

					Time (min)		
		5	10	15	20	30	60
DA 1	0.5	100	100	100	100	100	100
DA 3	0.5	10	30	60	90	100	100
DR 1	1.0		20	30	40	60	60
	0.5	100	100	100	100	100	100
DR 2	1.0	30	40	60	70	80	80
	0.5	50	70	80	85	90	100
DR 3	1.0	45	60	80	85	90	90
	0.5	60	70	85	90	100	100

between them show that the stability is in the following increasing order for both asphaltenes and resins.

$$\text{oil } D < \text{ oil } B < \text{ oil } G$$

However, asphaltenes form stable emulsions compared to resins in all the cases analysed. Furthermore, quantities of interfacially active components required to form emulsions of similar comparability are different for asphaltenes and resins. These facts clearly indicate that the interfacially active components that can form stable emulsions contain some special structural features in their components.

Table 3. Volume fraction of water seperated as a function of time from model emulsion stabilized by polar fractions of oil **D** at 50^0 C.

	% (w/w)	Separation of Water (%) T = 50° C					
		Separation Time (min.)					
		5	10	15	20	30	45
DA 1	1.0	20	50	60	70	80	100
	0.5	100	100	100	100	100	100
	1.0	20	50	60	80	80	100
DA 2	0.5	30	60	80	100	100	100
	0.25	70	100	100	100	100	100
DA 3	0.5	40	80	100	100	100	100
DR 1	1.0	40	80	100	100	100	100
	0.5	100	100	100	100	100	100
DR 2	1.0	80	100	100	100	100	100
	0.5	100	100	100	100	100	100
DR 3	1.0	80	100	100	100	100	100
	0.5	100	100	100	100	100	100

Asphaltenes and resin components of an oil are unique in their structural features. They are formed from the decomposition of a kerogen matrix. Therefore hydrocarbon, asphaltene and resin fractions are related to each other. Resins and asphaltenes of an oil contain some unique structural centres which help them to form intermolecular bondings for dispersing asphaltenes in the hydrocarbon medium. Location of these centres are different for resins and asphaltenes of different oils.
Resins and asphaltenes are structuraly similar but they differ in their molecular

weights. Both contain pericondensed polyaromatic rings bearing alkyl and alicyclic systems with heteroatoms [8,11] . Number of rings in resins are smaller compared to asphaltenes and molecular weight of resins are also smaller. However there is considerable overlap between resin and asphaltene fractions over molecular weight. The interfacially active components of total stability were found in all the asphaltenes and resin fractions. Wide spread of molecular weight, functionality distribution explains why these molecules are found in all the fractions. However, their quantities are different depending on the solvent medium used for their

Table 4. Volume fraction of water seperated as a function of time from model emulsion stabilized by polar fractions of oil G at 21⁰ C.

	% (w/w)	Separation of Water (%) T = 21° C					
		Separation Time (hour)					
		0.5	1.0	2.0	3.0	4.0	5.0
GA 1	0.05			5	5	5	5
	0.025	40	50	60	60	60	60
GA 2	0.05						0
	0.025		15	20	25	30	35
GA 3	0.05						0
	0.025			5	5	5	5
GR 1	0.5	90	100	100	100	100	100
GR 2	0.5	10	20	40	50	60	60
	0.025	85	100	100	100	100	100
GR 3	0.5						0
	0.025	40	60	80	100	100	100

extraction.

Asphaltenes are more aromatic in nature than resins. They have a low H/C ratio compared to resins. Their heteroelements composition is comparable [16]. It means that the hetero elements have high local concentration of functional groups in resins compared to asphaltenes. This is also evident from the work of Christy et al. [10].

However, resins form less stable emulsions suggesting that the local concentration of the functional groups alone is not a determining factor for emulsion stability.

Table 5. Volume fraction of water seperated as a function of time from model emulsion stabilized by polar fractions of oil G at 50⁰ C.

	%	Separation of water (%) T = 50⁰ C					
	(w/w)	Separation Time (hour)					
		0.5	1.0	2.0	3.0	4.0	5.0
GA 1	0.05		25	35	40	50	60
	0.025	60	90	100	100	100	100
GA 2	0.05			10	20	40	50
	0.025		30	60	85	95	100
GA 3	0.05			5	10	15	20
	0.025			10	20	40	50

Table 6. Volume fraction of water seperated as a function of time from model emulsion stabilized by polar fractions of oil B at 50⁰ C

	%	Separation of Water (%) T = 50˙ C					
	(w/w)	Separation Time (hour)					
		0.5	1.0	2.0	3.0	4.0	5.0
BA 1	0.25	15	15	30	35	40	40
	0.10	40	40	60	70	80	80
	0.25						10
BA 2	0.10				10	30	50
	0.05	25	50	75	85	85	85
	0.25						30
BA 3	0.10				10	40	70
	0.05	25	50	75	85	85	85

166

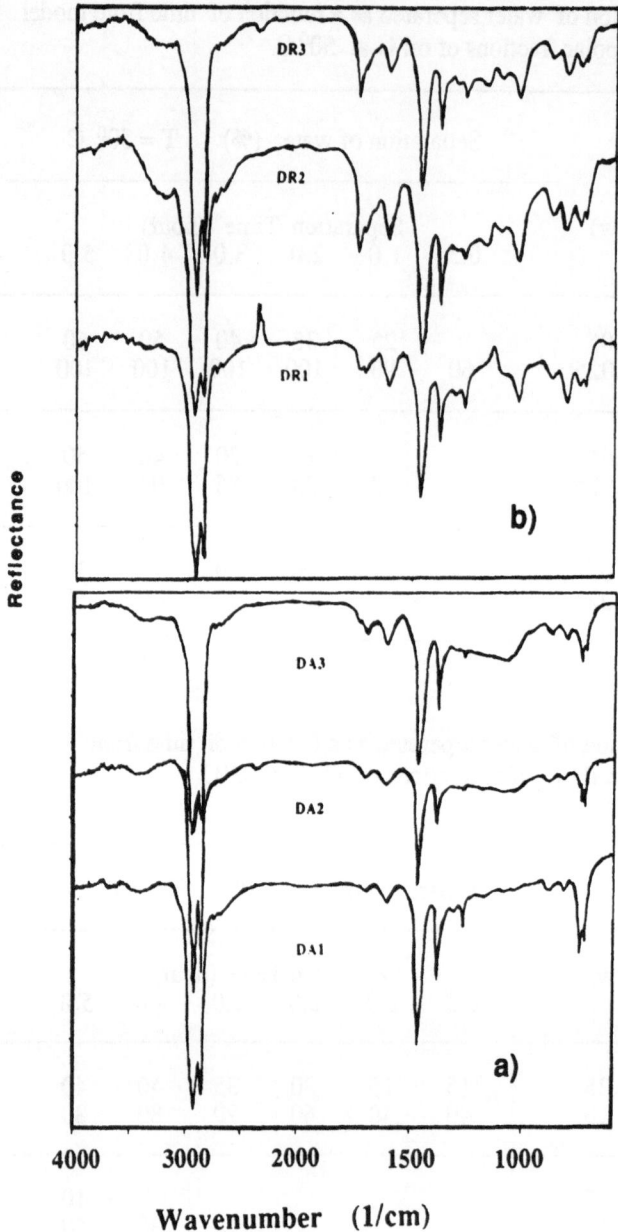

Fig. 2. Reflectance spectra of asphaltene and resin fractions of oil D.

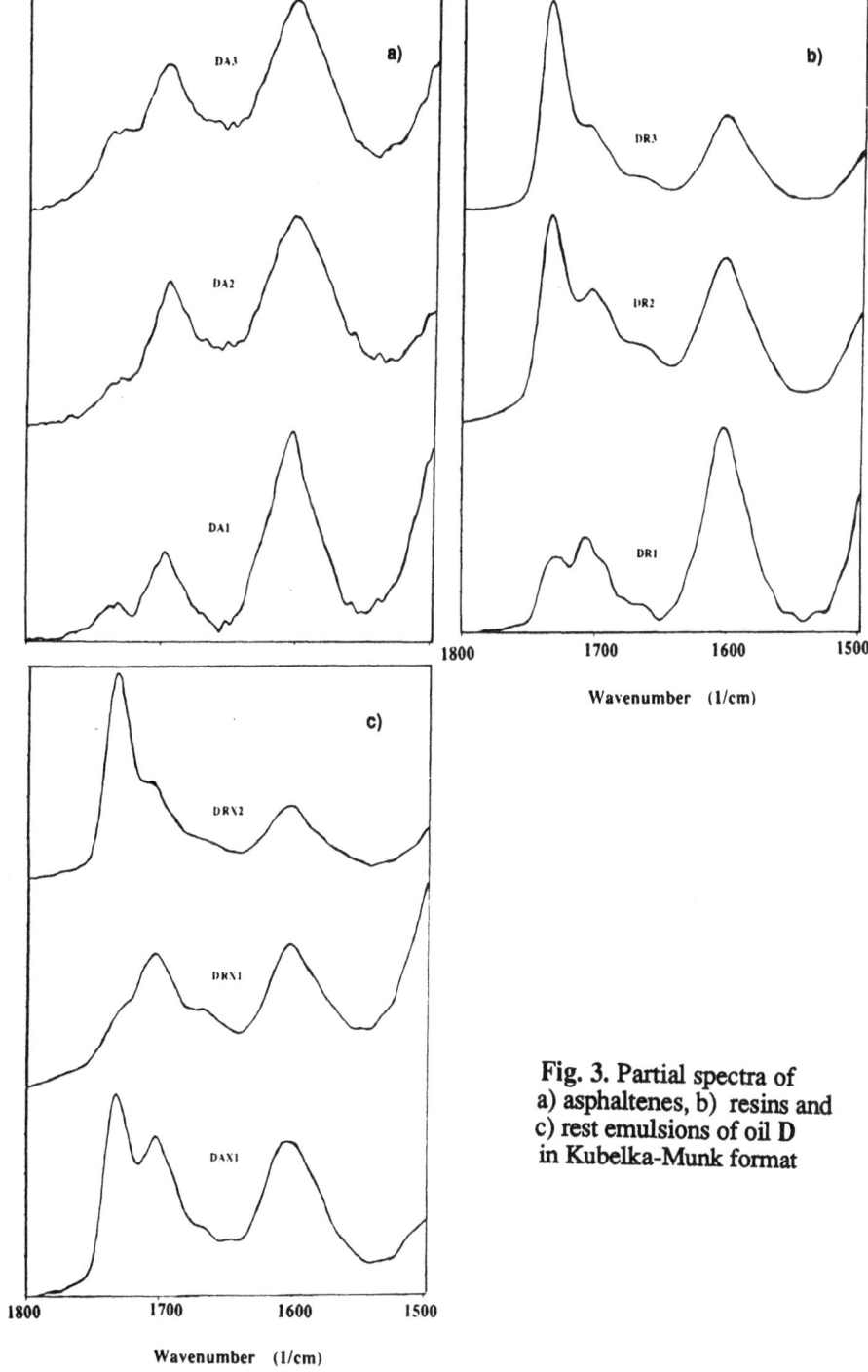

Fig. 3. Partial spectra of
a) asphaltenes, b) resins and
c) rest emulsions of oil D
in Kubelka-Munk format

168

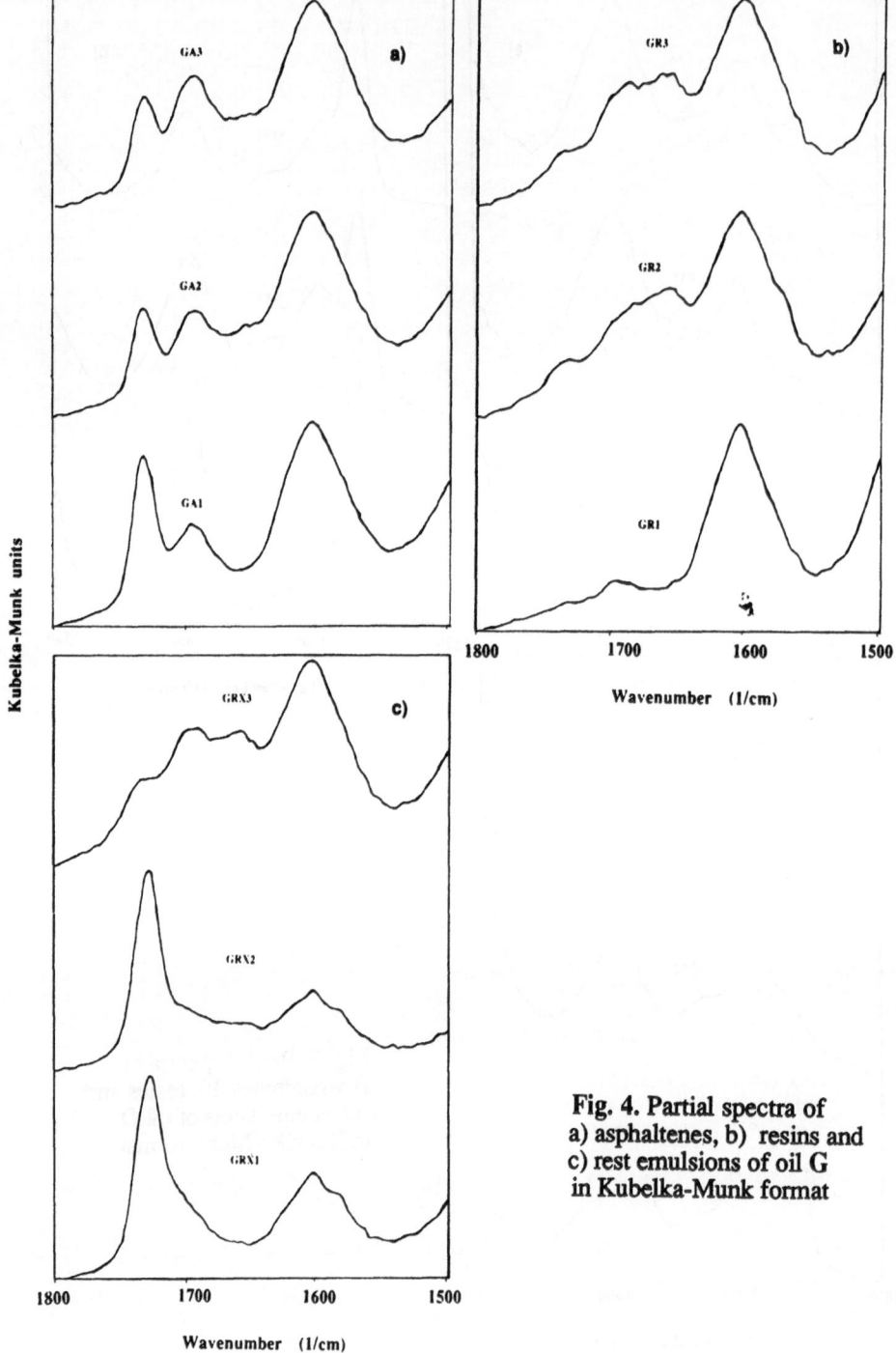

Fig. 4. Partial spectra of
a) asphaltenes, b) resins and
c) rest emulsions of oil G
in Kubelka-Munk format

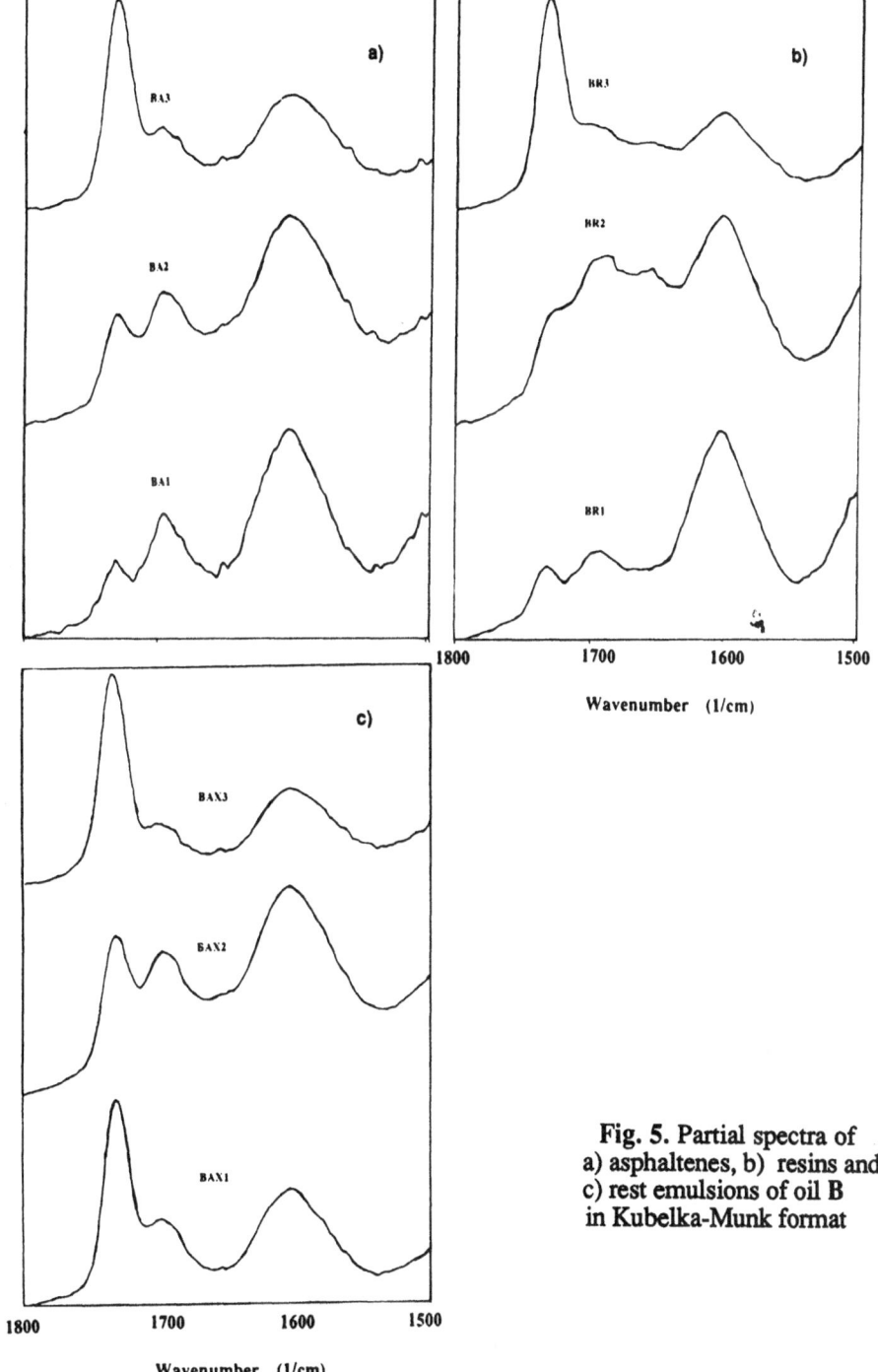

Fig. 5. Partial spectra of
a) asphaltenes, b) resins and
c) rest emulsions of oil B
in Kubelka-Munk format

Aromaticity and molecular weight may also be contributing factors.

3.2 FUNCTIONALITY DIFFERENCES

The functional group differences between the components of asphaltenes and resins can be compared in the reflectance format of the spectra shown in Figure 2. Partial spectra of the components in Kubelka-Munk format show differences in the carbonyl group concentrations. Asphaltene and resin components show some clear differences in functional group concentrations. Carbonyl absorptions arising from open chain carbonyl groups (1735 cm^{-1}), aromatic carbonyl groups (1700 cm^{-1}) and amide carbonyl groups (1650 cm^{-1}) vary in these components.

Absorption arising from NH and OH stretchings in the asphaltene components are exhibited by the broad band in between the region 3500 cm^{-1} and 3300 cm^{-1}. A contribution of these broad bands are due to carboxylic acis. Figures 3a, 4a and 5a show that the concentration of carbonyl groups increase from A1 to A3. This is true also for resins (Figures 3b, 4b and 5b).

The A2, A3 and R2, R3 fractions stabilize the emulsions in contrast to A1 and R1 fractions. These components contain more of the polar functional groups compared to 1 fractions. Selucky et al. [17] have concluded that the molecular weights of asphaltene fractions from silicagel decreased with increasing fraction of polar groups. This means that the fractions 1,2 and 3 of both asphaltenes and resins are in the order of decreasing molecular weight. Furthermore, the aromaticity of the fractions decrease from fractions 1 to 3. These observations clearly show that the molecular size of the interfacially active components, aromaticity and the relative concentrations of the functional groups present play an important role in the stabilization of emulsion.

Figures 3c, 4c and 5c show partial spectra of the interfacially active components of the rest emulsion. Asphaltenes and resins show that the fractions contain relatively high concentration of carbonyl groups, especially open chain carbonyl groups in most of the cases. This is evident when the corresponding parent asphaltene or resin fractions are compared. The interfacially active components of the rest emulsion extracted from resin fractions of oil D (Fig. 3c) show that they contain a high concentration of carbonyl groups. However, they form unstable emulsions. This shows that the number of molecules with suitable molecular size and functional groups distribution are few in these fractions.

All these observations indicate that there should be a compromise between the molecular size, polar functionality concentration and aromaticity to form stable water-in-crude oil emulsions.

The presence of carbonyl groups in the interfacially active components is a requirement for the emulsion stability. Instability of the emulsions formed by the resin fractions suggest that the presence of the carbonyl group alone is not sufficient to form strong hydrogen bonding at the w/o interface in order to prevent the coalescence of the water domains. Furthermore, the presence of OH and carbonyl groups may help the asphaltene molecules to form a mechanical barrier through hydrogen bonding around the water droplets and prevent coalescence. Among the carbonyl functionalities open chain carbonyl groups seem to play a very inportant role in the emulsion stability. This fact is evident in most of the fractions of total stability . Open chain carbonyl groups are suitable for hydrogen bonding compared to rigid carbonyl groups in the ring systems. The carbonyl groups on the open chains are considerably more elastic in nature and they can easily form intermolecular bondings to create the mechanical barrier required. Molecular size may be a crucial factor in forming this mechanical barrier. Aromatic rings in the asphaltene molecules can enhance the stability by forming π-type interaction with water molecules.

Asphaltenes and resin components of oils contain long aliphatic chains. The local concentration of aliphatic chains is higher in resins compared to asphaltenes. These chains can penetrate into the oil medium during emulsification.

Any model system of water-in-oil emulsion stabilized by asphaltenes or resins extracted from an oil is less stable than the emulsion based on the asphaltene and hydrocarbon portion of the original crude oil. This is because asphaltenes extracted from an oil cannot be dispersed or dissolved fully in a medium other than the original crude oil hydrocarbon portion [17]. This can be understood from the fact that the oil is produced from the thermal decomposition of kerogen and aliphatic and aromatic hydrocarbons are akins to the polar components released during the decomposition . The dispersal of asphaltene molecules in this medium is optimum.

4. Conclusion

Interfacially active components of oils are present both in resins and asphaltenes. Fractioning the resins and asphaltenes, and the behaviour of these fractions in stabilising the model water-in-oil emulsion, clearly shows that the molecular size and polar functionalities are important factors.

Infrared spectroscopy characterizes the different functionalities in the polar fractions and confirms the presence of especially open chain carbonyl groups in the interfacially active molecules. However size differentiation of the molecules cannot be achieved by the spectroscopic technique. The spectra of resins clearly show the high concentration of carbonyl groups in the molecules. The instability of the resin-stabilized emulsions is most likely due to deviation in molecular size. Furthermore, infrared spectroscopy shows that acidic and open chain carbonyl groups dominate in the interfacially active fractions. Aromaticity of the extracted fractions decreases from the benzene fraction to benzene/50% methanol fraction. A proper balance between aromaticity and polarity in the interfacially active molecules is necessary for emulsion stability.

Water-in-oil emulsion stability is a function of aromaticity, polarity, molecular size, and type of functionality in the stabilizing fraction.

5. Acknowledgement

The Norwegian National Multiphase Flow Research Programme (PROFF) is acknowledged for partial financial support.

172

6. References

1. Mackay, G.D.M., Mclean, A.Y., Betancourt, O.J. and Johnson, B.D. J. Inst. of Petr. 1973, **59**, 164

2. Sjøblom, J., Urdahl, O., Høiland. H., Christy, A.A. and Johansen, J. Progr. Colloid Polym. Sci., 1990, **82**, 131

3. Sjøblom, J., Søderlund, H., Lindblad, S., Johansen, E.J. and Skjårvø, I.M. Coll. Poly. Sci. , 1990, **268**, 389

4. Nordli, K.G., Sjøblom, J.,Kizling, J. and Stenius, P. Colloid Surfaces, in press.

5. Nordli, K.G., Sjøblom, J. and Stenius, P., Colloid Surfaces, in press.

6. Cason, J. and Graham, D.W. Tetrahedron, 1965, **21**, 471

7. Reisberg, J. and Dosher, T.M., Producers Monthly, 1956, **20**,46

8. Dunning, H.J., Moore, J.W. and Denekas, M.O. Ind. Eng. Chem., 1953, **45**, 1759

9. Johnson, E.C., Jr., Petrol Tech., 1976, 85

10. Christy, A.A., Dahl, B. and Kvalheim, O.M., Fuel, 1989, **68**, 430

11. Yen, T.F., Wu, W.H. and Chillinger, G.V., Energy Sources, 1984,**7** (3), 203

12. Layrisse, I. and Rivas, H., Dispersion Sci. & Technology, 1984, **5** (1), 1

13. Christy, A.A., Velapoldi, R.A., Karstang, T.V., Kvalheim, O.M., Sletten, E. and Telnaes, N., Chemomet. Int. Lab. Syst., 1987, **2**, 199

14. Fuller, M.P. and Griffiths, P.R., Anal. Chem., 1978, **50**, 1906

15. Christy, A.A., Hopland, A.L., Barth, T. and Kvalheim, O.M., Org. Geochem., 1989, **14**, (1), 77

16.Tissot, B.P., and Welte, D.H., 1978, Petroleum Formation and Occurrence, 2nd edn., 699pp. Springer, Berlin.

17. Selucky, M.L., Kim, S.S., Skinner, F. and Strauz, O.P., Chemistry of Asphaltenes, 1979, Advances in Chemistry series 195, 1979, 1981, 83

RHEOLOGICAL PROPERTIES OF EMULSION SYSTEMS

Th. F. TADROS
I.C.I. Agrochemicals, Jealott's Hill Research Station,
Bracknell, Berkshire RG12 6EY, U.K.

ABSTRACT. This overview will address a number of rheological properties of emulsion systems. It will start with the interfacial rheology of adsorbed emulsifier films. Both interfacial elasticity and viscosity play an important role in stabilisation of emulsions and in some cases a correlation can be obtained between the interfacial rheology and reduction of coalescence. The second part will deal with the rheological properties of concentrated emulsions that are stabilised by surfactants. When the surfactant layer thickness is small compared to the droplet size, the emulsion behaves as a hard sphere dispersion. The third part will deal with the viscoelastic properties of sterically stabilised emulsions, which were studied using dynamic measurements. Two systems will be described: oil-in-water emulsions stabilised by poly(ethylene oxide)-poly(propylene oxide) block copolymer and water-in-oil emulsions stabilised by a block copolymer of polyethylene oxide and polyhydroxystearic acid. In some cases, information could be obtained on the adsorbed layer thickness.

1. Introduction

It is fairly well established that emulsions, on standing, may undergo a number of breakdown processes, namely creaming or sedimentation, flocculation, Ostwald ripening, coalescence and phase inversion. Most of these processes are determined by the interaction forces between the droplets, i.e., electrostatic repulsion, steric interaction and van der Waals attraction. The balance of these forces determines the state of the emulsion and the structure of the units formed. These various breakdown processes affect the flow characteristics (rheology) of the emulsion. For example, any flocculation results in an increase in the relative viscosity of the emulsion, since the flocs tend to entrap liquid and the floc volume fraction can be several times larger than the volume fraction of the single droplets. In contrast, Ostwald ripening and coalescence may lead to a reduction in the

173

J. Sjöblom (ed.), Emulsions – A Fundamental and Practical Approach, 173–188.
© 1992 *Kluwer Academic Publishers*.

relative viscosity, particularly in high volume fraction emulsions. Phase inversion results in a dramatic decrease in the viscosity of the system, since the external medium volume fraction, which now forms the disperse phase is usually smaller than that of the disperse phase before inversion. Inspite of the large dependence of the rheology of the emulsion on its stability, only a few systematic investigations have been carried out to correlate the rheological parameters to the various breakdown processes. This is due to the difficulty in the interpretation of the complex rheological responses of various emulsion systems. In addition, the breakdown processes in many emulsion systems may take place simultaneously and variations in rheological parameters with system variables and conditions may not be simply related to stability. With emulsions the surfactant or polymer film at the liquid/liquid interface requires investigations of interfacial rheology (interfacial elasticity and viscosity), which may play an important role in its stability. Thus, studying the rheological properties of emulsion systems is by no means an easy task and it requires a combination of various techniques.

In this preliminary overview an attempt will be made to highlight the most important rheological properties of emulsions, giving examples to illustrate their value in evaluation of emulsion stability. The overview will start with a section on interfacial rheology and its correlation with emulsion stability. The next section will deal with the viscosity-volume fraction relationship and a comparison will be made with solid/liquid dispersions. The final section will deal with the viscoelastic properties of concentrated oil-in-water and water-in-oil emulsions which have been recently investigated in our laboratory.

2. Interfacial Rheology

It has long been argued that interfacial rheology, namely interfacial viscosity and elasticity, play an important role in emulsion stability. Most surfactants (and mixtures) and macromolecules adsorbed at the interface show a high surface-induced viscosity. This high surface viscosity (which can be orders of magnitude higher than the bulk viscosity) can be accounted for in terms of the orientation of the molecules at the oil/water interface. For example, orientation of surfactant molecules with the hydrophobic portion pointing towards to (or dissolved in) the oil and the polar groups pointing to the aqueous phase results in resistance to compression. This resistance is described by a two dimensional surface pressure π, which is given by the difference between the interfacial tension of the surfactant free interface γ_0 and that of the covered interface γ. Since

γ_0 is of the order of 30-50 mNm^{-1}, whereas γ may reach values that are a fraction of mNm^{-1}, surface pressures of the order 30-50 mNm^{-1} may be reached. These high surface pressures account for the high surface viscosity produced with many surfactant films. Macromolecular films also give high surface viscosities, since the macromolecular film (that consist of loops and tails) resists compression in two dimensions.

Interfacial films show both viscosity and elasticity. Films are elastic if they resist deformation in the plane of the interface and if the surface tends to recover its natural shape when the deforming forces are removed [1]. Similar to bulk materials, interfacial elasticity can be measured by static and dynamic methods. For example, a constant stress may be applied to the surface film, in two dimensions, and the deformation or compliance measured as a function of time. In dynamic measurements a sinusoidal oscillation of strain or stress is applied to the film and the phase lag between stress and strain is measured, allowing one to obtain the storage and loss modulus. This will be discussed below for bulk measurements of viscoelasticity. An important rheological parameter is the surface dilational elasticity ε, which is given by the following equation,

$$\varepsilon = \frac{d\gamma}{d\ln A} = A\frac{d\gamma}{dA} \qquad (1)$$

where A is the area of the interface and γ is the interfacial tension. The dilational elasticity can be measured using a Langmuir trough technique with two movable barriers. It has been argued that the dilational elasticity plays an important role in prevention of coalescence in emulsions. The latter process occurs as a result of thinning and disruption of the liquid film between the droplets. When the latter approach closely during a Brownian collision, or in flocculated or creamed (or sedimented) emulsions, surface or film fluctuations may occur. This leads to interfacial tension fluctuations due to the expansion of the interface. The surface dilational elasticity, described by equation (1), tends to dampen the surface fluctuations, thus preventing thinning and disruption of the liquid film between the droplets. As a result of interfacial tension gradients, surfactant molecules diffuse from the bulk to the interface to reduce these fluctuations by adsorption at areas deficient in surfactant molecules. These surfactant molecules will drag solvent to the liquid film making it thicker. The combined effect of surface elasticity and flow of solvent to the liquid film is usually referred to as the Gibbs-Marangoni effect.

Several examples may be quoted to illustrate the role of

interfacial rheology in maintaining emulsion stability. The addition of cosurfactants such as long chain alcohols is well known to enhance the stability against coalescence of oil/water emulsions. Although the alcohol is not particularly surface active, its presence at the interface, containing a surfactant such as sodium dodecyl sulphate, tends to lower the interfacial tension of the film. Prince et al [2] found that the dilational elasticity, ε, increased markedly in the presence of the alcohol and, therefore, attributed the enhanced stability as due to such a high surface elasticity. Other authors, attributed the enhanced stability to a high interfacial viscosity η_s [3], although Prince et al [2] argued against this since they found that the film stability was not very sensitive to either temperature changes (although of course η_s is) or the concentration of alcohol, which had a pronounced effect on η_s.

Another example that may be quoted to illustrate the role of surface rheology in emulsion stability was the work of Biswas and Haydon [4]. These authors have systematically investigated the rheological characteristics of various proteins, namely albumin, poly(ϕ-L-lysine) and arabinic acid at the O/W interface and correlated the measurements with the stability of the oil droplets at a planar oil/protein solution interface. The viscoelastic properties were studied using two dimensional creep and stress relaxation measurements, using a specially designed interfacial rheometer. In the creep experiments, a constant torque (expressed in mNm^{-1}) was applied and the resulting deformation γ (in radians) was recorded as a function of time. The creep recovery was measured by recording the deformation when the stress was removed. In the stress relaxation experiments, a certain deformation γ was produced in the film by applying an initial strain, the stress required to maintain a constant deformation was measured as a function of time. In this case, the stress decreases exponentially with time. Fig. 1 shows a typical creep curve for bovine serum albumin at the petroleum ether/water interface. The curve shows an initial, instantaneous deformation characteristic of an elastic body, followed by a nonlinear flow that gradually declines and approaches the steady state flow of a viscous body. After 30 minutes, when the external force was withdrawn, the film tended to revert to its original state, with an initial instantaneous recovery followed by a slow one. The original state was not obtained even after 20 hours and the film seemed to have undergone some flow. This behaviour illustrates the viscoelastic property of the bovine serum albumin.

Biswas and Haydon [4] also found a striking effect of pH on the rigidity of the protein film. This is illustrated in Fig. 2, where the shear modulus G and surface viscosity η_s are plotted as functions of pH. The elasticity of the film is

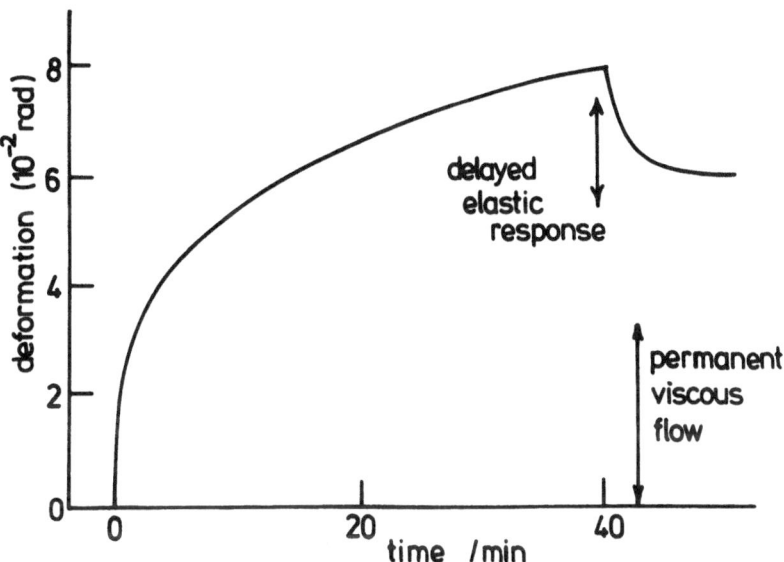

Fig. 1 Creep curve of an adsorbed bovine serum albumin film
(pH = 5.2) at a petroleum ether/water interface, at a constant
stress of 0.0116 Nm⁻¹.

seen to be at a maximum at the isoelectric point of the
protein. Biswas and Haydon [4] then measured the rate of
coalescence of petroleum ether drops at a planer O/W interface
by measuring the lifetime of a droplet resting beneath the
interface. The half-life of the droplets was plotted as a
function of pH, as shown in Fig. 2, which clearly illustrates
the correlation with G or η_s. Biswas and Haydon [4] derived
an equation relating the time of coalescence τ with the

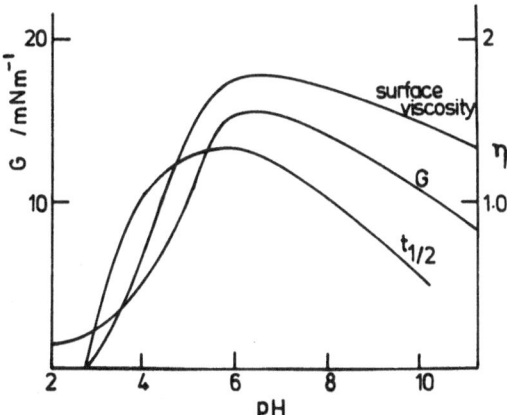

Fig. 2 Shear modulus, surface viscosity and half-life of
petroleum ether drops beneath a plane petroleum ether/0.1 mol
dm⁻³ aqueous KCl solution interface.

viscoelasticity of the film, the thickness of the adsorbed film h and the critical distortion of the plane interface under the weight of the drop, i.e.,

$$\tau \ - \ \eta_s \left[3C' \ \frac{h^2}{A} \ - \ \frac{1}{G} \ - \ \phi(t) \right] \tag{2}$$

where G is the (instantaneous) elasticity, η the long time viscosity (i.e., for infinite time of retardation), $\phi(t)$ elastic deformation per unit stress and 3C' is a critical deformation factor. Equation (2) predicts that: (i) the life time of the drop (τ) increases with increase of the viscosity of the protective film; (ii) the rate process of coalescence is not influenced by the instantaneous elasticity, but this quantity is likely to set a limit on the process through the critical deformation factor 3C'; (iii) the life time should depend on the film thickness and vary linearly with h^2 if the retarded elasticity $\phi(t)$ is neglected; (iv) τ should be a fixed but not a fluctuating quantity.

The results obtained by Biswas and Haydon [4] indicate clearly that no significant stabilisation occurred in the case of non-viscoelastic films. However, the presence of viscoelasticity is not sufficient to confer stability when drainage is rapid. For example, it was found that the highly viscoelastic films of bovine serum albumin or pepsin could not stabilise W/O emulsions; the same was found with pectin and gum arabic. It was clear in these cases that the drainage of the film was rapid even from rigid films, e.g., W/O droplets in the case of bovine serum albumin. In fact, as expected, it was only after solvent drainage had taken place, and the disperse phases were still separated by a film of high viscosity, that enhanced stability occurred. It was concluded from these investigations that, in agreement with the prediction of theory in this case, the requirements for stability to coalescence were the presence of a film of appreciable thickness. It is also necessary that the main part of the film should be located on the continuous side of the interface.

Several other examples may be found in the literature, in which a correlation between the interfacial viscosity of macromolecular stabilised interfaces with droplet stability may be found. However, there are also a number of cases where stable emulsions could be prepared without any significant interfacial viscosity or rigidity. Polyelectrolytes, such as poly(methacrylic acid) and carboxymethyl cellulose, were found to stabilise droplets better in their acid form than as salts.

This behaviour correlates with the interfacial viscosity behaviour, but not with the bulk solution viscosity.

It can be concluded, therefore, from the above discussion that interfacial rheological measurements should not be applied for the prediction of emulsion stability without consideration of other factors such as film thickness and film drainage. However, interfacial rheology still offers a powerful tool for understanding the properties of surfactant and macromolecular films at the liquid/liquid interface. In cases where a correlation exists between the viscosity and/or the elasticity of the film and emulsion stability, it is possible to modify the film properties by inclusion of additives, and/or alterations to the structures of the molecules to enhance the rheological parameters and hence emulsion stability.

3. Viscosity-Volume Fraction Relationship for Oil/Water and Water/Oil Emulsions.

The relative viscosity (η_r) - volume fraction (ϕ) curves for paraffin oil/water emulsions is shown in Fig. 3. Four emulsion systems were prepared using nonionic surfactants, namely Synperonic NPE 1800 and its analogues [5]. These surfactants have the following structural formula: $C_9H_{19}-C_6H_5-(CH_2-CH(CH_3)-O)_m-(CH_2-CH_2-O)_n-OH$. They all contain the same hydrophobic chain (nonyl phenyl and 13 moles propylene oxide, but have different moles of ethylene oxide: 27 for Synperonic NPE 1800, 48 for NPE A, 80 for NPE B and 174 for NPE C respectively. The average droplet size of each emulsion was determined using the Coulter counter. The volume mean diameter (VMD) of each emulsion is given in the legends of Fig. 3. The molecular weight and hydrodynamic thickness of these surfactant molecules was determined before [6] and the results are shown below:

	Synperonic NPE 1800	NPE A	NPE B	NPE C
Mw	2180	3080	4460	8650
δ_h (nm)	5.8	6.4	8.5	11.6

The above emulsion system is fairly simple, since it is likely that the nonyl phenyl and propyleneoxide chain are on the oil side of the interface, whereas the polyethylene oxide chain is on the aqueous side of the interface. The hydrodynamic thicknesses of the surfactants, given above, clearly show that they are much smaller than the droplet radius. Therefore, these sterically stabilised emulsions may be represented by hard spheres with an effective radius $R_{eff} = R + \delta_h$. This can be tested by fitting the data to the hard sphere model suggested by Dougherty and Krieger [7,8]. By application of the theory of corresponding states, these

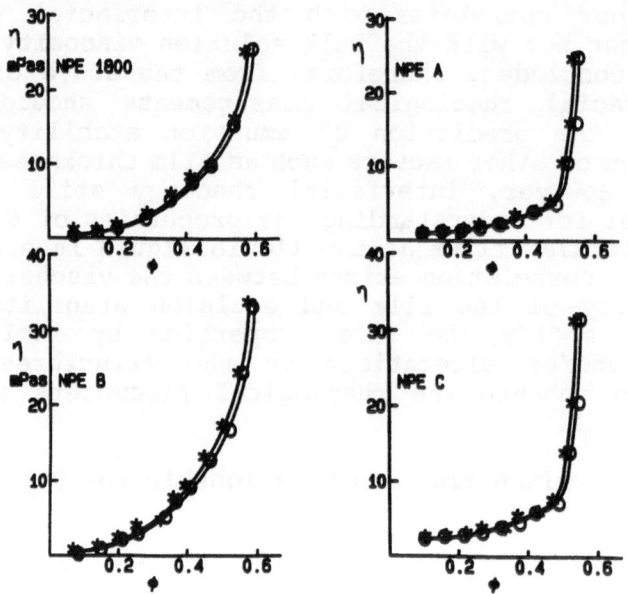

Fig. 3 η_r - ϕ curves for paraffin oil/water emulsions: (a) Synperonic NPE 1800; VMD = 3.5 µm; (b) NPE A; VMD = 4 µm; (c) NPE B; VMD = 4.5 µm; (d) NPE C; VMD = 5 µm. x, experimental results; o, theoretical values according to the Dougherty-Krieger equation.

authors derived the following equation for the relative viscosity,

$$\eta_r = \left[1 - \left(\frac{\phi}{\phi_p}\right)\right]^{-[\eta]\phi_p} \qquad (3)$$

where $[\eta]$ is the intrinsic viscosity which has a theoretical value of 2.5 for rigid spheres and ϕ_p is the maximum packing fraction which is equal to 0.64 for random packing and 0.74 for hexagonal packing of monodisperse spheres. However, Krieger [7] showed that with hard sphere dispersions ϕ_p is close to 0.6. Since the emulsions are polydisperse, a higher value for ϕ_p is to be expected. The value of ϕ_p for each emulsion was estimated from plots of $\eta^{-1/2}$ versus ϕ, which gave

straight lines. Extrapolation to $\eta^{-1/2} = 0$ (i.e., $\eta = \infty$) gave
the value of ϕ_p. The values obtained were 0.73, 0.73, 0.72
and 0.69 for Synperonic NPE 1800, NPE A, NPE B and NPE C
respectively. Using the calculated ϕ_p values, plots of η_r
versus ϕ were constructed and these are shown in Fig. 3. It
is clear from these results that the emulsions stabilised with
Synperonic NPE surfactants approximate very closely to the
behaviour of hard sphere dispersions.

The relative viscosity-volume fraction curve for water-in-oil
emulsions [9] is shown in Fig. 4. Isoparaffinic oil (Isopar
M) was used in this case and the emulsions were prepared using
an A-B-A block copolymer of PHS-PEO-PHS, where PHS refers to
poly-12-hydroxystearic   acid   and   PEO   refers   to
polyethyleneoxide. The weight average molecular weight of the
polymer is 6809, while its number average is 3499. The
emulsion had a narrow droplet size distribution with a z-
average radius R of 183 nm, as determined by photon

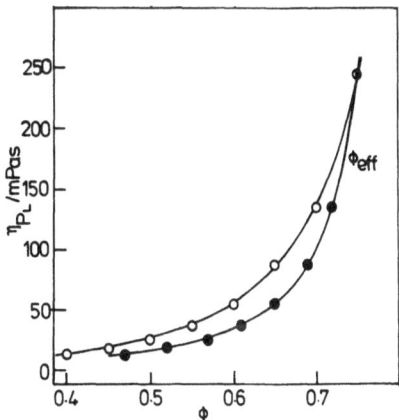

Fig. 4 Viscosity-volume fraction curves for w/o emulsions.

correlation   spectroscopy.   As   shown   before   [10],   the
viscosity-volume fraction curve may be used to obtain the
adsorbed layer thickness as a function of ϕ. Assuming that
the w/o emulsions behave as near hard sphere dispersions, it
is possible to apply the Dougherty-Krieger equation to obtain
the effective volume fraction. Using equation (3), ϕ_{eff} can
be calculated from η_r provided a reasonable estimate can be
made of both $[\eta]$ and ϕ_p. $[\eta]$ was taken to be equal to 2.5,
whereas ϕ_p was estimated from a plot of $1/\eta_r^{1/2}$ versus ϕ, as
described above. ϕ_p was found to be 0.84, which is a
reasonable   value   considering   the   polydispersity   of   the

emulsion. Fig. 4 shows the ϕ_{eff} values. From ϕ_{eff}, the adsorbed layer thickness δ was calculated using the following expression,

$$\phi_{eff} = \phi \left[1 + \left(\frac{\delta}{R} \right) \right]^3 \qquad (4)$$

A plot of δ versus ϕ is shown in Fig. 5. It can be seen that δ decreases linearly with increase in ϕ. The value at $\phi = 0.4$ is 10 nm, which is a measure of the fully extended PHS chain.

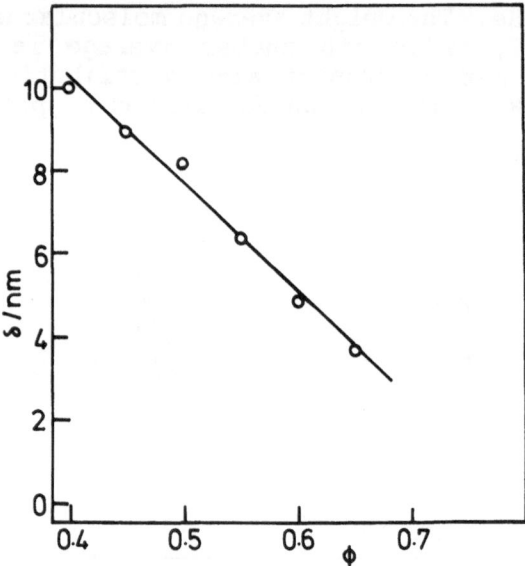

Fig. 5 Variation of δ with ϕ

At such relatively low ϕ, there will be no interpenetration of PHS chains since the distance between the droplets is fairly large. The value of δ obtained from rheology is in close agreement with the results recently obtained from thin liquid film measurements between two water droplets [11]. It is also in agreement with the results obtained by Ottewill and coworkers [12,13] using compression and small angle neutron scattering. The decrease in δ with increase in ϕ is similar to the results obtained using latex dispersions stabilised with grafted PEO chains. This reduction in δ with increase in ϕ may be attributed to interpenetration and/or compression of the chains on increasing ϕ. If complete interpenetration is possible, the δ value can be reduced by half that obtained in dilute dispersions. Indeed the results of Fig. 5 show that

δ is reduced to 4 nm at $\phi = 0.65$. The reduction in δ can also be attributed to compression of the chains on close approach, without need of invoking any interpenetration. It is also probable that a combination of interpenetration and compression may occur.

4. Viscoelastic Properties of Concentrated o/w and w/o Emulsions.

The viscoelastic properties of concentrated o/w and w/o emulsions were investigated using oscillatory measurements (a Bohlin VOR, Bohlin Reologie, Lund Sweden, was used for these measurements). Concentric cylinder platens were used for such measurements which were carried out at $25 \pm 0.1°C$. In oscillatory measurements, the response in stress of a viscoelastic material to a sinusoidally varying strain is monitored as a function of time. The stress response is also a sinusoidally varying function in time, but for a viscoelastic material it is shifted out of phase with the strain. This produces a phase angle shift δ ($\delta = \Delta t\, \omega$, where ω is the frequency in radians per second, i.e., $\omega = 2\pi\, v$, where v is the frequency in Hz). From measurement of the angular deflection and the resulting torque on the detector shaft used for monitoring the stress, the various viscoelastic parameters (complex modulus G^*, storage modulus G', loss modulus G'', phase angle shift δ and dynamic viscosity η') are obtained as functions of strain amplitude and frequency. The strain amplitude is gradually increased (at a fixed frequency) from the smallest possible value at which a measurement can be made and the rheological parameters are monitored as functions of strain amplitude γ_0. Initially the rheological parameters remain constant and independent of γ_0, but above a critical γ_0 these parameters begin to change with increase in strain amplitude. The region where the rheological parameters are independent of strain amplitude is the linear viscoelastic region, and frequency measurements are only made in this region.

The various rheological parameters are obtained from the measured stress and strain amplitudes and the phase angle shift,

$$G* \; - \; \tau_o / \gamma_o \qquad\qquad (5)$$

$$G' = G* \cos \delta \qquad (6)$$

$$G'' = G* \sin \delta \qquad (7)$$

$$\eta' = G''/\omega \qquad (8)$$

G' is a measure of the energy stored elastically in the system, whereas G'' is a measure of the energy dissipated as heat during viscous flow.

As an illustration Fig. 6 shows plots of G*, G', G'' and η' as functions of frequency in Hz for an isoparaffinic oil/water emulsion [14] with a volume fraction of 0.6. The emulsion was stabilised with an A-B-A block of PEO-PPO-PEO, namely Synperonic PE, with an average of 47.3 polypropylene oxide (PPO) units and 41.6 polyethylene oxide (PEO) units. The mean volume diameter of the droplets was 0.98 μm (as determined by the coulter counter). It can be seen from Fig. 6 that below a certain frequency G''>G', whereas above that frequency G'>G''. At high frequency (short time) most of the

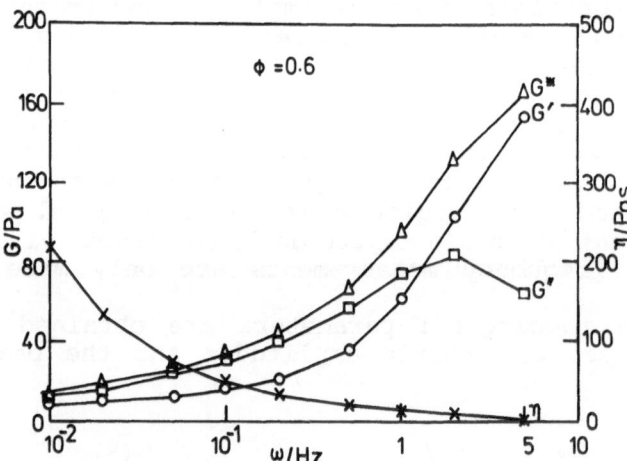

Fig. 6 Variation of G*, G', G'' and η' with frequency (Hz) for an o/w emulsion, φ = 0.60.

energy is stored elastically in the system and G'' tends to

zero. η′ is also shown to decrease rapidly with increase in frequency. Similar trends were obtained with other volume fractions, and the moduli values increased with increase in φ, whereas the cross frequency point at which G′ = G′′ was shifted to lower ω values as φ increases. This reflected the increase in relaxation time with increase in φ. The cross-over frequency may be related to the characteristic (relaxation) time for the concentrated emulsion, i.e., t_r = 1/ω. A plot of t_r versus φ is shown in Fig. 7, which clearly shows the rapid increase of t_r with φ, when the latter exceeds 0.54.

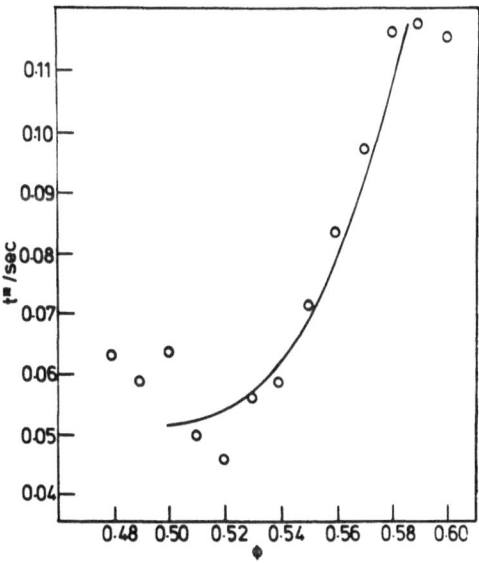

Fig. 7 Variation of t_r with φ for o/w emulsions.

Similar trends are obtained when G*, G′ and G′′ (at ω = 2 Hz) are plotted versus φ as shown in Fig. 8. At φ < 0.56, G′′> G′, whereas at φ > 0.56 G′ > G′′. This behaviour reflects the steric interaction in the emulsion system. At φ < 0.56, the droplet-droplet separation is probably larger then twice the adsorbed layer thickness and hence the steric interaction is not very strong. However, when φ > 0.56, the droplet-droplet separation may become smaller than twice the adsorbed layer thickness and the steric interaction becomes strong, increasing in magnitude as φ increases. This explains the predominantly more elastic behaviour of the emulsion when φ exceeds 0.56.

Similar results were obtained for the w/o emulsions stabilised by the A-B-A block of PHS-PEO-PHS. Fig. 9 shows the variation of relaxation time t* with φ, whereas Fig. 10 shows the variation of G′ and G′′ (at ω = 1 Hz) with φ. It can be seen from Fig. 9 that t* increases rapidly with increase of φ, when the latter is greater than o.67. Fig. 10 also shows that G' > G'', when φ exceeds o.67. This reflects the steric interaction between the PHS chains in these w/o emulsions. As mentioned above, at high volume fractions of

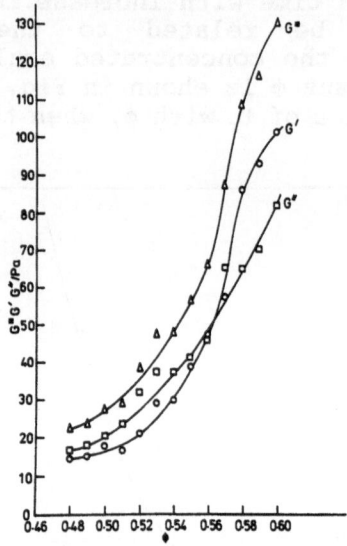

Fig. 8 Variation of G*, G' and G'' (at ω = 2 Hz) for o/w emulsions.

the emulsions, interpenetration and/or compression of the PHS chains may occur and this leads to a predominantly elastic response. Due to the polydisperse nature of the emulsion such

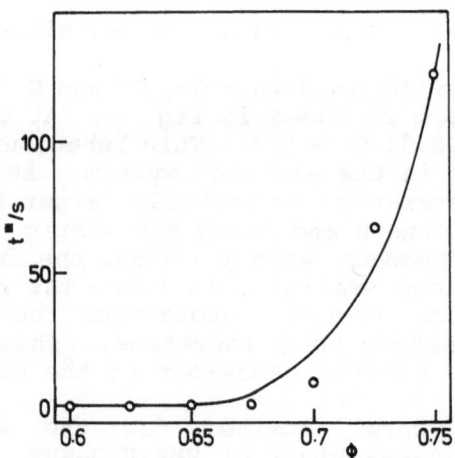

Fig. 9 Variation of t* with φ for w/o emulsions.

interaction is felt at relatively high volume fractions of the emulsions. With monodisperse systems such as latex dispersions, the transition from a predominantly viscous to

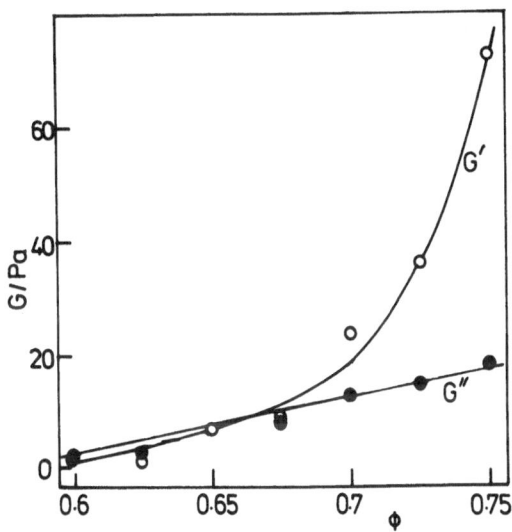

Fig. 10 Variation of G' and G'' with φ for w/o emulsions.

a predominantly elastic response occurs at lower volume fractions. For example, with polystyrene latex dispersions (with a radius of 175 nm) containing grafted PEO chains of molecular weight of 2000, the transition from a viscous to an elastic response occurred at φ = 0.48. This results from the higher ratio of δ/R and the relatively more monodisperse nature of the latex when compared with the emulsion systems.

The above results on the viscoelastic behaviour of emulsions clearly demonstrate their value in studying interaction between emulsion droplets that are sterically stabilised. Inspite of the polydisperse nature of the emulsion systems, the results obtained follow the same as were trends obtained with model monodisperse latex dispersions. Further investigations are required to study the viscoelastic behaviour of emulsions that are flocculated, e.g., by addition of free (non-adsorbing) polymers or by reduction of solvency of the medium for the chains.

ACKNOWLEDGEMENTS

The results on rheology of oil/water and water/oil emulsions were obtained by my coworkers: Mr. P. Winn, Miss J.L. Cutler and Mr. P.K. Thomas and these will be published in more detail in separate papers. I am grateful to my coworkers for allowing me to present some of their data in this review. I am also grateful to Mr. P.K. Thomas for improving the text.

References

1. Criddle, D.W. (1960), "The Viscosity and Elasticity of Interfaces", in "Rheology, Theory and Applications", Editor, Eirich, F.R., Vol. 3, Chapter 11, p. 429.

2. Prince, A., Arcuri, C., and Van den Tempel, M. (1967), J. Colloid Interface Sci., **24**, 811.

3. Miles, G.D., Ross, J., and Shedlovsky, L. (1950), J. Amer. Oil Chemists' Soc., **27**, 268.

4. Biswas, B., and Haydon, D.A. (1963), Proc. Royal Soc., **A271**, 296, 317.

5. Tadros, Th. F., and Winn, P., to be published.

6. Van den Boomgaard, A. (1985), Thesis, Wageningen University, The Netherlands.

7. Krieger, I.M., and Dougherty, M. (1959), Trans. Soc. Rheol., **3**, 137.

8. Krieger, I.M. (1972), Adv. Colloid Interface Sci., **3**, 111.

9. Tadros, Th.F., and Thomas, P.K., to be published.

10. Prstidge, C., and Tadros, Th.F. (1988), J. Colloid Interface Sci., **124**, 660.

11. Aston, M.S., Herrington, T.M., and Tadros, Th.F. (1989), Colloids and Surfaces, **40**, 49.

12. Cairns, R.J., and Ottewill, R.H. (1976), J. Colloid Interface Sci., **54**, 45.

13. Ottewill, R.H., Personal communication.

14. Cutler, J.L., and Tadros, Th.F., to be published.

PHARMACEUTICAL EMULSIONS AND CREAMS

H.E. JUNGINGER
Division of Pharmaceutical Technology
Center for Bio-Pharmaceutical Sciences
P.O. Box 9502
Leiden University
2300 RA Leiden
The Netherlands

ABSTRACT. O/W-creams with crystalline gel structures in general are four phase systems consisting of a hydrophilic gel phase and a lipophilic gel phase. The first one is able depending on the polarity of surfactants to incorporate a certain amount of water in between the lamellae and to further fix the bulk water (third phase) as coherent (outer) phase. The lipophilic gel phase stabilizes the dispersed (inner) phase (fourth phase). This concept showed its validity for gel structures in O/W-creams built up by ionic and non-ionic surfactants.

Amphiphilic creams show the properties that they can be transformed either to O/W-creams by water addition or to W/O-creams by addition of oil. Their essential gel structures consist of an amphiphilic gel structure with a limited swellability with water and a hydrophilic gel structure (comparable to that of O/W-creams) with unlimited water swellability. The amphiphilic gel structures with limited swellability are built up by mixed crystals consisting of e.g. glycerolmonostearates or non-ionic (EO)$_n$-ether-surfactants (n=2-3) and cetostearyl alcohol. At a moderate water content the swollen amphiphilic gel structure represents the basic features of such an amphiphilic cream. On dilution with oil a W/O-cream results and on dilution with water the hydrophilic gel phase swells additionally resulting in an O/W-cream.

W/O-creams represent the most simple systems with respect to colloidal gel structures: water droplets are stabilized by surfactant layer(s). The coherent oily phase is strengthened by a three-dimensional gel-network consisting of crystalline paraffins or crystalline fatty substances.

1. Introduction

Ointments and creams have been used for many centuries to improve the healing conditions of wounds and to treat empirically skin diseases as well as to retard the aging process of the skin and to preserve natural beauty. Dermatological preparations and cosmetics as the modern generations of these systems are also named "semisolids" due to their unique properties of being in the solid state under ambient conditions and getting transformed to the liquid state when mechanically stressed during the application on the skin. These properties allow the systems to spread easily on the surface of the skin. Depending on the formulation ointments are able to remain on the surface of the skin and show skin protection and occlusion whereas creams may be able to penetrate into the different layers of the skin, especially the stratum corneum (horny layer) and to exert interactions with both the keratines in the horny cells (corneocytes) and the lipid bilayers in which the horny cells are embedded.

Ointments are defined as water free systems ranging from systems with extreme lipophilic properties (e.g. liquid paraffins, white petrolatum and paraffin waxes) until systems with good hydrophilic properties (e.g. polyethylenglycol ointments). Ointments with medium polar character are W/O- or O/W-absorption bases (mostly lipophilic vehicles) in which O/W- or W/O-emulsifiers are incorporated. Addition of water to these absorption bases results in the formation of O/W- or W/O-creams, which are generally defined as water containing systems. We speak of O/W-creams when water forms the outer (continuous) phase and of W/O-creams when the lipid phase is the outer phase. Creams may be simplified as O/W- or W/O-emulsions which are predominantly stabilized by a colloidal gel structure. In his pioneering work Münzel (1953) described semisolids as "plastic gels" for cutaneous application.

189

J. Sjöblom (ed.), Emulsions – A Fundamental and Practical Approach, 189–205.
© *1992 Kluwer Academic Publishers.*

The dominant colloidal structural elements of semisolid preparations are three-dimensional colloidal solid networks in which a liquid is incorporated. Such a bi-coherent (sponge-like) structure may be referred to as a gel. These gel structures may be either in a crystalline or in a liquid-crystalline state. The type of the gel structure mainly determines the systems'
* rheological properties
* stability
* possible interactions with the skin
* drug release
The knowledge of the colloidal structure of these semisolid systems is of essential importance for their proper application on the skin.

2. Colloidal Gel Structures of O/W-Creams

2.1 Water Containing Hydrophilic Ointment

The formulation of the water containing hydrophilic ointment DAB 9 (German Pharmacopoeia, 9th edition) is as follows:
- Emulsifying wax   9,0   % wt/wt
- Liquid paraffins   10,5   % wt/wt
- White petrolatum   10,5   % wt/wt
- Water   70,0   % wt/wt

X-ray investigations by the Kratky low angle technique and by goniometry in combination with quantitative differential scanning calorimetry (DSC) led to the following structure model of the water containing hydrophilic ointment DAB 9 (Junginger, et al. (1979), (1984a), (1984b)). It was found that such O/W creams may be regarded as four-phase systems (Fig. 1). The dominant matrices are the hydrophilic and the liphophilic gel phases. Both phases build up layers of mixed crystals. The bilayers are oriented in such a way that the hydrocarbon tails are directed towards each other, as are the polar groups (Fig. 1, region a). The hydrophilic gel phase consists of cetostearyl alcohol and the whole amount of the ionic sodium-n-alkylsulfates; thus strong hydrophilic moieties and hydrophobic cores counteract each other (Fig. 1, region a).

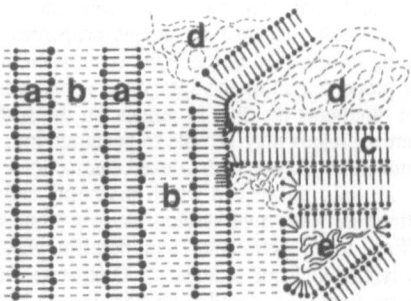

Fig. 1.   Gel structures of the Water Containing Hydrophilic Ointment DAB 8, a: mixed crystal bilayer of cetostearylalcohol and cetostearylalcoholsulfate, b: interlamellarly fixed water layer, a+ b: hydrophilic gel phase, c: lipophilic gel phase (cetostearylalcohol-semihydrate), d: bulk water phase, e: lipophilic components (dispersed phase)

Water molecules if present, are inserted between the polar sulfate and alcohol groups of the surfactant's molecules (Fig. 1, region b). The regions a and b together form the hydrophilic gel phase. The water molecules interlamellarly fixed in the hydrophilic gel phases are in equilibrium with the molecules of the bulk water phase (Fig. 1, region d). It is assumed that the interlamellary fixed water molecules exhibit other physicochemical properties than those of the bulk water phase.

The surplus of cetostearyl alcohol not incorporated in the hydrophilic gel phase, builds up a matrix with lipophilic properties (Fig. 1, region c) called lipophilic gel phase. The inner or dispersed phase (Fig. 1, region e) is mainly immobilized mechanically by this lipophilic gel phase. The lipophilic gel phase consisting of pure cetostearyl alcohol is only able to form a semihydrate with water.

Freeze fracture electron microscopy (FFEM) has added a new dimension to the studies of O/W-cream organization. This technique allows the visualization of the previously mentioned structural elements (Junginger et al. (1981), Junginger et al. (1983), Junginger et al. (1990)).

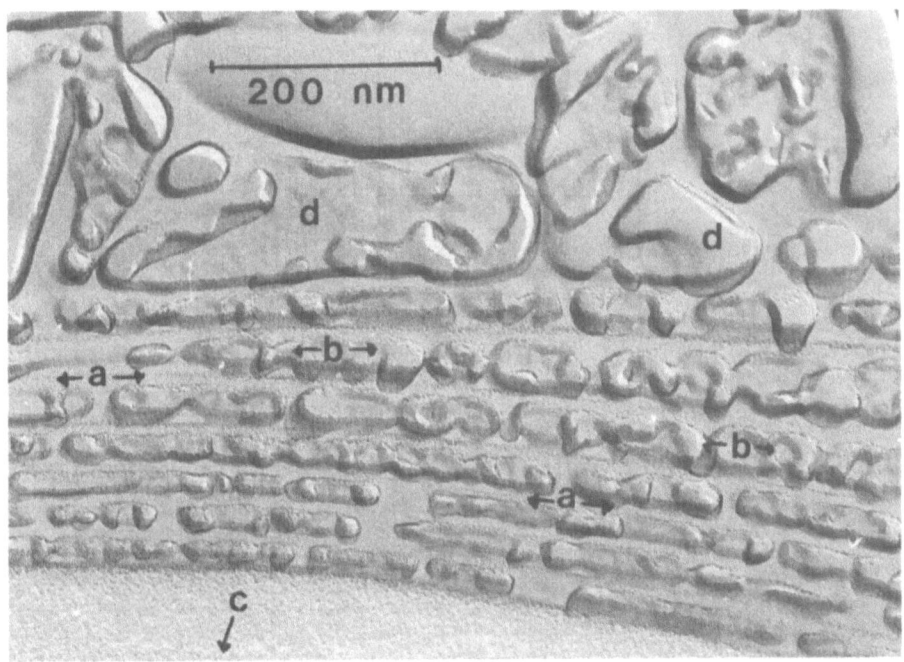

Fig. 2.   Freeze fracture micrograph of Emulsifying Wax DAB 9 (main structural component of the Water Containing Hydrophilic Ointment DAB 9) with 70% (wt/wt) of water.
a: mixed crystal bilayer; b: interlamellarly fixed water; a + b: hydrophilic gel phase; c: fracture edge of a lipophilic plane; d: bulk water phase;

From Fig. 2 the hydrophilic gel phase can be recognized very clearly. In this photograph the alternating layers of the hydrophilic gel phase are nearly at right angles to the fracture face. Together with areas of bulk water (d) entrapped in the hydrophilic gel phase, the interlamellarly bound layers of water (b) and the bilayers of the surfactant molecules (a) are visible. Together, (a) and (b) form the hydrophilic gel phase.

Investigation about the swelling ability of the Emulsifying Wax DAB 9 (main surfactant component of the Water Containing Hydrophilic Ointment DAB 9) with water (Fig. 3) show a swelling of the lamellar gel structure

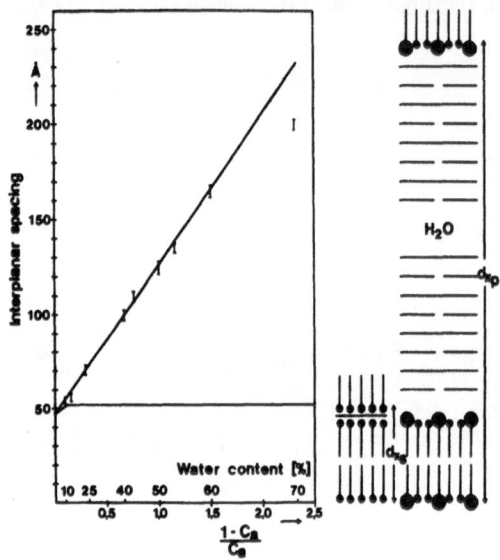

Fig. 3.   Swelling behaviour of Emulsifying Wax DAB 9 with water. C_s: weight fraction of surfactant; $1-C_s$: weight fraction of water; d_{sg}: interplanar spacings of cetostearylalcohol semihydrate (lipophilic) gel; d_{xp}: interplanar spacings of the hydrophilic gel phase

when the long spacings as obtained from small angle X-ray diffraction (SAXD) are plotted versus the ratio water/surfactant (wt/wt) (C_s is the weight fraction of surfactants; $1-C_s$ is the weight fraction of water). At a water content of 70% (wt/wt) the thickness of the interlamellar fixed water layer is about 15 nm (long spacing = 20.0 ± 0.5 nm). In comparison, the sizes of molecules for the semihydrate of cetostearylalcohol (lipophilic gel phase) and for the hydrophilic gel phase are given on the right side of Fig. 3.

It must be emphasized that the degree of swelling of the hydrophilic gel phase depends on the total water content of the cream. Thus a dynamic equilibrium exists between the bulk water and the interlamellarly fixed water. Both water phases form the continuous phase of the system. The capacity of the hydrophilic gel phase to incorporate interlamellar water is high enough to get clearly defined melting and recrystallization peaks by means of DSC which vary strongly from the water-free systems.

Once this swelling point of the water layer is reached the water molecules in the middle of the layer have the same mobility as water molecules of the bulk phase. Consequently, the lamellar gel structure of the hydrophilic gel phase breaks down and the system behaves as a liquid. The transition of the systems into this state means the transition of the cream into the (unstable) state of a suspension (emulsion). Undergoing this transition, the plastic flow behaviour properties of the cream are lost and the system exhibits the pseudo-plastic flow behaviour of an emulsion or a suspension.

The lipophilic gel phase (Fig. 1, region c), however, is only able to form a semihydrate, independent from the total water present in the system. After the transition from a cream into an emulsion (suspension) state the lipophilic gel phase is still surrounding the dispersed inner phase (Fig. 1, region e).

To further characterize these O/W creams for a technique was searched that allows a more quantitative differentiation between the different types of water existing in these bases. It can be expected that the swelling capacity of the hydrophilic gel phase may influence not only the water release to the skin and the diffusion rate of drugs across the vehicle but also the penetration ability of drugs through the hydrated horny layer of the skin. By means of a dynamic thermogravimetric analysis (TGA), a method was developed which enables us to differentiate between interlamellarly fixed water and the bulk water fraction. The results from the water containing ointment DAB 9 using the TGA are summarized in Fig. 4. The total water contents of the different creams are plotted against the total amount of interlamellarly fixed water fractions. At a total water amount of 70% (wt/wt) about 40% (wt/wt) is present interlamellarly in the hydrophilic gel phase. If the total water amount exceeds 80% (wt/wt) the hydrophilic gel phase reaches a saturation state, in which the water molecules in the middle of the water layer in between the bilayers are supposed to have the same free energy in comparison with that of the bulk water molecules at the same temperature. This saturated state of the hydrophilic gel phase represents the transition point from a cream into an emulsion Simultaneously the fraction of interlamellarly fixed water markedly decreases in favour of the fraction of the bulk water phase (Fig. 4).(Junginger et al. (1984b)).

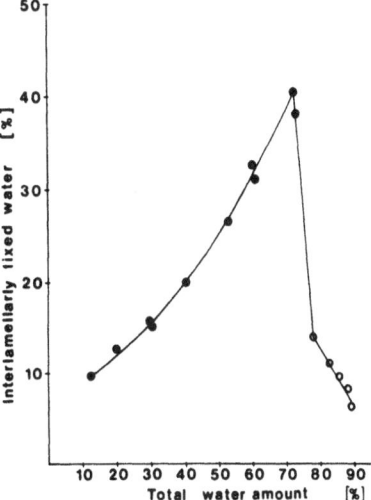

Fig. 4.   Amount of interlamellarly fixed water in Water Containing Hydrophilic Ointment DAB 9 depending on the total water amount of the system (0 = unstable systems)

194

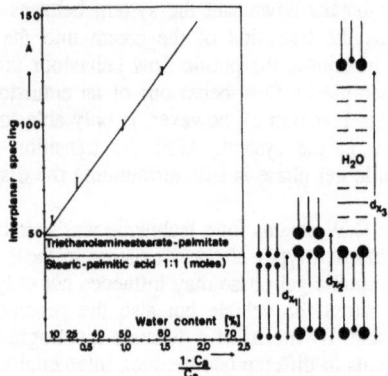

Fig. 5.   Swelling behaviour of the gel forming components (triethanolamine-stearate-palmitate) of a stearate cream with water. C_s: weight fraction of surfactants; $1-C_s$: weight fraction of water; d_{x1}: interplanar spacing of stearic-palmitic acid mixture (1:1 moles); lipophilic gel phase; d_{x2}: interplanar spacing of triethanolamine-stearate-palmitate (water free); d_{x3}: interplanar spacing of triethanolamine-stearate-palmitate swollen with water (hydrophilic gel phase of the stearate cream).

2.2   STEARATE CREAMS

For these creams a formulation proposed by Tronnier was chosen as a model system (Tronnier (1964)). The stearate cream investigated had the following composition:
-   Stearic acid   12   % (wt/wt)
-   Palmitic acid   12   % (wt/wt)
-   Triethanolamine   1.2   % (wt/wt)
-   Gycerol   13.5   % (wt/wt)
-   Water   10-61.3% (wt/wt)
Addition of water to the water free system which consists of stearic acid, palmitic acid and triethanolamine, results in lamellar mixed crystals in which incorporated water is in between the hydrophilic moieties of the lamellae. At increasing water concentration (10-61.3% wt/wt) a swelling capacity of the formed hydrophilic gel phase could be stated (Fig. 5). The thickness of the interlamellar water layer is about 6.5 nm at a water content of 60% (wt/wt) (long spacing = 12.6 ± 0.2 nm). (Junginger (1984d)).

The results obtained by means of SAXD, DSC and TGA led to a structure model for stearate creams as given in Fig. 6 (Junginger (1984a), (1984d)). One part of the lamellar mixed crystals, which consists of free fatty acids and their triethanolamine salts, are able to form the hydrophilic gel phase. Between the polar moieties of the mixed crystals (Fig. 6, region a) water molecules are present (Fig. 6, region b). This water of the hydrophilic gel phase is in equilibrium with the bulk water of the continuous phase (Fig. 6, region d).

The second part of the gel network, consisting of mixed crystals of palmitic and stearic acid, which is not able to retain water interlamellarly, forms a lipophilic gel phase (Fig. 6, region c). If a dispersed (lipophilic) phase is present in such a system it is mainly immobilized by the lipophilic gel phase. Stearate creams show a special pearl effect due to the crystallization of very small isolated platelets

(Fig. 6, region e) depending on the amount of added triethanolamine as well on as the manufacturing conditions. Platelets are formed preferably in place of a coherent lipophilic gel phase especially in the absence of a lipophilic phase.

The results of thermogravimetry of the stearate cream are depicted in Fig. 7. It becomes quite clear that above all at a high water content only one third is fixed interlamellarly, but two thirds of the water is present as a bulk water phase which is directly available for skin hydration. These facts explain the results of Tronnier who stated that the hydration rate of the skin is much higher by stearate creams than by other O/W creams (Tronnier (1964)).

These facts could also be confirmed by isothermal TGA comparing systems with different ratios of interlamellarly fixed water in their hydrophilic gel phases.

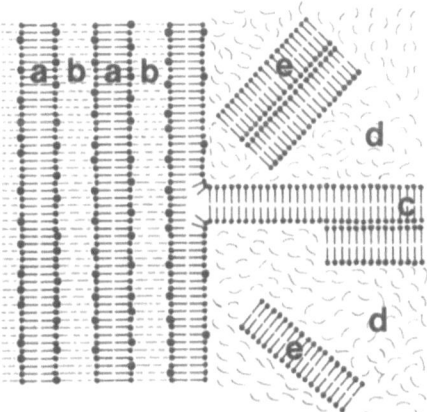

Fig. 6.   Gel structures of stearate creams. a: mixed crystal bilayer of triethanolamine-palmitate-stearate; b: interlamellarly fixed water; a + b: hydrophilic gel phase; d: bulk water phase; c: lipophilic gel phase ("stearate"); e: isolated "stearate" platelets, dispersed in the bulk water phase.

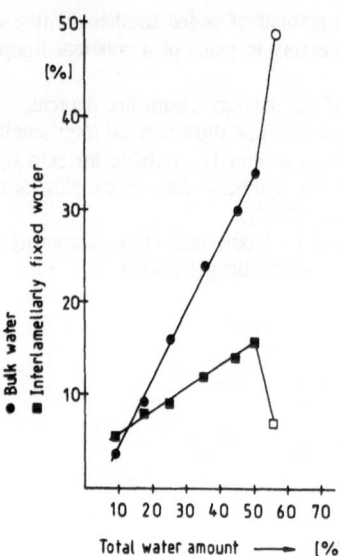

Fig. 7.   Ratio of bulk water and interlamellarly fixed water of a stearate cream
depending on the total water amount of the systems (, 0 = unstable
systems)

At a total water amount higher than 55% these systems become unstable (Fig. 7) and a transition takes
place from a cream with a coherent three-dimensional hydrophilic gel network to an emulsion without
these structural elements (Junginger et al. (1984b)).

2.3   NON-IONIC HYDROPHILIC CREAMS

Dermatological preparations containing ionic surfactants may show different disadvantages such as a
higher irritation potential of the skin, strong effects when salts are added, interactions with drugs etc.
For these reasons preparations containing only non-ionics will be more suitable as a general applicable
base. As one of the most representative formulation of non-ionic hydrophilic creams the following
formula was studied:

- Poly(oxyethylene) 20 glycerolmonostearate (PGM 20)	5.0	% (wt/wt)
- Liquid paraffin	7.5	% (wt/wt)
- Cetylalcohol	5.0	% (wt/wt)
- Stearylalcohol	5.0	% (wt/wt)
- Glycerol	8.5	% (wt/wt)
- White soft paraffin	17.5	% (wt/wt)
- Water	51.5	% (wt/wt)

This particular system is known as "Unguentum Hydrophilicum Non-ionicum Aquosum, DAC"*. Similar
formulations appear in the Swiss Pharmacopoeia, 6th edition and in the Formulary of the Dutch
Pharmacists (FNA).

*   Deutscher Arzneimittel Codex (1979).

Investigations by means of SAXD about the structure of the elementary units of the gel present in the above mentioned formulation show that PGM 20 and cetostearyl alcohol form mixed crystals if the water free melt recrystallizes (Fig. 8). In the water free system also the diffractions of pure PGM 20 and cetostearyl alcohol are to be found.

With SAXD no reproducible distances are found if water is added to the water free system up to 20% (wt/wt) (Fig. 8). The samples remain solid-like and polarization microscpy shows anisotropic structures. It is assumed that the hydrophilic polyoxyethylene chains together with the water molecules and the hydroxyl groups of cetostearyl alcohol form the hydrophilic layers between the lipophilic sheets. At a water content of 20% (wt/wt) the polyoxyethylene chains are surrounded with the minimum of hydration water needed to build up a homogenous lamellar structure. Thus reduction of the water content leads to formation of mixed crystals containing partially hydrated polyoxyethylene chains as well as cetostearyl alcohol and PGM 20. This picture (Fig. 8) is reinforced by the observation that samples containing less than 20% (wy/wt) water show several endothermal peaks with DTA and no electrical conductivity; thus, the coherency of the water layer is not reached yet (De Vringer et al. (1984), (1986)).

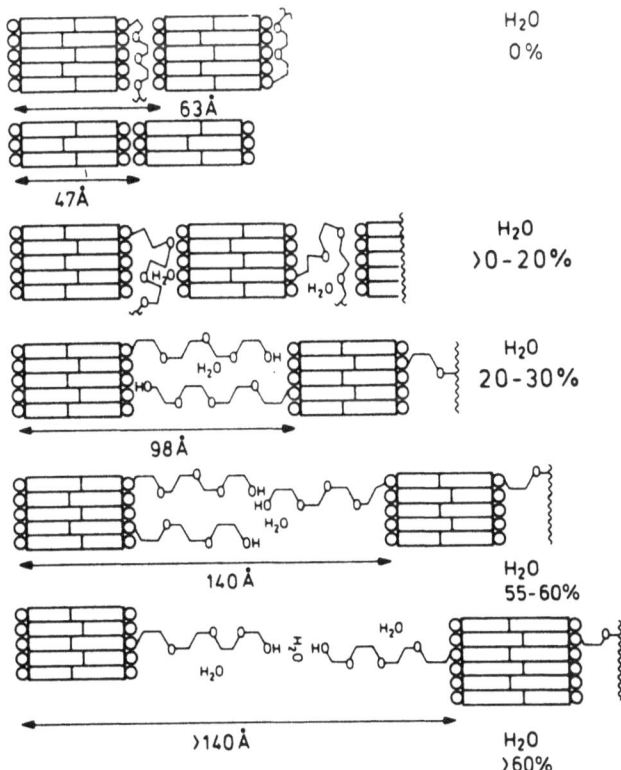

Fig. 8.   Swelling characteristics of cetostearylalcohol, polyoxyethylene-glycerolmonostearate and water mixtures. In the water-free state two diffraction peaks are found for cetostearylalcohol (4,7 nm) and cetostearylalcohol-polyoxyethylene-glycerolmonostearate mixed crystals (6,3 nm). Up to 20% (wt/wt) water unreproducible diffraction peaks are

198

found due to unsufficient hydration of the polyoxyethylengroups. Between 25-60% (wt/wt) water continuous swelling of the hydrophilic gel phase takes place.

Also in this non-ionic cream system polyoxyethylene-glycerolmonostearate crystallizes together with cetostearylalcohol as mixed crystals (Fig. 9a), the swelling degree of which is depending on total water content. The length of the polyoxyethylene unit determines the maximum swelling capacity of the systems.

Together with polyoxyethylene bound water the lamellar mixed crystals are forming the hydrophilic gel phase (Fig. 9a and b) in which partly bulk water is incorporated, too. The hydrophilic gel phase together with (part of) the bulk water (Fig. 9d) builds up a three-dimensional gel-network.

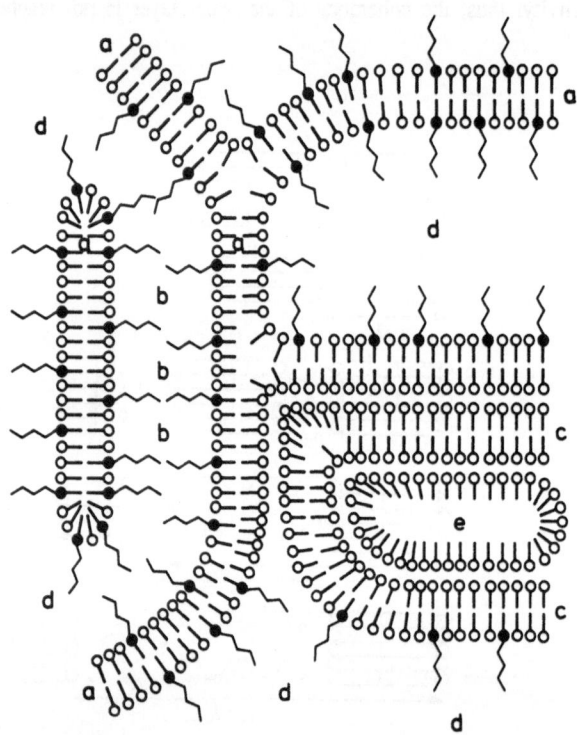

Fig. 9.   Schematic presentation of the gel structures of non-ionic hydrophilic cream (DAC 79). a: mixed crystal bilayer of cetostearylalcohol and poly(oxyethylene) 20 glycerol-monostrearate (PGM 20), b: interlamellarly fixed water, a + b: hydrophilic gel phase, c: lipophilic gel phase (cetostearylalcohol-semihydrate), d: bulk water phase, e: lipophilic components (dispersed phase)

The surplus of cetostearylalcohol forms again the lipophilic gel phase (Fig. 9c) which immobilizes the lipophilic dispersed phase (Fig. 9e) consisting mainly of white petrolatum and liquid paraffins.

As a result of these investigations it is concluded that non-ionic O/W-creams also may be regarded as four phase systems consisting of the same structural elements as the ionic O/W-creams (De Vringer et al.(1987a)).

Many non-ionic surfactants are obtainable with various PEG chain lengths. So it becomes feasible to develop non-ionic O/W creams with a desired ratio of interlamellarly fixed water and bulk water fraction. It will be clear that knowledge about these gel structures is of fundamental importance for developing a formulation with desired properties such as controlled water release especially considering the interactions between the vehicles and the skin (Junginger et al. (1985), De Vringer et al.(1987b)).

3.   Colloidal Gel Structures of W/O-Creams

According to the definition of DAB 9 W/O-creams are hydrophobic systems, the continuous phase of which is lipophilic. The general formula of an absorption ointment with incorporated water is as follows (according to DAB 9):

Anhydrous lanolin (wool fat)*	3,00	% (wt/wt)
Cetostearylalcohol	0,25	% (wt/wt)
White petrolatum	46,75	% (wt/wt)
Water	50,00	% (wt/wt)

A schematic presentation of the W/O-cream gel structures is given in Fig. 10:

*   with cholesterol as the most important ingredient.

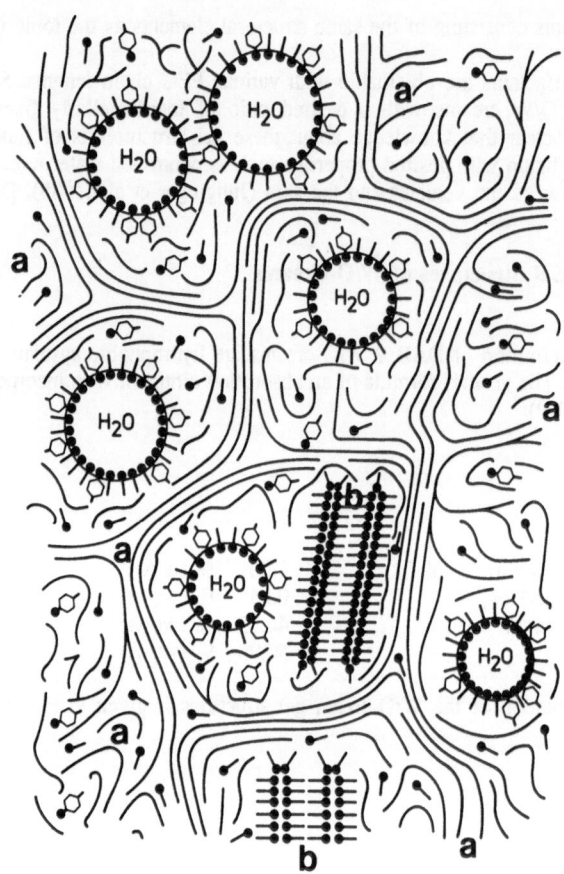

Fig. 10.   Schematic presentation of the gel structures of a W/O cream. a: long paraffin
chains are forming the solid gel in which liquid paraffin chains are immobilized
by lyosorption. Both cetostearylalcohol (   ) and cholesterol (derivatives)
(   ) are accummulated at the water-paraffin interface and are both molecularly
dispersed in the paraffin gel according to their solubilities. b: a surplus of
cetostearylalcohol may crystallize as separate   lamellar crystals.

The W/O-surfactants (cholesterol and other sterols as well as cetostearylalcohol) mainly accumulate at the interface between the water droplets and the oily phase (white petrolatum), forming a monomolecular mixed layer of surfactants at the water/oil interface. Experimental work has proven, that the capacity of water uptake strongly increases when mixtures of fatty alcohols and sterols are used. It sems to be important to create a liquid crystalline monolayer at the oil/water interface. Crystallization of the surfactant film at the interface drastically reduces the water uptake capacity of the system. According to their solubilities the sterols and fatty alcohols are dissolved in the paraffin mixture (white petrolatum), too. A surplus of these O/W-emulsifiers may crystallize separately in the lipophilic phase and may additionally enforce the (para)crystalline gel structures of white petrolatum. W/O-creams represent simple W/O-emulsions which are stabilized by a high viscous gel of paraffins. The long chain paraffins are able to build up a three-dimensional solid gel-network in which the short chain liquid paraffins are mainly immobilized by lyosorption.

4.   Colloidal Gel Structures of Amphiphilic Creams

Amphiphilic creams are colloidal systems which transform by addition of an oily phase to a W/O-cream and by water addition to an O/W-cream. They therefore represent a special transition state between the other two cream types. This colloidal state is obligatory to the existence of special gel structures. A prerequisite for the existence of a cream with amphiphilic properties are lamellar mixed crystals which show upon water addition only a limited swellability. Examples for suitable compounds which fulfill these requirements are glycerolmonostearate and esters or ethers of PEG with fatty alcohols or fatty acids, respectively. PEG should have a low polymerisation degree of n = 2-3. All these compounds are surfactants of the W/O-type. The collodial structure of an amphiphilic cream shall be illustrated with the "basis cream" of DAC having the following formulation:

Glycerolmonostearate	4,0	% (wt/wt)
Cetylalcohol	6,0	% (wt/wt)
Medium chain triglycerides	7,5	% 9wt/wt)
White petrolatum	25,5	% (wt/wt)
Poly(oxyethylene)20 glycerolmonostearate		
(PGM 20)	7,0	% (wt/wt)
Propyleneglycol	10,0	% (wt/wt)
Water	40,0	% (wt/wt)

The limited swellability of glycerolmonostearate is well documented in literature (Larson (1967), Krog (1973)). Melted glycerolmonostearate together with water continuously swells by interlamellarly incorporating water molecules in between the hydrophilic glycerol-rest until a water content of 30% (wt/wt) is reached. At increased water amounts the degree of swelling remains constant and the exceeding amount of water is incorporated mechanically as droplets (bulk water phase) in the glycerolmonostearate gel structure.

Conductivity measurements on mixtures of glycerolmonostearate, liquid paraffins, and water with a constant amount of glycerolmonostearate of 30% (wt/wt) revealed (Fig. 11), that up to a water content of 20% (wt/wt) no conductivity could be measured. Therefore, up to this water amount the systems show W/O-character.

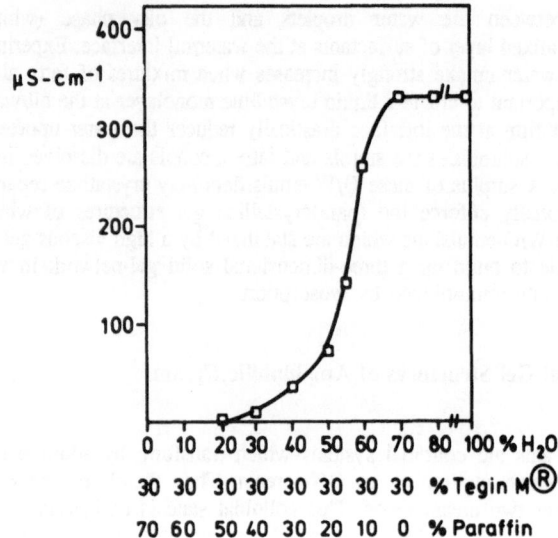

Fig. 11.   Conductivity of paraffin-water mixtures with a constant amount glycerolmonostearate (Tegin M^R) of 30% (wt/wt).

On the other hand mixtures with a water content between 60 and 70% (wt/wt) show O/W-characteristics. In the range between 20 and 50% (wt/wt) of water the systems behave as amphiphiles, i.e. addition of water results in O/W-systems, additions of liquid paraffins directs to W/O systems.

However, all these simple systems show no favourable properties with respect to the requirements of well developed creams for pharmaceutical use. Furthermore, especially the resulting O/W-systems are unstable and phase separation of the only mechanically stabilized bulk water phase occurs. Therefore, well developed formulations of these amphiphilic systems additionally contain non-ionic O/W-emulsifiers with a high water binding capacity. In the above mentioned formulation of DAC poly(oxyethylene) 20 glycerolmonostearate is added, which is able to form stable O/W-creams at high water contents, i.e. exceeding 50% (wt/wt). Amphiphilic creams contain alltogether a relatively high amount of O/W- and W/O-surfactants. Oily and water phase are approximately in the same order of weight range.

Amphilic creams are tri-coherent systems:

-        the dominant coherent gel phase is built up by the monoglycerolstearate lamellae.
-        the continuous water phase is mainly performed by the interlamellarly bound water in between the monoglycerolstearate lamellae.
-        the lipophilic phase is coherent, too.

The liquid paraffins are predominantly mechanically fixed by the monoglycerolstearate lamellae.

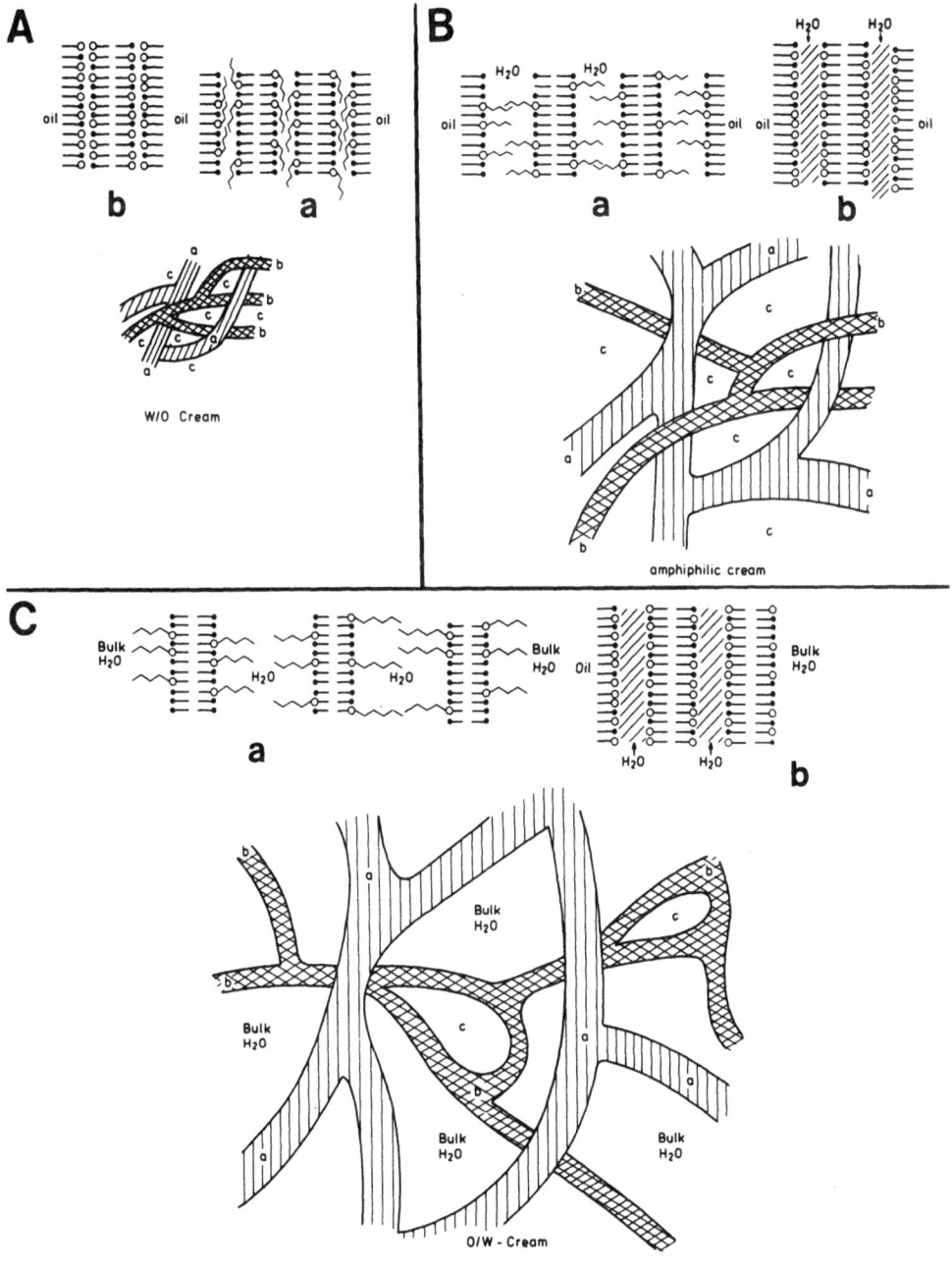

Fig. 12.   Schematic presentation of gel structures existing during the transition of a O/W-
cream (A) to an amphiphilic cream (B) and to a W/O-cream (C).
Poly(oxyethylene) 20 glycerolmonostearate
Glycerolmonostearate
Cetostearyl Alcohol

a)   Mixed crystals consisting of poly(oxyethylene) 20 glycerolmonostearate and cetostearyl alcohol (Fig. A water-free state, Fig. B partly swollen state, Fig. C swollen state).

b)   Mixed crystals of glycerolmonostearate and cetostearyl alcohol with limited swellability (Fig. A water-free state, Fig. B and Fig. C state of limited swelling).

c)   Lipophilic phase (Fig. A coherent continuous phase, Fig. B coherent phase, Fig. C dispersed inner phase).

A schematic representation of the proposed gel structures of an amphilic creams and the transitions to an O/W-system, respectively, are given in Fig. 12 A, B, C.

Originating with the above described amphiphilic system (Fig. 12 B) and adding an oily phase, the water swelling degree of the monoglycerolstearate lamellae is strongly reduced. Due to the favourable phase-volume-ratio regarding the lipophilic phase a W/O-cream automatically results (Fig. 12 A).

Increasing the water amount of the amphiphilic basic cream especially the poly(oxyethylene) 20 glycerolmonostearate surfactant system (stabilized with cetylalcohol) strongly swells by interlamerly incorporating the added water and fixing the resulting bulk water phase, thus increasing the conductivity tremendously (Fig. 12 C). As a result of water addition a stable O/W-cream results.

Literature

De Vringer, T., Joosten, J.G.H. and Junginger, H.E., (1984)," Characterization of the gel structure in a nonionic ointment by small angle X-ray diffraction" Coll. Polym. Sci. 262, 56-60

De Vringer, T., Joosten, J.G.H. and Junginger, H.E., (1986), "A study of the gel structure in a nonionic O/W cream by differential scanning calorimetry" Coll. Polym. Sci. 264, 691-700

De Vringer, T., Joosten, J.G.H. and Junginger, H.E., (1987a) "A study of the gel structure in a nonionic O/W cream by X-ray diffraction and microscopic methods" Coll. Polym. Sci 265, 167-179

De Vringer, T., Joosten, J.G.H. and Junginger, H.E., (1987b), "A study of the ageing of the gel structure in a nonionic O/W cream by X-ray diffraction, differential scanning calorimetry and spin lattice relaxation measurements" Coll. Polym. Sci 265, 448-457.

Junginger, H.E., Führer, C., Ziegenmeyer, J. and Friberg S.E. (1979) "Strukturuntersuchungen an der Wasserhaltigen Hydrophilen Salbe DAB 7" J. Soc. Cosmet. Chem. 30, 9-23

Junginger, H.E., Heering, W., Führer, C. and Geffers, I. (1981) "Elektronenmikroskopische Untersuchungen über den kolloid-chemischen Aufbau von Salben und Cremes" Coll. Polym. Sci., 259, 561-567

Junginger, H.E. and Heering, W. (1983) "Darstellung kolloider Strukturen von Salben, Cremes, Emulsionen und Mikroemulsionen mittels Gefrierbruch-Ätztechnik und TEM" Acta Pharm. Technol. 29, 85-96

Junginger, H.E. (1984a) "Colloidal structurtes of O/W creams", Pharm. Weekbl. Sci. Ed. 6, 141-149

Junginger, H.E., Ackermans, A.A.M.D. and Heering, W. (1984b) "The ratio of interlamellarly bound water to bulk water in O/W creams" J. Soc. Cosmet. Chem. 35, 45-57

Junginger H.E. (1984c) "The ratio of interlamellarly fixed water to bulk water used as quality criterion in O/W creams" Ber. Bunsen Ges. Phys. Chem. 88, 1070-1074

Junginger, H.E. (1984d) "Strukturuntersuchungen an Stearatcremes" Pharm. Ind. 46, 758-762

Junginger, H.E. and H.E. Boddé, (1985) Gelstructures of dermatological vehicles and their influence on drug release, in Topics in Pharmaceutical Sciences, D.D. Breimer and P. Speiser eds., Elseviers Science Publ., Amsterdam, New York, Oxford, p. 329-343

Junginger, H.E. and Heering, W. (1990) "Gelstrukturen in Cremes" Dtsch Apoth. Ztg. 130, 684-685

Krog, N. and Borup, A.P. (1973) "Swelling behaviour of lamellar phases of saturated monoglycerides in aqueous systems, J. Sci. Food Agric. 24, 691-701

Larsson, K. (1967) "The structure of mesomorphic phases and micelles in aqueous glyceride systems" Z. Phys. Chem. (Neue Folge) 56, 173-198

Münzel, K. (1953) "Versuch einer Systematik der Salben nach galenischen Gesichtspunkten" Pharm. Acta Helv. 28, 320-336

Tronnier, H. (1964) "Über die Wirkungsweise indifferenter Salben- und Emulsionssysteme an der Haut in Abhängigkeit von ihrer Zusammensetzung" Editio Cantor, Aulendorf

PERFLUOROCHEMICAL EMULSIONS AS BLOOD SUBSTITUTES

Robert J. Kaufman
HemaGen/PFC
11444 Lackland Ave.
St. Louis, MO 63146

ABSTRACT. Perfluorochemicals (PFC's) dissolve large volumes of oxygen (35-52 vol. %) and carbon dioxide (135-160 vol.%) and are chemically and biochemically inert. These properties have made them candidates for blood substitutes and other therapeutic applications requiring oxygen delivery to tissues. In this paper, the history of PFC's in medicine and the requirements for a successful commercial emulsion are briefly reviewed. The practical aspects and current status of the production of a safe, shelf stable, small particle size, high PFC content emulsion are reviewed and recent advances in the field such as new PFC's, novel surfactants, new clinical trials and recent regulatory approvals are presented.

1. Introduction

Whole blood serves a number of critical functions including delivery of oxygen to tissues, hemostasis, host defense, maintenance of ionic and protein balances, transport of nutrients and hormones, and the removal of metabolic waste products. However, since oxygen deprivation rapidly and irreversibly degrades the function of cells, tissues, and organs, leading to death, the most important function of blood is oxygen delivery to the tissues.

Historically, donated human blood has been the therapeutic of choice for acutely and chronically anemic patients. However, the use of donated blood is not without risks. Blood can be contaminated with an array of infectious agents including HIV, hepatitis virus, cytomegalovirus (CMV), Epstein- Barr virus, and *Brucella abortus*. While blood is currently screened for many pathogens including HIV, it is not possible to identify all pathogens nor to eliminate them with heat sterilization. Recent estimates place the risk of contracting non-A, non-B hepatitis at 7-10% of all patients transfused in the USA, with 50% of these patients developing chronic, active hepatitis (Haljamae and Rosenberg 1988).

Immunosuppression after transfusion of homologous blood is also a significant problem and has been linked with increased recurrence of cancer (Blumberg et al. 1985) and with an increased incidence of postoperative infection in certain types of surgical patients (Maetani et al. 1986). In addition to the health related risks inherent in blood usage, there are other limitations to the use of donated blood. Donor blood must be typed and cross-matched for each patient, resulting in a transfusion delay of at least 20 to 30 min. The use of type O negative blood to circumvent the typing and crossmatching time delay is not without risks.

Blood has a storage lifetime of only 42 days and it must be refrigerated. This makes blood unavailable in many of the situations in which it is most needed, such as in rural trauma incidents, in ambulances and helicopters, on battlefields and during civilian disasters. Complicating the storage lifetime is the fact that after blood has been stored for a few days, the erythrocyte depletes its 2,3-DPG and oxygen bound to hemoglobin in these cells becomes relatively unavailable. Complete restoration of 2,3-DPG content and oxygen delivery function requires about 12 hours in the circulation. Thus, transfusion of stored red cells does not immediately fulfill the oxygen transport function for which it was administered (Huetis et al. 1981).

The search for agents to replace the oxygen transport function of blood has been underway for over 50 years and has centered largely on three approaches: purified, hemoglobin derivatives, synthetic heme complexes and perfluorochemical (PFC)

J. Sjöblom (ed.), Emulsions – A Fundamental and Practical Approach, 207–226.
© 1992 *Kluwer Academic Publishers.*

emulsions. Synthetic hemes have remained an academic curiosity while hemoglobin and PFC emulsions have been extensively studied. The goals of both PFC and hemoglobin research have been to develop a readily available, disease-free, shelf stable, safe, cost-effective, universal donor product that would be available when and where it was needed. While neither approach has yet achieved broad clinical success, enormous progress has been made with PFC emulsions over the last five years in understanding and overcoming obstacles to successful product introduction. New medical uses for oxygen transport agents have been identified and explored. Several PFC products are in various stages of preclinical and clinical evaluation. In late 1989, Fluosol™, developed and manufactured by the Green Cross Co. (Osaka, Japan), was approved as an oxygen transport agent for distal oxygenation of the myocardium during high risk coronary balloon angioplasty by the US FDA (Anonymous 1990).

The purpose of this chapter is to review the use of PFCs and their emulsions in medical oxygen transport and to highlight recent developments in the field.

2. History of Perfluorochemicals in Oxygen Transport

Fluorine, which is the most reactive and corrosive of all the halogens, forms the most stable single bonds to carbon known to chemists, with bond dissociation energies on the order of 116-122 Kcal/Mol compared to 102 Kcal/Mol for a typical carbon - hydrogen bond. This high bond strength translates into enormous molecular stability and chemical inertness, particularly when the carbon skeleton is completely fluorinated.

PFC's were first prepared in the 1930's in small quantities for research purposes from direct fluorination of carbon. However, the fact that fluorine was only available in gram quantities and could not be purchased at all until after World War II severely limited the scope of investigations of preparative methods for PFC's. During World War II, the Manhattan Project required the preparation of extremely inert solvents for separation of uranium isotopes via diffusion as their hexafluorides. This need served as the impetus to develop commercially useful methods of producing fluorine. Shortly thereafter fluorine became readily available electrochemically and methods of synthesis of perfluorinated compounds were developed resulting in the preparation of a wide variety of structurally diverse and commercially useful PFC's.

PFC's were found to have a number of interesting properties including chemical inertness, high gas solubility and diffusivity, and low toxicity. The only significant reactions of PFC's are pyrolysis and aromatization at temperatures around 350 to 550°C. Early studies of the fundamental physical properties of PFC's included measurements of gas solubilities, performed by Gjaldbaek and Hildebrand. They found that PFC's dissolved gases in accordance with Henry's law and remarkably, they could dissolve 10 to 20 times the gas dissolved by water (Gjaldbaek and Hildebrand, 1949).

Leland Clark galvanized medical research in the perfluorochemical field when he demonstrated that mice and young puppies submerged in liquid perfluorobutyltetrahydrofuran saturated with oxygen could derive their physiological oxygen requirements via the PFC in their lungs without toxicity (Clark and Gollan 1966). Within a year of Clark's publication, Henry Sloviter reported that PFCs could be rendered into a plasma compatible form by emulsification with bovine serum albumin in Kreb's Ringer bicarbonate (Sloviter and Kamimoto 1967). Using this emulsion, he was able to extend the electrical activity in isolated, perfused rat brains far longer when compared to rat erythrocytes. Geyer exchanged the blood of rats with an emulsion made from perfluorotributylamine, the surfactant pluronic F-68, and physiological salts. Animals survived on PFC "blood" at high oxygen tensions until sufficient regeneration of RBCs had occurred to support life on room air (Geyer et al. 1968, 1973). The rats developed normally and survived in apparent good health to the end of their normal life expectancy. These total exchange experiments were a graphic demonstration of both the efficacy and the safety of PFCs. The fact that the rats survived with hematocrits as low as 3% demonstrated the physiological gas transport capabilities of PFCs. The safety of the emulsion was evident because serum albumin, immunoglobulins, clotting factors, platelets, leucocytes, and erythrocytes, which had all been removed as a consequence of the total exchange, all regenerated, indicating no damage to the liver or the marrow.

Subsequently, clinical research groups initiated animal studies and identified potential clinical applications for PFC oxygen transport agents in fields ranging from shock resuscitation and wound healing to heart attack and cancer therapy. Industrial research resulted in the preparation of the first clinically acceptable emulsion, Fluosol, by the Green Cross Co, (Osaka, Japan).

3. Characteristics Required of Perfluorochemicals for Medical Use

The pioneering work of Clark, Sloviter and Geyer allowed a clear definition of the requirements for a successful commercial oxygen transport agent based on PFC emulsions.

The PFC required two seemingly simple characteristics:

(A) Readily available in high purity - At the time interest arose in their medical applications, PFC's were exclusively used in industrial applications including vapor phase resoldering and as fire resistant lubricants. The purity requirements for industrial grade materials are significantly different than the specifications for a fluid for intravenous use. Many potential impurities in PFC's, particularly compounds bearing residual hydrogens and double bonds, are too reactive to be used intravenously.

(B) Satisfactory tissue residence times without toxicity - Geyer's work had shown that perfluorotributyl amine was sequestered in the liver and spleen of the treated rats for the duration of their lifetime. Lifetime sequestration of xenobiotics, particularly in the quantities required for oxygen transport agents, is rightfully considered medically unacceptable.

Using these guidelines, Clark (Clark et al. 1974), in the US, and Yokoyama (Yokoyama et al. 1975) in Japan screened over 100 PFC's searching for compounds which had the right combination of properties to form a stable emulsion, possess minimal side effects and which would leave the body in a reasonable amount of time. Of these only a handful received detailed study. Reiss's excellent 1978 review of PFC blood substitutes contains citations for each study for the reader interested in more historical detail (Reiss and LeBlanc, 1978).

The results of this research were the development of rather general, empirical rules for the physical properties of PFC's suitable for intravenous usage. In general, the rate at which a PFC is eliminated from the body was found to be related to its molecular weight, structure, vapor pressure and lipophilicity, with vapor pressure being perhaps the major determinant. Generally, the higher the vapor pressure, the more rapidly the PFC is eliminated from the body. However, the relationship of vapor pressure to organ clearance rate is not perfect, and structural changes and heteroatoms have a significant impact on tissue clearance rate regardless of vapor pressure. For example, perfluorodecalin, with a vapor pressure of 12 mmHg, was found to leave most organs in which it was sequestered with a $T_{1/2}$ of seven days. However, perfluoromethyldecalin, with a vapor pressure of 5-6 mm/hg, has an organ $T_{1/2}$ of 105 days while perfluorotripropylamine, with a vapor pressure of 20 mmHg, has an organ $T_{1/2}$ of 65 days; quite a difference in half-life for such minor changes in structure, molecular weight or vapor pressure! Addition of a heteroatom to a structure generally increases its organ $T_{1/2}$. For example, perfluoro-4-methyl-4-isopropylpentane has an organ $T_{1/2}$ of 11 days while a similar C-9 ether, perfluoro-4,4-dimethylbutylpropyl ether has a $T_{1/2}$ of 35 days (Reiss, 1984). Molecular weight can also be misleading as a predictor of tissue clearance time. For example, perfluorooctyl bromide has a molecular weight of 499 and an organ $T_{1/2}$ of 7 days (Liu, 1973) compared to perfluoromethyldecalin with a molecular weight of 512 and an organ $T_{1/2}$ of 105 days. It is unlikely that this small difference in molecular weight could have such a profound impact on organ half-life, and suggests that molecular weight is not the key determinant of organ half-life *within* the structural range embraced by C8 to C12 PFC's. The critical solution temperature (CST), an index of PFC lipophilicity has been proposed as a predictor of tissue clearance rate (Moore and Clark, 1985). The CST is a measure of the temperature required for a PFC to dissolve in hexane. The data support the hypothesis that, within the range of C8 to C12, the molecules with a lower CST and thus a higher lipophilicity have lower tissue residence times. In the final analysis, the clearance rate depends upon a poorly understood and almost totally empirical relationship between structure, molecular weight range, vapor pressure and lipophilicity.

At the same time it was found that PFC's with vapor pressures above 30 mmHg were unacceptable because they caused pulmonary hyperinflation (an emphysema like condition without any visible tissue lesions) and subsequent lethality due to pulmonary complications (Yokoyama, Suyama and Naito, 1982). This result places an upper ceiling on the acceptable range of PFC vapor pressures. When this screening work was completed, it was clear that the structures of medically acceptable PFC's were fairly restricted between 8-12 carbons, were preferably more lipophilic and required vapor pressures between about 10-20 mmHg in order to be non-toxic and yet clear the body within a reasonable timeframe. While these studies narrowed the range of structures that might be physiologically acceptable, all structures within that range still had to be synthesized and tested individually for toxicity and clearance rate before conclusions about their medical utility could be drawn.

The net result of these studies was the recognition that only four of the PFC's synthesized and tested had the correct balance of organ retention time and safety necessary to be further considered for commercial development as medical oxygen transport agents (Table 1). These were perfluorodecalin, perfluoromethyladamantane, 1,2 bis(perfluorobutyl)-ethylene and perfluorooctyl bromide. Predictably, Clark was the first to discover the rapid transpiration rate of both perfluorodecalin and perfluoromethyladamantane (Clark, Wesseler, Miller and Kaplan, 1974). The organ clearance data for the liver and the spleen for perfluorodecalin compared to perfluorotributylamine shows that PFD leaves the organs in a dose dependent fashion and is largely gone from rat livers and spleens at the doses of two to eight g/kg in about two weeks, while perfluorotributylamine content does not decrease at all during the same time frame.

Table 1. Properties of Perfluorochemicals tested for medical oxygen transport

Perfluorochemical	MW	Vapor Pressure	Organ $T_{1/2}$(days)	CST (°C)	Toxicity
Perfluorodecalin	462	12	7	22	non-toxic
Perfluoromethyldecalin	512	6	105	50	non-toxic
Perfluorobutyltetrahydro-furan	416	58	na	28	toxic
Perfluorooctahydoindan	412	34	na	na	toxic
Perfluoro-4-methyl4-isopropylpentane	488	na	11	na	na
Perfluoro-4,4-dimethylbutyl propyl ether	504	na	35	na	na
Perfluorotributylamine	671	2	400	61	non-toxic
Perfluorotripropylamine	521	20	40 to 60	43	non-toxic
Perfluorodimethyladamantane	524	na	10 to 20	32	non-toxic
Perfluoro-1,3-dimethylcyclohexane	400	67	na	na	toxic
Perfluorooctyl Bromide	499	12	7	<0	non-toxic

4. Characteristics Required for Medical PFC Emulsions

The properties required for the PFC emulsion are much more extensive:
(A) Surfactant safety - The surfactant used in PFC emulsions must be biocompatible and extremely non-toxic.
(B) Good shelf-stability of the PFC emulsion - Quarantine times post manufacture coupled with distribution and shipping times mandate as long a shelf life as possible for an emulsion with a minimum shelf life of six months and a preferred shelf life of 18 months. Storage temperature is flexible for a pharmaceutical product but room temperature storage is cheapest and most convenient.
(C) *In vivo* emulsion stability - An emulsion must be stable to the ions, proteins and enzymes of the plasma in order not to undergo particle size growth. Growth of particles in the plasma could result in PFC emboli forming in the microcirculation, causing widespread tissue ischemia.
(D) Stable during terminal sterilization at 121°C - All large volume parenteral products sold in the US today are terminally sterilized at 121°C to insure complete reduction of any bioburden which may have been acquired in the processing of the product. Historically, this requirement has been a serious obstacle for PFC emulsions.
(E) High PFC content - In contrast to hemoglobin which has a sigmoidal oxygen delivery curve that is 97% saturated at normal ambient pO_2's of about 150 mmHg of oxygen, PFC emulsions carry oxygen in direct proportion to both the PFC content and the oxygen tension of the gaseous environment with which the emulsion is in contact. Since the

emulsion will be diluted upon infusion into the patient, the concentration of PFC in the emulsion needs to be as high as possible to insure sufficient circulating concentration of PFC in the patient.

(F) Acceptable particle size distribution (PSD) - The particle size of an emulsion plays an important role in its toxicity and organ distribution. Early investigators in the intravenous fat emulsion field discovered that large particles, in the size range 5-15μm, were sieved by the pulmonary circulation resulting in significant pulmonary toxicity.

(G) Viscosity compatible with whole blood - It is well established that perfusion and tissue oxygenation are inversely related to the viscosity of blood. As blood viscosity increases, the rate of perfusion falls and so does the effectiveness of tissue oxygenation. The viscosity of an emulsion must not increase the viscosity of the circulating blood to a point where the flow characteristics become significantly impaired and reduce perfusion.

(H) Intravenous half-life - The emulsion must remain in the circulation long enough to insure a therapeutically useful period of oxygen delivery.

5. Medical Oxygen Transport with Perfluorochemicals

The fact that neat PFC's dissolve large amounts of oxygen when in equilibrium with pure oxygen gas is well documented and the number of measurements made on PFC's since interest arose in their use as blood substitutes has lead to a large and structurally diverse database on their oxygen solubility properties. There have been a number of mechanisms proposed to explain the enormous solubility of gases in PFC's, including charge transfer interactions, structural sites in the PFC molecule itself and cavities in the bulk solution phase as a consequence of the very weak Van der Waal's attractions between molecules. The cavity proposal has received some support from NMR studies of the relaxation times of carbon and fluorine nuclei when oxygen is dissolved in PFC's (Hamza et al. 1981). Regardless of mechanisms, the oxygen content of PFC's is significant compared to other fluids.

Oxygen dissociation curves for PFC emulsions and hemoglobin are distinctly different. PFC oxygen content is linearly dependent upon the oxygen tension with which it is in equilibrium. Gases dissolved in PFC's behave like gases dissolved in water; that is, they are driven only by the environmental gradients. Oxygen moves into the PFC phase when environmental tensions are high and out when they are low. There is no chemical affinity of PFC's for oxygen. In contrast, hemoglobin exhibits a sigmoidal dependence upon oxygen tension and is 97% saturated at atmospheric oxygen tensions. Thus, oxygen tension has two important ramifications for PFC emulsions. First, they require supplemental oxygen to be achieve maximum oxygen content and second, they can exploit oxygen therapy far better than hemoglobin which is saturated at about 150 mm Hg of oxygen.

The oxygen content in PFC emulsions on thus depends directly on PFC content and oxygen tension: the more PFC in the emulsion, the greater the oxygen content of the resulting solution. This underscores the need for a high PFC content emulsion to insure an adequate circulating concentration of PFC in the patient resulting in adequate oxygen transport after dilution of the infused emulsion by the patient's systemic circulation.

The starting point for assessment of a PFC emulsion's oxygen transport capability is its oxygen content at various oxygen tensions *in vitro*. However, the ultimate utility of a PFC emulsion to transport oxygen therapeutically depends upon its contribution to oxygen delivery and ultimately to oxygen consumption. Oxygen delivery is the product of arterial oxygen content (CaO_2) and cardiac output (CO) (Equation 1):

$$QO_2 = CaO_2 \times CO \qquad (1)$$

The arterial oxygen content for blood containing PFC is the sum of the three phases carrying oxygen: [O2]aHb, the oxygen content due to hemoglobin, [O2]aPFC, the oxygen content due to PFC and [O2]aH2O (Equation 2).

$$CaO_2 = [O_2]aHb + [O_2]aPFC + [O_2]aH_2O \qquad (2)$$

To assess the fractional contribution of an emulsion to oxygen delivery requires that the arterial and venous contents of all three phases be known or calculated. The oxygen content due to hemoglobin can be measured using a co-oximeter or calculated using the hematocrit and the well accepted value of 1.34 cc O_2 per gram of hemoglobin. The contribution of the aqueous and PFC phases can be calculated from the blood gases, the Bunsen solubility coefficients for water and PFC and the fluorocrit [the fluorocrit (Fct) is the volume percent of PFC, <u>not</u> emulsion, in the whole blood]. By Henry's law, the arterial or venous oxygen content of the PFC phase depends upon the solubility coefficient (aPFC) of the PFC and the arterial or venous blood gases as shown for arterial content in equation 3:

$$[O_2]aPFC = Fct \times PaO_2/760 \times aPFC \qquad (3)$$

The arterial or venous content of the aqueous phase can be calculated from the arterial and venous blood gases, the solubility coefficient for water and for the fluorocrit and hematocrit, as shown for the arterial aqueous content according to equation 4:

$$[O_2]aH_2O = PaO_2/760 \times (1-Hct-Fct) \times aH_2O \qquad (4)$$

These equations coupled with known solubility coefficients and a few measurements, available from the patient's chart, allow the investigator to determine the relative contribution of PFC's to the overall oxygen transport of a given patient. The equations may also be coupled with direct oxygen content measurements to determine solubility coefficients for any emulsion at any given temperature as Moss's group has done for Fluosol® (Rosen et al. 1985).

6. Medical Applications of Perfluorochemicals and Their Emulsions

Perfluorochemical emulsions have been investigated in a number of therapeutic roles including their use as blood substitutes, as therapy for ischemic heart and cerebral tissue, as a adjunct to balloon angioplasty, as cardioplegia for coronary bypass surgery, as imaging agents, as an adjunct to cancer radiation therapy and as a therapy for the bends. Most of these applications have been investigated in animal models and at least four indications have been the subject of clinical studies and one indication has received FDA clearance. This section will focus on the major indications in which sufficient animal and/or clinical data exist to evaluate the therapeutic value of PFCs.

6.1 BLOOD SUBSTITUTES

The most frequently investigated therapeutic application for PFC emulsions has been as an oxygen transport substitute for red blood cells. Red blood cell substitutes cover many possible, diverse indications including hemorrhagic shock resuscitation, restoration of blood oxygen content during elective and emergency surgical hemorrhage, priming pumps during coronary artery bypass and as an oxygen transport diluent for autologous donation preceding elective surgery.

Geyer's experiments with total exchanged rats provided the first substantial evidence that PFC emulsions could take the place of blood in providing the total oxygen requirements of an animal. Subsequently, exchange transfusions have been reported in dogs (Suyama et al. 1975) and monkeys (Ohyanagi et al. 1978; Rosenblum et al. 1985). Monkeys were exchanged transfused with either Fluosol or the plasma expander Hespan (HES) to a hematocrit of <2%. After six hours, the survivors were infused with autologous blood. All of the HES-exchanged monkeys died before the blood transfusion, while 8 of 10 monkeys given Fluosol survived to receive the blood transfusion, and six of the eight that received blood after six hours survived.

Okada and coworkers studied the use of PFCs in resuscitation of hemorrhagic shock in dogs (Okada et al. 1975). Arterial, venous and skeletal muscle oxygen tensions were significantly higher animals resuscitated with Fluosol DC than in the group resuscitated with Lactated Ringer's. Remarkably, the mixed venous oxygen tension of the Fluosol DC group was 78 mm Hg indicating that, in anesthetized, intubated dogs breathing 100% oxygen, the dog's oxygen requirements were largely being met by the PFC with almost no oxygen off-loading by hemoglobin (hemoglobin is greater than 90% saturated at 78 mm Hg).

Ohyanagi and Mitsuno (1975) compared shock resuscitation with PFC or low-molecular-weight dextran (LMD). In all cases, the PFC emulsion produced superior survival to the LMD resuscitation, and when combined with blood transfusions given six hours post resuscitation, 100% survival was observed. The data from these studies offer strong support that interim resuscitation by PFC emulsions in environments where blood is not immediately available followed by blood transfusion could make an enormous difference in survival of shock victims.

Makowski compared Fluosol and Fluosol DA-35 with 6% HES and whole blood in shock resuscitation (Makowski 1978). All four treatments restored mean arterial pressure and cardiac index and increased mean pulmonary artery pressure. Blood and both PFC emulsions restored arterial and venous oxygen contents, but HES did not. All treatments restored oxygen consumption immediately after infusion, however, the PFC emulsions increased oxygen consumption compared to HES or blood. Mixed venous oxygen tensions were also higher immediately after perfusion in the PFC-treated animals. In addition, Makowski found that while 80% of the shocked animals reinfused with whole blood or HES died, only 20% of the Fluosol treated animals succumbed. The increase in oxygen consumption indicated that PFC emulsions are superior at reducing the oxygen debt that accumulate as a consequence of hemorrhagic shock.

A more recent study by Elliott and coworkers compared Fluosol to lactated Ringer's (LR) in an dog resuscitation model. The authors concluded that Fluosol was effective in producing volume expansion, oxygen delivery, and oxygen consumption. They observed an increasing oxygen consumption at 60 min. post-infusion similar to previous work and which was statistically significantly greater at 24 hours post infusion than LR. The authors also found that the PFC contributed as much as 40% to the animals' overall oxygen consumption in the one-hour period post shock (Elliott et al. 1989).

In summary, the data from the rat, dog, and primate studies on hemodiluted or shocked animals indicate that PFC emulsions can effectively replace volume, deliver adequate oxygen, increase oxygen consumption in animals with an oxygen debt, and markedly increase overall survival rate. The data indicate that PFC emulsions can provide a major portion of an animal's oxygen consumption. The ability of PFCs to increase oxygen consumption in shocked animals may also reflect the superior ability of the small, oxygen-laden particles in the emulsion to traverse and oxygenate previously ischemic microcirculation.

Fluosol was tested for efficacy as a blood substitute in clinical trials in Japan and the U.S.A. In the Japanese trials (Ohyanagi et al. 1984; Mitsuno, Ohyanagi et al. 1982; Mitsuno and Ohyanagi 1985), which ran from 1979 to 1982, 270 patients received Fluosol in dosages of 20 to 30 ml/kg. Instead of blood transfusion for replacement of surgical blood loss or for improvement of acute hemorrhagic anemia. These studies do not report on survival nor were the results of Fluosol-treated patients compared to control groups, leaving the question of efficacy unanswered. The authors determined that Fluosol provided about 17% of the tissue oxygen consumption at an fractional inspired oxygen (FiO2) of 0.5 to 0.6, an amount equal to the plasma contribution. No acute or chronic adverse reactions were observed, except for a transient decline in neutrophils and platelets shortly after infusion. Hemodynamic parameters of patients were either maintained or recovered to normal after treatment with Fluosol. In patients who died of their diseases, there were no organ abnormalities that could be attributed to Fluosol. Although PFC could still be detected in some tissues 7 weeks post-infusion, organs analyzed 7 months after treatment were free of PFC.

In the USA, Fluosol was tested first on a humanitarian protocol. The objectives of this study were to determine clinical safety, hemodynamic, and oxygen transport profiles of Fluosol. In contrast to earlier reports, fully one-third of the patients experienced an acute reaction to infusion of a 0.5-ml test dose of Fluosol which was controlled by treatment with corticosteroids (Tremper et al. 1985). The treated patients had about 3% PFC in their systemic circulation which contributed about 0.7 volume % oxygen to the total oxygen transported, or about 25% of the oxygen consumption of the patient.

A subsequent clinical trial by Gould reported similar results (Gould et al. 1986). The patients in this study, who received up to 40 ml Fluosol/kg, were severely anemic with pretreatment hemoglobin levels of about 3 g/dl. Fluosol made a significant contribution to overall oxygen consumption, providing 28% of the oxygen consumed and demonstrating the ready availability of the oxygen carried by the PFC phase. Despite the significant contribution to oxygen consumption noted in the Gould paper, and in contrast to most of the previously published animal studies, the overall level of oxygen delivery and consumption declined in Fluosol treated patients . Gould concluded that Fluosol failed to make greater contributions to overall oxygen delivery because of the low volume of PFC in the Fluosol formulation and its short intravascular half-life of 24 hours. He speculated that second- generation emulsions with a higher initial concentration of PFC might lead to higher circulating PFC concentrations and better efficacy. He also indicated that a clinical trial where the efficacy of Fluosol was evaluated as a short term replacement until red blood cells became available might have a more favorable outcome. Perhaps the most important aspect of this particular clinical trial was the fact that there were no adverse patient reactions to Fluosol. Gould concluded that Fluosol was safe but not effective, largely because of low PFC content. Despite this outcome, Fluosol continues to be utilized on a humanitarian protocol at certain institutions (Spence et al. 1989).

The fact that Fluosol failed in clinical trials should not obscure the fact that PFCs did make a significant and reproducibly measurable contribution to oxygen consumption in these severely anemic patients, that side reactions to the product were minimal and controllable and that the overall safety of Fluosol was adequate to insure its testing in other clinical indications such as balloon angioplasty and cancer therapy.

6.2 MYOCARDIAL INFARCTION

Numerous studies have demonstrated that mortality and morbidity in acute myocardial infarct (MI) are directly related to the degree of destruction of the myocardial tissue due to ischemia post infarct. The timeframe of effective response to coronary artery occlusion has been shown to be short, on the order of six hours. Thereafter, tissue damage is largely irreversible. Thus, the more rapidly the infarcted myocardium is reoxygenated and reperfused, the more myocardial tissue will be preserved and the final outcome improved for the patient.

PFC emulsions have been proposed as therapeutic agents for preservation of myocardium after an infarct because of their ability to transport oxygen in significant quantities and their small particle size. The small particle size and lowered viscosity of

PFC emulsions suggest that they might flow more readily through the long and thin intercapillary connections that make up the collateral circulation in humans, and thus help to oxygenate the myocardium distal to the occlusion (Faithful et al. 1986). These same properties suggest that PFC particles might reperfuse the edematous "no reflow" vessels resulting from ischemia, bringing in oxygen and removing carbon dioxide in the acidotic tissues functioning under conditions of anaerobic metabolism (Nunn et al. 1983).

Glogar and coworkers assessed the efficacy of PFCs in preservation of ischemic myocardium in dog models (Glogar et al. 1981). After infusion of the PFC emulsion or a comparable amount of LR, the animals breathed 100% oxygen for 15 min. prior to permanent occlusion of the left anterior descending (LAD) artery. After six hours of occlusion, the animals were sacrificed and the area of necrosis (A_N), the area at risk (A_R) and the ratio of A_N/A_R were measured. The area of necrosis (A_N) and the ratio of A_N/A_R were both significantly reduced by about 30% (p<.01) in the PFC-treated group compared to the LR and the control groups.

Nunn and coworkers have examined the effect of Fluosol on myocardial salvage in a similar model (Nunn et al. 1983). One hour after ligation of the LAD, Fluosol or 0.9% saline was infused with simultaneous withdrawal of blood to a dose of 30 ml emulsion/kg. The ratio of the A_N/A_R is statistically significantly reduced by about 33% in the Fluosol treatment group.

In studies done in pigs, whose collateral circulation is most similar to humans and lacks the variation found in dogs,
Faithful et al. (1986) evaluated the ability of Fluosol to preserve myocardium during infarct by measuring the change in myocardial oxygen tension polarographically in the most hypoxic region of the ischemic myocardium for five hours post infarct. Oxygen tension of the ischemic myocardium in the PFC-treated animals increased greatly relative to controls.

Minimization of reperfusion damage by Fluosol post MI has been reported by Forman, using both an intracoronary (Kolodgie et al. 1986), and a systemic infusion model (Bajaj et al. 1989). In the earlier intracoronary model, reperfusion with Fluosol after 90 min. of ischemia resulted in a 60% reduction in infarct size and improved ventricular function two weeks post infarct. There was no evidence of increased myocardial oxygen tensions in the Fluosol group compared to control despite ventilation with 100% oxygen during and for three hours after reperfusion. Histologically, there was reduced neutrophil deposition in the microcirculation of the A_R of Fluosol-treated dogs compared to controls. Electron microscopy revealed capillary obstruction involving endothelial cell protrusions and neutrophil and red cell plugging of the obstructed capillaries in control dogs but not in Fluosol treated animals. The authors interpreted this as evidence for improved reflow after the occlusion was eliminated.

In the systemic infusion model (Bajaj et al. 1989), A_N/A_R was significantly reduced after 90 minutes of ischemia in the Fluosol group (24% compared to the LR control) while A_R was similar in both treatment groups at 48-50%. Epicardial blood flow measurements in the central ischemic zone indicate that collateral blood flow variability did not bias the A_N/A_R. Control animals and Fluosol-treated animals had significantly reduced blood flow in the central ischemic zone during occlusion. One hour post reperfusion, the control animals had significantly reduced myocardial blood flow in the epicardial and endocardial ischemic zone compared to Fluosol. There were two- to fourfold increases in neutrophil infiltration into the epicardium, midmyocardium, and endocardium ischemic regions in control dogs compared to Fluosol treated dogs. Non-ischemic regions of the myocardium had similar levels of neutrophils in the LR and Fluosol treated groups. Venous neutrophil counts were lowered within one hour of Fluosol infusion and remained below control counts throughout the reperfusion period. Neutrophils from the Fluosol group exhibited lowered *ex vivo* chemotaxis one hour after reperfusion than control groups. (Bajaj et al. 1989) concluded that the primary mechanism by which Fluosol preserves myocardium is by reducing neutrophil chemotaxis and adhesion in the ischemic tissue during reperfusion.

In summary, PFC emulsions have been shown to reduce relative infarct size, increase myocardial oxygen tension and neutralize inflammatory cells during infarct and after reperfusion. Despite the overwhelming evidence of myocardial preservation during MI in animal studies of PFC emulsions from at least four independent labs, there have no published clinical studies on the use of PFC emulsions to treat acute MI.

6.3 PERCUTANEOUS TRANSLUMINAL CORONARY ANGIOPLASTY (PTCA)

Percutaneous transluminal coronary angioplasty (PTCA) is a procedure that is used to remodel arterial plaque and allow increased coronary artery blood flow. It has gained wide acceptance with over 300,000 angioplasties being performed in the USA annually. In this procedure, a catheter with an inflatable balloon at its tip is threaded into the coronary artery experiencing partial blockage. Once centered in the lesion, the balloon is inflated, compressing the plaque against the vessel wall and increasing the effective diameter of the coronary artery. Current medical practice requires balloon inflation times of 45 seconds or longer to insure optimal results, and there is an interest in increasing the inflation time to determine if (re-occlusion of the artery) rates can be reduced. During the period of balloon inflation, the coronary artery is occluded and part of the left ventricle experiences temporary ischemia. Recent studies have shown that balloon inflation times of 20 seconds are enough to produce ECG changes, induce ventricular wall motion irregularities, and reduce the left ventricular ejection fraction. Such responses are exacerbated in patients with multivessel disease. In order to increase the time of inflation or to reduce risk in patients with multivessel disease, physicians have attempted to perfuse blood through the central lumen of the angioplasty catheter. Blood perfusion was found to have a number of disadvantages including high viscosity, hemolysis, and the necessity of another arterial access point to provide blood for the perfusion.

In the early 1980s, Spears piloted the use of oxygenated Fluosol perfused through the central lumen of the catheter to alleviate symptoms of ischemia during balloon inflation in dogs. The results showed that 8 of 10 dogs perfused with oxygenated Fluosol during prolonged balloon inflation times (>5 min.) had normal ECGs while controls perfused with oxygenated saline had an ECG injury pattern and increased ventricular ectopy (Spears et al. 1983).

Both Spears (Spears et al. 1984) and Anderson (Anderson et al. 1985) issued preliminary reports on the use of Fluosol in humans but the definitive clinical studies were performed by Jaffe and coworkers (Cleman et al. 1986; Jaffe et al. 1988). They used two-dimensional echocardiography to detect the location, extent and evolution of ischemic contractile dysfunction and quantitatively assessed regional wall motion and ejection fraction by computer analysis of the echocardiograms.

In the clinical trial, 42 symptomatic patients with single lesions of 70%-or-greater stenosis of the coronary artery undergoing PTCA were studied. All patients underwent a preliminary inflation without perfusion. Subsequent inflations were done with either oxygenated LR, oxygenated Fluosol or nonoxygenated Fluosol. The data showed that oxygenated Fluosol maintained an ejection fraction identical with the baseline value 45 seconds post balloon inflation, while there was 35% reduction in ejection fraction in the controls. These studies were the basis of the recent FDA approval granted for the use of Fluosol in PTCA on high-risk patients (Anonymous 1990).

6.4 CANCER THERAPY

Solid tumors possess vascular insufficiencies and blood flow irregularities resulting in significant areas of hypoxic cells, frequently amounting to 10-20% of the total viable tumor cell population (Thomlison and Gray 1955). Hypoxia has long been known to protect cells from the cytotoxic effects of radiation and chemotherapy. The surviving hypoxic cell fraction is generally recognized as capable of reestablishing the tumor and limiting the therapeutic effectiveness of these modalities. There have been numerous efforts to render the hypoxic cell fraction susceptible to radiation and chemotherapy including the development of radiosensitizers such as misonidazole and the use of hyperbaric oxygen (HBO) chambers.

In 1983, Teicher and Rockwell demonstrated that infusion of PFC emulsions in advance of radiation coupled with carbogen (95% oxygen and 5% carbon dioxide) breathing during radiation therapy significantly reduced the surviving fraction of hypoxic cells and the growth rate of implanted rodent tumors (Teicher and Rockwell 1983). Subsequently, Teicher and Rose found that mice bearing tumors from Lewis lung tumor or FSa-II fibrosarcoma treated with Fluosol and breathing carbogen had tumor growth delays of up to 30 days compared to rats receiving either radiation alone, radiation and carbogen alone, or radiation and Fluosol with air breathing (Teicher and Rose 1984). Others have reported similar results (Song et al. 1986; Rockwell 1985; Lustig and McIntosh 1986).

The mechanism of enhancement of radiation therapy appears to be due to increased oxygenation of the tumor. Song et al. (1987) measured the pO_2 of tumors in mice and verified that tumor oxygenation is markedly increased by the combination of Fluosol and carbogen breathing. The results show that tumor oxygenation is increased six-seven-fold in mice breathing carbogen and infused with Fluosol over mice breathing room air and two-fold over mice breathing carbogen without Fluosol. Klubes and coworkers (1987) found that Fluosol does not increase blood flow through solid tumors in the model they studied. Long has documented by X-ray and histological studies that most implanted and spontaneous animal tumors preferentially accumulated PFC emulsion in macrophages associated with the tumors (Long et al. 1978). What, if any, role this phenomenon plays in radiation therapy enhancement is not clear.

Clinical trials of Fluosol in cancer radiation therapy were first reported in 1986 (Rose et al. 1986). These trials were conducted on 15 patients with Stage III/IV head and neck cancer, who were not candidates for surgical therapy. Radiation was fractionated into about 25 doses of 1.8 Gy over a five-week period. Fluosol was infused at a rate of eight to nine ml/kg on the first day of each week for a total dose of 40-45 ml/kg and patients breathed 100% oxygen before and during all 25 radiation fractions. Of the 15 patients, 10 had primary and nodal clearance with the longest follow-up post treatment being eight months. The authors noted four cases of acute reactions controllable with diphenhydramine and eight of 15 patients exhibited serum enzyme elevations of two to three times normal which returned to normal three months post therapy. Coagulation times, BUN, creatinine, serum albumin and bilirubin were unaffected by the treatment. White blood cell counts and hematocrit were slightly depressed but the changes were typical of normal responses to radiation therapy and could not be attributed to Fluosol. There was some evidence for acceleration of the onset of the mucositis normally caused by the radiation.

Lusting et al. (1989a) have also reported on clinical trials with Fluosol in patients with head and neck cancer. In this study, 37 patients were enrolled and 28 (76%) had complete response (no evidence of primary disease) two months post treatment. The determinant survival, excluding those who died from other causes, was 78% one year post therapy compared to the Radiation Therapy Oncology-Group reported survival rate of 53-62%. Patient side effects were reported to be mild and reversible and of about the same frequency, extent, and duration as observed in the first clinical trial (Rose et al. 1986). While cancer patients are not considered cured until after five years of complete remission, the initial response to Fluosol has been very promising. In addition to the head and neck trials discussed above, clinical trials using Fluosol and radiation therapy for treatment of non-small-cell lung cancer (Lustig et al. 1989b) and brain glioma are underway (Rockwell 1988).

The effects of Fluosol and oxygen breathing were also evaluated in conjunction with chemotherapy. Ohyanagi and coworkers found that animals treated with vincristine or spadicomycin and Fluosol with oxygen breathing had smaller tumor masses compared to animals treated with the chemotherapeutics alone (Ohyanagi et al. 1983). Subsequently, a number of investigators have reported on the potentiating effect of Fluosol on chemotherapy induced tumor growth delay (TGD) (Teicher and Holden 1987). Fluosol with oxygen breathing enhanced the effectiveness of all three major classes of chemotherapeutics: alkylating agents, antibiotics and alkaloids and antimetabolites. Despite the large number of papers on the use of Fluosol with chemotherapeutics, there have been no reports of clinical studies.

6.5 LIQUID BREATHING

Liquid breathing using PFCs was the catalyst for an explosion in research on the use of PFCs as oxygen transport agents. While liquid breathing research was rapidly overshadowed by research on emulsions, the field remained active and now promises to make clinical impact. The list of potential applications of "liquid breathing" can be divided into two major groups: liquid ventilation therapies designed to oxygenate a patient with compromised pulmonary function; and lavage therapies which strive to maintain oxygenation while the lung is purged of obstructive material. Liquid ventilation is being investigated as a therapy for premature infant respiratory distress syndrome (RDS), adult respiratory distress syndrome (ARDS), and "respirator lung"-lungs stressed due to prolonged mechanical ventilation. Lavage therapies under investigation include removal of meconium, fibrotic material in cystic fibrosis, and proteinaceous matter in proteinemia diseases.

After Clark and Gollan's famous demonstration of liquid breathing, Modell and coworkers evaluated the feasibility of long-term maintenance of dogs and primates via liquid ventilation (Modell et al. 1970b). First they determined that, unlike saline lavages, PFC lavages did not extract surfactant from the lungs of dogs nor did they modify their surface tension properties (Modell et al 1970a). Liquid ventilation of dogs and primates was accomplished with PFC FX-80 perfluorobutyltetrahydrofuran for 30 min. to eight hours. Both species became hypercarbic (arterial CO_2 of 40 to 80) and acidotic (arterial pH of 7.05 to 7.2) as the liquid ventilation progressed but were otherwise well oxygenated. After the PFC had been drained from their lungs, the animals resumed normal air breathing but with significantly depressed arterial oxygen levels of 45-55 mm Hg. Measurement of the PaO_2 over time indicated a return to normal values about 10 days after liquid ventilation. Gross examination of lungs from dogs serially sacrificed from one hour to 10 days after termination of liquid ventilation showed a translucent sheen suggestive of PFC in the dependent alveolar sacs of the one-hour lungs. Three days after liquid ventilation the translucent areas were less extensive and numerous, and after 10 days only a few lobes had a translucent sheen. Microscopically, the lungs from the three-hour post liquid ventilation were hyperemic, contained neutrophil exudate in the bronchioles and had some congested alveolar septa filled with numerous intra-alveolar vacuolated macrophages. Ten days after treatment the lungs were virtually normal microscopically. The authors suggested that the hyperemia and inflammatory reactions were due to the alveolar distention of liquid ventilation or to an irritant effect of the PFC itself or both. They concluded that the reduced PaO_2, which was readily improved by breathing $FiO_2 = 0.4$, was due to the PFC in the alveolar septa forming a diffusion barrier to oxygen transfer. On balance, the data support the conclusion that liquid ventilation does not cause any adverse morphological, biochemical, or histological effects.

The problems of acidosis and hypercarbia were probably related to inadequate removal of metabolic CO_2. To determine if the hypercarbia and acidosis limited the duration of liquid breathing, four dogs were liquid ventilated for eight hours with buffer administered to prevent acidosis. All four dogs survived, the acidosis was controlled by the buffer, and the hypercarbia leveled off at a 60-80 mm Hg. While all four dogs survived, two who were given higher-pressure ventilations suffered lung tissue damage, suggesting that pressure control during liquid breathing would be necessary. Moskowitz and Schaffer have subsequently developed ventilators for liquid breathing to solve the problems of CO_2 removal and pressure induced tissue damage (Schaffer and Moskowitz 1974).

Modell also studied tissue distribution in dogs and primates several years after liquid breathing. The tissues were examined grossly, microscopically, and chromatographically. The results were unremarkable except for the presence of trace amounts (a few milligrams per 100 g of tissue) of PFC in the lungs, liver, and fatty tissue (Modell et al. 1976).

In summary, liquid ventilation studies with adult dogs and primates resulted in adequate oxygenation during liquid breathing accompanied by acidosis and hypercarbia. After liquid ventilation, PaO_2 levels were depressed for three to 10 days, probably due to PFC in the alveolar sacs. After 10 days and for periods thereafter up to three-years post liquid ventilation, lungs were normal grossly, microscopically, and biochemically, with the exception of traces of PFC in various tissues

In studies with immature and premature animals, the results were markedly different. Rufer and Spitzer studied the use of liquid ventilation with PFC in immature pigs (Rufer and Spitzer 1974). In pigs of 95-days gestation, air ventilation was difficult due to atelectasis, and 75% of the animals died within 15 min. At 100 days gestation, air ventilation was maintained for 90 min. and at 110 days gestation (full term), air ventilation was easily achieved and survival was high. Rufer and Spitzer found that liquid ventilation of the 95-day immature mini-pig could be sustained for over three hours and that the animals did not become hypercarbic. A more exciting finding was that compliance in the immature lung after liquid ventilation was improved almost to that of a gestationally mature mini-pig in spite of the fact that lung lavage of the 95-day gestational animals failed to recover any surfactant. It appeared that PFC assumed the role of surfactant in the immature lung, decreased the surface tension of the lung, and restored compliance towards normal during subsequent air breathing. The results suggest that periods of liquid breathing need not be terribly long to have lasting benefit. Schaffer and coworkers found that premature sheep were readily ventilated with PFCs and that peak intratracheal pressures were significantly reduced after liquid ventilation (Schaffer et al. 1976). Schaffer's work also confirmed that hypercarbia and acidosis were not as extreme in the immature lung undergoing liquid ventilation. In a subsequent paper, Schaffer and coworkers found that pre-term lambs could be liquid ventilated for three hours and that their gas exchange and lung compliance were similar to mature lambs undergoing gaseous ventilation. They believe that this result extends the viability of the pre-term lamb to the limit of the pulmonary capillary development rather than that of the pulmonary surfactant system (Schaffer et al. 1983).

Schaffer and coworkers also found that cardiac output is decreased (Lowe et al. 1979) and pulmonary vascular resistance is increased during liquid breathing in pre-term lambs (Lowe and Schaffer 1986). Modell and coworkers on the other hand, found no change in cardiac output in adult dogs (Modell et al. 1970b). The differences in these reports may be due to differential response in premature animals compared to adults, species differences, or methodological differences. If cardiac output decrease and PVR increase are confirmed during liquid breathing, it may have consequences for long-term maintenance via liquid breathing. However, effects would have no impact on the short-term usage required for the PFC to serve as a temporary surfactant and open the alveoli.

Waldrop (1989) conducted a clinical trial on a 24 to 28-week-old human infant, with 15 min. of liquid breathing. Although the baby died, the lungs functioned for 19 hours after the liquid ventilation ceased. Further trials are planned in premature infants.

7. Second Generation Products

At least five companies are in the development stage with second generation products. HemaGen/PFC (San Francisco, CA.), Green Cross Co. (Osaka, Japan), Alliance Pharmaceutical (San Diego, CA.), Dupont (Wilmington, De.) and Affinity Biotech (Linwood, PA.). HemaGen has developed an emulsion based on the use of triglyceride oils to improve surfactant interaction with the PFC, resulting in emulsions with improved stability at PFC contents as high as 70 vol%. Invented by R.F. Shaw, a physician and medical entrepreneur and L.C. Clark, the original PFC pioneer, these high-PFC emulsions can be sterilized at 121°C for as long as 30 min. without degradation or cracking (Shaw and Clark 1987). The mean particle size after sterilization is 0.2 μm by laser light scattering, and the emulsions can be stored for over a year at 4 and 25°C. Particle size does not increase during storage. The viscosity of the emulsion is 7 cP at 37°C, and it does not impair flow through the microcirculation. Animal safety is several times greater than Fluosol DA-20 on a ml PFC/kg basis.

Green Cross Co. (Osaka, Japan) recently published information on a second-generation emulsion based on lecithin and perfluoro-N-methyldecahydroisoquinoline (Tsuda and Yokoyama 1989). The emulsion is reported to be only a 10 vol% emulsion and is stable at "cold room temperature." Perfluoro-N-methyldecahydroiso- quinoline reportedly has a short tissue half-life.

Alliance Pharmaceutical (San Diego, CA.) has developed an emulsion based on high perfluorooctyl bromide content and using high concentrations of lecithin as the surfactant (Burgan et al. 1987). The formulation was published as a 6% egg yolk lecithin and 100 (wt/vol)% perfluorooctyl bromide (about 52 vol%) perhaps including coadditives such as tocopherol. The LD_{50} in mice was reported to be 45 g emulsion/kg or about 32 ml emulsion/kg. Perfluorooctyl bromide was widely investigated as an oxygen transport agent in the 1970s by other investigators including Clark and Yokoyama but was not developed. Long, who had been investigating the utility of bromoperfluorochemicals as radiographic contrast agents for many years, developed the emulsion which is now being evaluated in clinical trials. Total exchanges using the same formulation have been reported. Mean particle size is reported to be 0.2 μm, and the emulsion is reported to be shelf stable at room temperature (Long et al. 1987). Hemodynamic responses in dogs were negligible with this emulsion compared to the well-known hemodynamic collapse in dogs when given Fluosol DA-20® (Mattrey et al. 1989).

Dupont has developed an emulsion called Therox® using lecithin and containing 40 volume% 1, 2-bis(perfluorobutyl)ethylene. The emulsion is shelf stable at 4°C for over one year, has a PSD mean of about 0.25μ, has a viscosity compatible with whole blood and is well tolerated in animals. 1, 2-bis(perfluorobutyl)ethylene is expired from the liver and spleen with a half-life of 3-18 days. The clinical status of this product is unknown.

These second-generation products have solved two of the more serious problems with Fluosol DA-20: the acute complement activation caused by pluronic F-68 and the necessity to be stored frozen.

8. Conclusion

PFCs have come a long way since Clark's first experiment and the ensuing euphoric period when too much was expected and too little was known. Over the intervening years, the complexities of formulating, manufacturing, testing, and utilizing PFC emulsions have been unraveled and the first PFC product, Fluosol DA-20, is readily available in the USA. Unfortunately, Fluosol DA-20 proves inadequate for the goals set for it. Nonetheless, it served to define the next level of hurdles that the second-generation of oxygen transport agents will have to clear to achieve widespread use. Fluosol DA-20 was also responsible for the development of a body of knowledge based on the preclinical and clinical efficacy testing of PFC based products. Medical understanding of PFCs and of their use have been brought into focus by the Fluosol DA-20 experience. The finding that pluronic F-68 was responsible for the undesired acute reactions has led to its removal from second- generation products. The benign and reversible side effects attributable to PFCs now are fairly well understood and are not a threat to organ function at therapeutic doses. Several new formulations have been developed that will expand the uses of PFC emulsions into more applications. Research on third- generation products that will persist significantly longer in the bloodstream is underway, and the first new classes of safe intravenous surfactants have been identified. After a somewhat- long induction period, the use of PFC oxygen transport agents in medicine is poised to enter an exponential growth phase.

REFERENCES

Anonymous (1990) 'Alpha Therapeutic's Fluosol oxygen transport fluid approved for use in angioplasty; launch will be initiated at cardiologist conference in March', FDC Reports 52, 8.

Bajaj, A.K., Cobb, M.A., Virmani, R., et al. (1989) 'Limitation of myocardial reperfusion injury by intravenous perfluorochemicals', Circulation 79, 645-656.

Blumberg, N., Agaral, M. and Chuang, C. (1985) 'A possible association between survival time and transfusion in patients with cervical cancer', Blood 66 (suppl. 1), 274a.

Burgan, A.R., Long, D.M., Mattrey, R.F., et al. (1987) 'Results of pharmacokinetic and toxicologic studies with PFOB emulsions', Biomaterials, Artificial Cells, and Artificial Organs 15, 367.

Clark, L.C. and Gollan, F. (1966) 'Survival of mammals breathing organic liquids equilibrated with oxygen at atmospheric pressure', Science 152, 1755-1756.

Clark, L.C., Wesseler, E., Miller, M. and Kaplan, S. (1974) 'Ring versus straight chain perfluorocarbon emulsions for perfusion media', Microvascular Research, 8, 320-340.

Cleman, M., Jaffe, C.C., and Wohlgelernter, D. (1986) 'Prevention of ischemia during percutaneous transluminal coronary angioplasty by trans-catheter infusion of oxygenated Fluosol DA-20%', Circulation 74, 555-562.

Elliott, L.A., Ledgerwood, A.M., Lucas, C.E., et al. (1989) 'Role of Fluosol-DA 20% in prehospital resuscitation', Critical Care Med. 17, 166-172.

Faithful, N.S., Erdmann, W., Fennema, M. and Kok, A. (1986) 'Effects of haemodilution with fluorocarbons or dextran on oxygen tensions in the acutely ischemic myocardium', Brit. J. Anesth. 58, 1031-1040.

Geyer, R.P., Monroe, R.C., and Taylor, K. (1968) 'Survival of rats totally perfused with a fluorocarbon-detergent preparation', in Organ Perfusion and Preservation (Norman, J., ed.), pp. 85-97, Appleton Century and Crofts, New York.

Geyer, R.P.,Taylor, K., Duffett, E., et al. (1973) 'Successful complete replacement of the blood of living rats with artificial substitutes', Fed. Proc. 32, 927.

Gjaldbaek, J., and Hildebrand, J.H. (1949) 'The solubility of nitrogen in carbon disulfide, benzene, normal and cyclo-hexane and in three fluorocarbons', J. Am. Chem. Soc. 71, 3147.

Glogar, D.H., Kloner, R.A., Muller, J., et al. (1981) 'Fluorocarbons reduce myocardial ischemic damage after coronary occlusion', Science 211, 1439-1441.

Gould, S.A, Rosen, A.L, Sehgal, L.R., et al. (1986) 'Transfusion trigger literature', New England J. Med. 314, 1653-1656.

Haljamae, H., and Rosenberg, P.H. (1988) 'Perioperative transfusion practice-An update', Acta Anaesthesiol. Scand. 32 (suppl. 89), 1-3.

Hamza, M.A., Serratrice, G., Stebe, M-J., Delpuech, J-J. (1981) 'Solute-solvent interactions in perfluorocarbon solutions of oxygen. An NMR study', J. Am. Chem. Soc. 103, 3733-3738.

Huestis, D.W., Bove, J.R., and Busch, S. (1981) Practical Blood Transfusion, Little Brown and Company, Boston.

Jaffe, C.C., Wohlgelernter, D., and Cabin, H., et al. (1988) 'Preservation of left ventricular ejection fraction during percutaneous transluminal coronary angioplasty by distal transcatheter coronary perfusion of oxygenated Fluosol DA 20%', Am. Heart J. 6, 1156-1164.

Klubes, P.S., Hiraga, S., Richard, L.C., Owens, E.S. and Blasberg, R.G. (1987) 'Attempts to increase intratumoral blood flow in the rat solid Walker 256 tumor by the use of perfluorocarbon emulsion Fluosol-DA', Eur. J. Cancer Clin. 23, 1859-1867.

Kolodgie, F.D., Dawson, A.K., Roden, D.M., et al. (1986) 'Effect of Fluosol-DA on infarct morphology and vulnerability to ventricular arrhythmia', Am. Heart J. 112, 1192-1201.

Liu, M.S., Rosen, A., and Long, D.M. (1973) 'Biological disposition of perfluorooctylbromide-A new contrast agent', Fed. Proc. 32, Abstr. No. 3982, 927.

Long, D.M., Multer, F.K., Greenburg, A.G., et al. (1978) 'Tumor imaging with X-rays using macrophage uptake of radiopaque fluorocarbon emulsions', Surgery 84, 104-112.

Long, D.C., Fallano, R., Reiss, J.G., et al. (1987) ' Preparation and applications of highly concentrated PFOB emulsions', Biomaterials, Artificial Cells, and Artificial Organs 15, 417.

Lowe, C.A., and Schaffer, T.H. (1986) 'Pulmonary vascular resistance in the fluorocarbon-filled lung', J. Appl. Physiol. 60, 154-159.

Lowe, C., Tuma, R.F., Sevieri, E.M. and Schaffer, T.H. (1979) 'Liquid ventilation: cardiovascular adjustments with secondary hyperlactatemia and acidosis.', J. Appl. Physiol.:Respirat. Environ. Exercise Physiol. 47, 1051-1057.

Lustig, R.A., and McIntosh, N. (1986) 'Fluosol-DA in radiation therapy', Prog. Clin. and Biol. Res. 211, 29-38.

Lustig, R., McIntosh-Lowe, N., Rose, C., et al. (1989a) 'Phase I/II study of Fluosol-DA and 100% oxygen as an adjuvant to radiation in the treatment of advanced squamous cell tumors of the head and neck.', Int. J. Radiation Oncology Biol., and Phys. 16, 1587-1593.

Lustig R.A., Lowe, N., Prosnitz, L., et al. (1989b) 'Phase I/II study of Fluosol and 100% oxygen breathing as an adjuvant to radiation in the treatment of unresectabel non small cell carcinoma of the lung', Int. J. Radiation Oncology, Bio., and Phys 17, 202.

224

Makowski, H. (1978) 'The properties of Fluosol-DA infusion in the treatment of hemorrhagic shock', in Proc. 4th International Symposium on Perfluorochemical Blood Substitutes, Oct. 21-22, 1978, Kyoto Japan 439-448, Excerpta Medica, Amsterdam.

Mattrey, R.F., Hilpert, P.L., Long, C.D. et al. (1989) 'Hemodynamic effects of intravenous lecithin-based perfluorocarbon emulsions in dogs', Critical Care Medicine 17, 652-656.

Mitsuno, T., Ohyanagi, H., and Naito, R. (1982) 'Clinical studies of a perfluorochemical whole blood substitute (Fluosol-DA): Summary of 186 cases', Ann. Surg. 195, 60-69.

Mitsuno, T., and Ohyanagi, H. (1985) 'Present status of clinical studies of Fluosol-DA (20%) in Japan in Tremper, K. (ed), International Anesthisiology Clinics. Perfluorochemical Oxygen Transport, Little, Brown and Company Publishers, Boston pp. 169-184.

Modell, J.H., Gollan, F., Giammona, S.T. and Parker, D. (1970a) 'Effect of fluorocarbon liquid on surface tension properties of pulmonary surfactant', Chest 57, 263-265

Modell, J.H., Newby, E.J. and Ruiz, B.C. (1970b) 'Long-term survival of dogs after breathing oxygenated fluorocarbon liquid', Fed. Proc. 29, 1731-1736.

Modell, J.H., Hood, C.I., Kuck, E.J. and Ruiz, B.C. (1971) 'Oxygenation by ventilation with fluorocarbon liquid (FX-80)', Fed. Proc. 34, 312-320.

Modell, J.H., Calderwood, H.W., Ruiz, B.C., Tham, M.K., and Hood, C.I. (1976) 'Liquid ventilation of primates', Chest 69, 79-81.

Moore, R.E., and Clark, L.C. (1985) 'Chemistry of fluorocarbons in biomedical use', in Tremper, K. (ed), International Anesthisiology Clinics. Perfluorochemical Oxygen Transport, Little Brown and Company Publishers, Boston pp. 11-24.

Nunn, G.R., Dance, G., Peters, J., and Cohn, L.H. (1983) 'Effect of fluorocarbon exchange transfusion on myocardial infarction size in dogs', Am. J. Cardiol. 52, 203-205.

Ohyanagi, H., and Mitsuno, T. (1975) 'Biophysiological effects of perfluorochemicals as artificial blood', in Proc. of the Post Congress Symposium, Xth International Congress of Nutrition on Perfluorochemical Blood, Kyoto, Japan, pp.21-54.

Ohyanagi, H., Sekita, M., Yokoyama, K., et al. (1978) 'Studies on perfluorochemical emulsion (Fluosol DA) as artificial blood in exchange transfusion and emergency use in monkeys', in Proc. of the 4th International Symposium on Perfluorochemical Blood Substitutes, Oct. 21-22, 1978, Kyoto Japan, pp.373-389. Excerpta Medica, Amsterdam

Ohyanagi, H., Nishijima, M., Usami, M., et al. (1983) 'Experimental studies on the possible combined chemotherapy to neoplasms with Fluosol-DA infusion', Prog. Clin. Biol. Res. 211, 315-320.

Ohyanagi, H., Nakaya, S., Okumura, S., and Saitoh, Y. (1984) 'Surgical use of Fluosol-DA in Jehovah's Witness patients', Artificial Organs 8, 110-118.

Okada, K., Kosugi, I., Kawashima, Y. et al. (1975) 'Effect of Fluosol-DC on tissue pO_2 and pCO_2 in treatment of hemorrhagic hypotension', in Proc. of the Post Congress Symposium, Xth International Congress of Nutrition on Perfluorochemical Blood, Kyoto, Japan, pp.215-224.

Reiss, J.G. and LeBlanc, M. (1978) 'Perfluoro compounds as blood substitutes' Angew. Chem. Int. Ed. Engl., 17, 621-634.

Reiss, J.G. (1984) 'Reassessment of Criteria for the selection of perfluorochemicals for second-generation blood substitutes: Aanalysis of structure/property relationships', Artif. Organs, 8, 44-56.

Rockwell, S. (1985) 'Use of perfluorochemical emulsion to improve oxygenation in a solid tumor', Int. J. Radiation Oncology, Biol. and Phys. 11, pp.97-103.

Rockwell, S. (1988) Personal Communication, Yale Medical Center, Hunter Radiation Center, New Haven, CT.

Rose, C. M., Lustig, R., McIntosh, N., and Teicher, B. (1986) 'A clinical trial of Fluosol DA 20% in advanced squamous cell carcinoma of the head and neck', Int. J. Radiation Oncology, Biol. and Phys. 12, 1325-1327.

Rosen, A.L., Sehgal, L.R., Gould, S.S. Sehgal, H.L., Moss, G.S. (1985) 'Transport of oxygen by perfluorochemical emulsions', in Tremper, K. (ed), International Anesthisiology Clinics. Perfluorochemical Oxygen Transport, Little Brown and Company Publishers, Boston pp. 95-103.

Rosenblum, W.I., Moncure, C.W. and Behm, F.G. (1985) 'Some long-term effects of exchange transfusion with fluorocarbon emulsions in macaque monkeys', Arch. Pathol. Lab Med. 109, 340-344.

Rufer, R., and Spitzer, H. L. (1974) 'Liquid ventilation in the respiratory distress syndrome', Chest 66(suppl.) 29S-30S.

Schaffer, T.H., and Moskowitz, G.D. (1974) 'Demand-controlled liquid ventilation of the lungs', J. Applied Physiology 36, 208-213.

Schaffer, T.H., Rubenstein, D., Moskowitz, G.D. and Delivoria-Papadopoulos, M. (1976) 'Gaseous exchange and acid-base balance in premature lambs during liquid ventilation since birth', Pediat. Res. 10, 227-231.

Schaffer, T.H., Tran, N., Bhutani, V. K., and Sivieri, E.M. (1983) 'Cardiopulmonary function in very preterm lambs during liquid ventilation', Pediat. Res. 17, 680-684.

Shaw, R.F. and Clark, L.C. (1987) 'Stable emulsions of highly fluorinated organic compounds' European patent application no. 0 231 091.

Sloviter, H., and Kamimoto, T. (1967) 'Erythrocyte substitute for perfusion of brain', Nature 216, 458.

Song, C.W., Zhang, W.L., Pence, D.M., et al. (1986)' Increased radiosensivity of tumors by perfluorochemicals and carbogen', Int. J. Radiation Oncology, Biol. and Phys. 12, 934-936.

Song, C.W., Lee, I., Hasegawa, T., et al. (1987) 'Increase in pO2 and radiosensitivity of tumors by Fluosol-DA (20%) and carbogen', Cancer Res. 47, 442-446.

Spears, J.R., Serur, J., Baim, D.S. et al. (1983) 'Myocardial protection with Fluosol-DA during prolonged coronary balloon occlusion in the dog', Circulation 68 (suppl. III), 317.

Spence, R.K., McCoy, S., Costabile, J., et al. (1989) 'Fluosol DA-20% in treatment of severe anemia: A randomized, controlled study of 47 patients', Critical Care Med. 17, S144.

Suyama, T., Matsumoto, T., Watanabe, M., Hamano, T., and Naito , R. (1975) 'Exchange transfusion with perfluorodecalin emulsion (Fluosol-DC) in severely hemodiluted dogs', in Proc. of the Post Congress Symposium, Xth International Congress of Nutrition on Perfluorochemical Blood, Kyoto, Japan, pp. 225-235.

Teicher, B.A., and Holden, S.A. (1987) 'Survey of the effect of adding Fluosol-DA 20% O2 to treatment with various chemotherapeutic agents', Cancer Treatment Reports 71,

Teicher, B., and Rockwell, S. (1983) 'Increased efficacy of radiotherapy in mice treated with perfluorochemical emulsions plus oxygen', Amer. Assoc. of Cancer Res. Abst. 25-28, May 1983.

Teicher, B.A., and Rose, C.M. (1984) 'Perfluorochemical emulsions can increase tumor radiosensitivity', Science 223, 934-936.

Thomlinson, R.H., and Gray, L.H. (1955) 'The histological structure of some human lung cancers and the possible implications for radiotherapy', Brit. J. Cancer 9, 539-549.

Tremper, K.K., Levine, E.M. and Waxman, K. (1985) 'Clinical Experience with Fluosol-DA (20%) in the United States', in Perfluorochemical Oxygen Transport (Tremper, K., ed.), Little Brown and Company, Boston, pp 185-197.

Tsuda, Y. and Yokoyama, K. (1989) 'Perfluorochemical emulsions: The industrial view', Abstract no. 698 International Chemical Congress of Pacific Basin, Dec. 17-22, 1989, Honolulu, Hawaii.

Waldrop, M.M. (1989) 'The (liquid) breath of life', Science 245, 1043-1045.

Yokoyama, K., Yamanouchi, K., and Watanahi, M., et al. (1975) 'Preparation of perfluorodecalin emulsion, an approach to the red cells substitute', Fed. Proc. 34, 1478-1483.

Yokoyama, K., Suyama, T. and Naito, R. (1982) 'Selection of 53 PFC substances for better stability', Proc. Int. Sym., San Francisco, Ca. Sept. 29-Oct. 1, pp 189-196.

CHARACTERIZATION OF NIOSOMES

J.A. BOUWSTRA, H.E.J. HOFLAND, G.S. GOORIS, H.E. JUNGINGER
Center for Bio-Pharmaceutical Sciences
University of Leiden
P.O. Box 9502
2300 RA Leiden
The Netherlands

ABSTRACT. In this paper the forming of vesicles from single chain surfactants, i.e. polyoxyethylene alkyl ethers, has been discussed. The alkyl chain of the surfactants varied between 10 and 18, while the number of oxyethylene units varied between 3 and 7. Stable vesicles could only be prepared from those surfactants which from liquid state bilayers and which possess relative small hydrophilic head group compared to the alkyl chain. No vesicles could be prepared from surfactants which form gel-state bilayers. However after addition of cholesterol it was possible to prepare vesicles from all the surfactants under study. This is probably due to a decrease in interfacial area per "surfactant/cholesterol" molecule and to an increase in the mobility of the alkyl chains present in bilayers in the gel-state. The former has been confirmed by phase diagram studies of the system H_2O + $C_{12}EO_6$ (60% m/m)/cholesterol (40% m/m).

1. Introduction

Niosomes are vesicles mainly consisting of nonionic surfactants. These kind of vesicles have been introduced by Handjani-Vila (1) in 1978. One of the reasons for preparing niosomes is the assumed higher chemical stability of the surfactants compared to phospholipids, which are being used in preparing liposomes. Vesicles consisting of only one spherical bilayer are referred to as unilamellar, while multilamellar vesicles consist of more than one concentric bilayer. In general vesicles have been studied for two reasons. On one hand the vesicles are being used as model systems for membranes. On the other hand vesicles have been applied as drug carrier systems. Lipophilic drugs are incorporated in the bilayer regions, while hydrophilic drugs are solubilized in the water rich regions of the vesicles.

Niosomes have been prepared from several types of nonionic surfactants, e.g. polyglycerol alkylethers (1, 2), glycosyl dialkylethers (3) and crown ethers (4). Often a charged surfactant has been intercallated in the bilayers in order to introduce electrostatic repulsion between the vesicles which may increase their stability. This is especially necessary at higher surfactant concentrations. In our group mainly technical grade polyoxyethylene alkylethers (C_nEO_m) have been used in which the number of oxyethylene units (m) varied between 3 and 7. The alkyl chain length varied between 10 and 18 carbon atoms. In addition studies have been performed using Brij 96 consisting of an oleic tail and a mean number of 10 oxyethylene units.

One of the aims in developing delivery systems is controlling the release of the drug from

J. Sjöblom (ed.), Emulsions – A Fundamental and Practical Approach, 227–238.
© 1992 *Kluwer Academic Publishers.*

228

the carrier system. The release can be controlled by various parameters. In figure 1 the release of a model substance, 5,6 carboxyfluorescein (CF) is shown. The vesicles were prepared using various molar ratios Brij 96 and cholesterol. From this figure it is clear that the release has been influenced by the composition of the bilayers, the size and the number of bilayers of the vesicles. That means that these parameters are important to controll drug release in using vesicles for pharmaceutical applications.

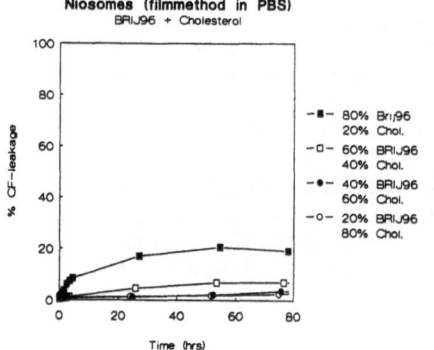 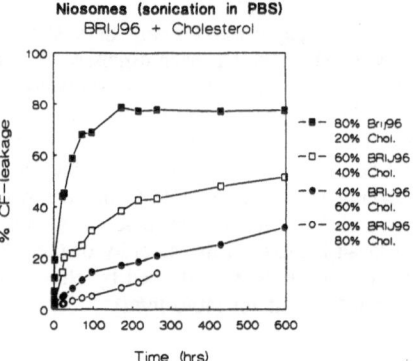

Fig. 1a

CF release from multilamellar vesicles consisting of cholesterol nd Brij 96. The composition of the bilayers varied between 20% and 80% m/m Brij 96.

Fig. 1b

CF release from unilamellar vesicles consisting of cholesterol and Brij 96. The composition of the bilayers varied between 20% and 80% m/m Brij 96.

In the first part of this paper the ability of technical grade polyoxyethylene alkyl ethers to form vesicles will be discussed. The forming of vesicles will be related to the presence of liquid crystalline phases in the water rich regions of the binary $H_2O + C_nEO_m$ phase diagrams. The influence of addition of cholesterol to the surfactants on the ability in forming vesicles will be discussed.

In the second part the influence of heating, vigorous shaking and NaCl on the structure of the vesicles will be presented.

Finally the speudo binary phase diagram of $H_2O + C_{12}EO_6$ (60% m/m)/cholesterol (40% m/m) will be discussed and compared to that of $H_2O + C_{12}EO_6$. This composition is chosen, since it is possible to prepare vesicles from the above mentioned cholesterol\surfactant ratio.

2. Materials and Methods

2.1 Materials

Technical grade polyoxyethylene surfactants were purchased from SERVO (The Netherlands), except pure $C_{12}EO_6$ which was obtained from Nikko (Japan). The surfactants varied in alkyl chain length between 10 and 18 carbon atoms. The number of oxyethylene units varied between 3 and 7. Cholesterol with a purity higher than 99% was obtained from J.T. Baker Chemicals (Deventer, The Netherlands).

2.2 Preparation of Vesicles

The vesicles were prepared either by the sonication method or by the film method (5). Using both methods 150 μmol surfactant/cholesterol was solubilized in a mixture of chloroform + methanol (weight ratio 3:1). The organic solution was evaporated which resulted in a film on the inner wall of a tube. The temperature of the tube was raised to 60-80°C, which is above the gel liquid transition of the surfactants. At this temperature the vesicles were formed by hydrating the film. Hydration with 5 ml water solution was performed either under vigorously shaking (sonication method) or under gently rotating (film method) tube. The vesicles were prepared in H_2O or in a NaCl solution in which the concentrations varied between 1 g/l to 288 g/l. The influence of NaCl on the repeat distance of the multilamellar vesicles was determined with small-angle X-ray scattering (SAXS).
A part of the vesicles prepared by the film method was shaking using a bath sonicator. The effect on size and the effect on the mean number of concentric bilayers of the shaking procedure were investigated using photon correlation spectroscopy and SAXS.

2.3 Preparation of Liquid Crystalline Phases

The liquid crystalline phases were prepared by mixing cholesterol (40 mol %) and $C_{12}EO_6$ (60 mol %). To this mixture several amounts of water were added, after which the mixture was increased in temperature to approximately 80°C. Sealed ampules were used to avoid water loss. The mixtures were carefully shaken and cooled down to room temperature. These mixtures were examined by SAXS and by Freeze-fracture electron microscopy (FFEM).

2.4 Small Angle X-Ray Measurements

The scattering experiments on the niosomes have been carried out at the synchrotron radiation source in Daresbury using station 8.2. This station has been built as a part of a NWO/SERC agreement. A description of the camera has been given elsewhere (6). The sample was measured in a sample cell with two mica windows, the X-Ray pathlength through the sample was about 1 mm. The scattering intensities are plotted as a function of the scattering factor Q defined as $Q = (4 \pi \sin\vartheta)/\lambda$, in which ϑ is the scattering angle and λ is the wavelength.
The phase behaviour of the surfactant/cholesterol mixtures to which different amounts of water were added was measured using a Kratky camera (A. Paar, Austria) using slit collimation. The camera was connected to a position sensitive detector (M. Braun, Germany). The resolution of the detector was approximately 0.05 mm. Ni-filtered Cu-Kα radiation was used (wave length = 0.154 nm). The distance between sample holder and detector was 0.30 m. The samples were pressed in a capillary with a diameter of 1.0 mm. During this procedure, the lamellae of the D-phase aligned along the longest axis of the capillary, which reduced the smearing during the measurements to a large extent. For this

reason no desmearing of the curves were carried out. The scattering was measured as a function of Q.

2.5 Freeze Fracture Electron Microscopy

The vesicles or lamellar mesophases were visualized with freeze fracture electron microscopy. The samples were cryofixed in liquid propane using the plunching method (KF 80, Reichert Jung). After fixation the samples were fractured in a Balzers BAF400 at -140°C and at a pressure of 10^{-2} Pa. The freshly obtained surface was shadowed by evaporation of Pt/C from an angle of 45°. The obtained replica was strengthened by C evaporation. After cleaning with a mixture of ether/methanol the replica was examined in a transmission electron microscope.

3. Results

3.1 The Ability of Surfactants in forming Vesicles

3.1.2   *Theoretical considerations*
Considering the geometry of the possible structures of micelles, the interfacial area per surfactant molecule (A) plays an important role. Hereby it is postulated that the hydrocarbon interior is a continuum. This means that the shortest distance between the center of the micel and the interface should be smaller or equal to the maximally attainable length 1_c of the hydrocarbon tail. With V_c being the hydrocarbon volume simple geometric conditions can be derived for the forming of micelles and vesicles (7).
These are

sphere micelles	: $A > 3 \, V_c / 1_c$
rod micelles	: $A > 2 \, V_c / 1_c$ and
bilayer micelles, vesicles	: $A > \quad V_c / 1_c$

If in accordance to Mitchell et al. (8) an increase in surfactant concentration results in an ordering of the surfactants the following disorder to order transitions are expected: bilayer micelles will be packed into a lamellar phase, rod micelles will be packed into an hexagonal phase and sphere micelles will be packed into a cubic or an hexagonal closed packed phase. From this simplified theory it is expected that vesicles can only be prepared from those surfactants which form a lamellar phase upon concentrating its micellar solution and that those surfactants exhibit a small interfacial area per molecule (see equations). Surfactants exhibiting this phase behaviour possess a small hydrophilic head group in comparison with the alkyl chain length.

3.1.3   *Forming of vesicles*
The forming of vesicles has been studied by freeze electron microscopy and polarization microscopy. In the case of polarization microscopy large multilamellar vesicles are recognized by the so called maltezer crosses. In table I the ability of C_nEO_m surfactant in forming vesicles are presented.

Table I

The ability of polyoxyethylene ethers in forming vesicles; + : forming of vesicles; ± forming of vesicles, but many aggregates have also been detected; - no forming of vesicles.

number of carbon atoms	number of oxyethylene units		
	EO_3	EO_5	EO_7
10	+	-	-
12	+	-	-
14	+	+	-
16	±	±	±
18	±	±	-

It appears that $C_nEO_{<3>}$ surfactants are able to form vesicles. This group of surfactants possesses a rather small hydrophilic head group in comparison with the alkyl chain length. The phase diagram of H_2O + $C_{12}EO_3$ exhibits a lamellar phase in the water rich regions (8), which confirms the relationship between the ability of the surfactants to form vesicles and the presence of a lamellar phase. The comparison between the existance of lamellar phases and the forming of vesicles has the shortcoming that the phase diagram studies have been carried out using pure surfactants while the studies on the vesicle forming has been carried out using technical grade surfactants. In previous studies we found that the phase diagram of the technical grade surfactant C_{12} $EO_{<7>}$ is similar to that of pure $C_{12}EO_6$ (9). From this study it was concluded that the purity influences the phase behaviour only to a minor extent. Although the phase diagram of H_2O + $C_{14}EO_3$ is not known, again a lamellar phase is expected in the water rich regions, since H_2O + $C_{14}EO_4$ only form a lamellar liquid crystalline phase (8). It seems that these surfactants obey the theory described above. However in the case of $C_{18}EO_{<3>}$ and $C_{16}EO_{<3>}$ many large aggregates have been formed. This might be due to the gel state bilayers of these vesicles (10), which are known to destabilize uncharged vesicles (11). The gel-liquid transition of hydrated $C_{16}EO_{<3>}$ and $C_{18}EO_{<3>}$ surfactants took place in the temperature range of 40 - 48°C and 50 - 58°C respectively (10). That the instability of the vesicle suspension is due to the gel state bilayers has been confirmed by the observation that aggregates have been formed upon cooling the suspension and not during the hydration step, which takes place at elevated temperatures.
In the case of $C_nEO_{<5>}$ surfactants niosomes could only be prepared from $C_{14}EO_{<5>}$, $C_{16}EO_{<5>}$ and $C_{18}EO_{<5>}$, although vesicles prepared from $C_{16}EO_{<5>}$ and $C_{18}EO_{<5>}$ were not stable and formed large aggregates. It seems that the head groups of $C_{10}EO_{<5>}$ and $C_{12}EO_{<5>}$ are too large for forming bilayer structures. This is confirmed by the phase diagram of H_2O + $C_{12}EO_5$ and $C_{12}EO_6$ which exhibits an hexagonal phase in the water rich regions of the phase diagram (8). This phase behaviour favours the forming of micelles. In the case of $C_nEO_{<7>}$ only $C_{16}EO_{<7>}$ shows the ability to form vesicles, although also large aggregates are formed which again points in the direction of destabilized vesicles. The gel-liquid transition of hydrated $C_{16}EO_{<7>}$ and C_{18} $EO_{<7>}$ occurred in a temperature range of 30 - 40°C and 35 - 55°C respectively (10). In the case of 10 to 14 C atoms in the alkyl chain the hydrophilic head group is to bulky for forming bilayered structures.
It appears that the nonionic surfactants behave in accordance with the theory described

above, although the longer alkyl chains are destabilized by the formation of gel state bilayers.

A possibility for forming vesicles from single chain surfactants is the addition of cholesterol, which possesses only a hydroxyl group as hydrophilic moiety. On the one hand cholesterol compensates for the large hydrophilic head groups of the surfactants and it is thus possible to prepare vesicles from C_nEO_m surfactant with relatively large headgroups as e.g. $C_{10}EO_5$, $C_{12}EO_5$ and $C_{12}EO_7$ by adding cholesterol to these surfactants. On the other hand cholesterol is known to decrease the chain order below T_c (12). The latter mechanism may explain the stabilisation of vesicles prepared from $C_{16}EO_m$ and $C_{18}EO_m$ after adding cholesterol to the solution. It seems that cholesterol also increases the membrane undulations of DMPC liposomes which results in larger repulsion forces between the vesicles (13). For these reasons cholesterol favours the forming of vesicles when using surfactants with large head groups or when using surfactants with alkyl chains which form gel-state bilayers. Using technical grade C_nEO_m and cholesterol in a molar ratio of 3/2 it is possible to prepare vesicles from the series of C_nEO_m surfactants, in which n varies between 10 and 18 and the number of oxyethylene groups varies between 3 and 7.

3.2 Vesicle Morphology

3.2.1 *Influence of vigour shaking and temperature on the structure of vesicles*
The structure of the vesicles depends on the preparation method. Vesicles prepared using the film method are often large and multilamellar (MLVs), see figure 2, while vesicles prepared by sonication are often small and unilamellar (SUVs), see figure 3.

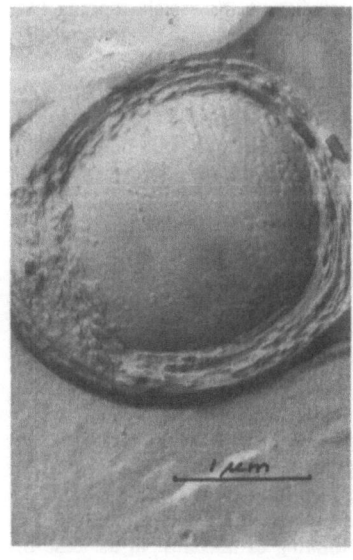
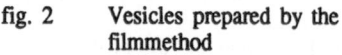
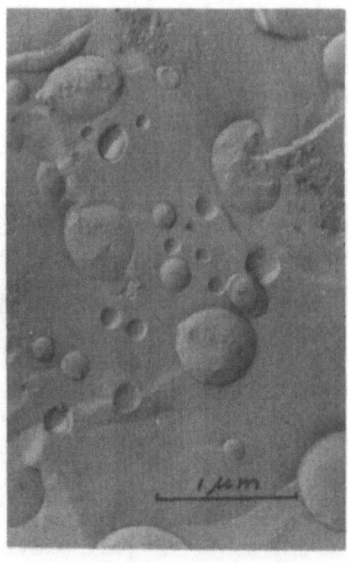

fig. 2   Vesicles prepared by the filmmethod

fig. 3   Vesicles prepared by sonication

The structure of niosomes prepared from $C_nEO_{<m>}$ can be changed from MLVs to SUVs by vigorously shaking. An example is given in figure 4, in which the mean diameter of the vesicles has been plotted as function of the time during which shaking took place. The measurements have been carried out by dynamic light scattering (DLS). Due to the z-average mean values for the diameters and the high polydispersity of the mixtures the absolute values are under debate, therefore this figure is only used to present a trend. In this figure the z-average mean diameter decreases tremendously after shaking during only 10 minutes. The influence of the shaking process on the structure of a vesicle preparation was also measured by SAXS.

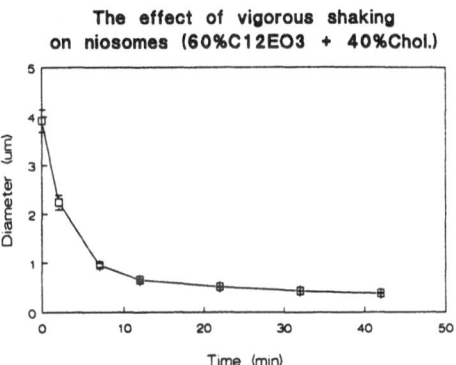

Fig. 4   Means diameter of vesicles prepared from $C_{12}EO_{<3>}$ as a function of vigorous shaking.

In figure 5a the scattering curves are shown before and after the shaking process. A multilamellar vesicle suspension is characterized by various diffraction peaks, from which the repeat distance (d) of the bilayers can be calculated by the equation:

$$Q_n = 2n\pi/d$$

in which n is the order of the diffraction peak.
The curve obtained before shaking exhibits a first and second order diffraction peak accounting for a large number of bilayers.

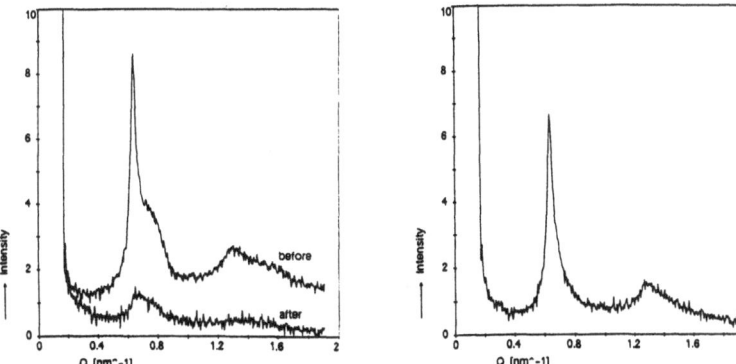

Fig. 5   The scattering curve before and after the shaking proces (a). The scattering curve after heating to 60°C (b). After heating and cooling down the vesicles changed from unilamellar to multilamellar ones.

234

The repeat distance calculated from the peak position at maximum intensity appeared to be 9.9 nm. After the shaking procedure almost no first order diffraction peak could be detected, which strongly indicates that during the shaking process the vesicles changed from multilamellar to unilamellar ones. Therefore this shaking procedure offers a very elegant way to manipulate the size of the niosomes. After shaking the vesicles were heated to 60°C (the temperature at which the vesicles have been prepared) and cooled down to room temperature. During this procedure the vesicles changed from unilamellar ones to multilamellar vesicles as is shown in figure 5b. The scattering curve exhibits a first and a second order diffraction peak. Following the Scherrer equation (14) the mean number of bilayers is 11. The exact mechanism of the transition from unilamellar to multilamelar vesicles is not yet known, but it might be due to demixing of the niosome solution upon passing the cloud point, which results in a phase with a high surfactant concentration and in a phase with a low surfactant concentration. Upon cooling the surfactants will be stabilized in water and the phase with the high surfactant concentration may form multilamellar vesicles. The change in the structure upon shaking or heating was not only observed in case of $C_{12}EO_{<3>}$ vesicles but also using other members of the $C_nEO_{<m>}$ group, with and without addition of cholesterol.

3.2.2.   *The influence of NaCl on the swelling of the bilayers*
The influence of NaCl on the swelling of the bilayers of the vesicles was determined by SAXS (Daresbury Synchrotron Radiation Station). In these experiments pure $C_{12}EO_6$ and cholesterol were used in a molar ratio of 3/2. The results are shown in figure 6.

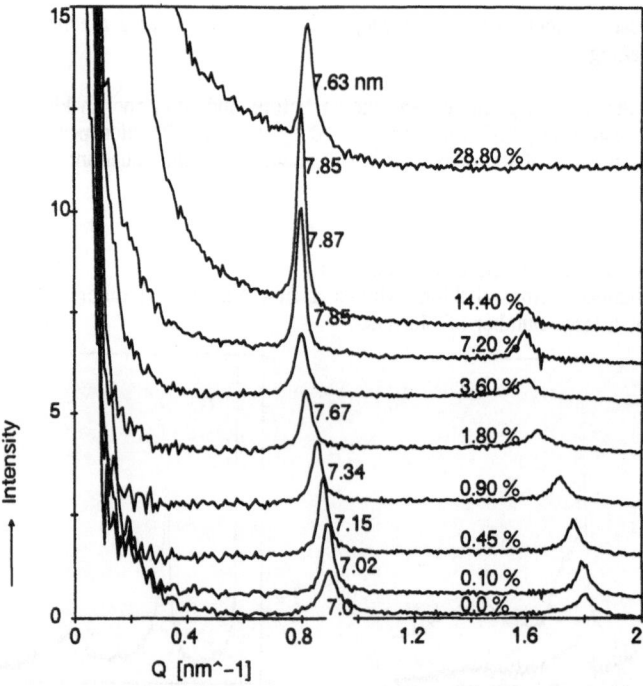

Fig. 6   Scattering curves of vesicles prepared from $C_{12}EO_6$ (60 % m/m) and cholesterol (40 % m/m). The vesicles have been prepared in NaCl solutions. NaCl has only a slightly influence on the repeat distance of the multilamellar vesicles.

The NaCl concentration varied between 0 and 288 g/l (7 molar). The repeat distance of the multilamellar vesicles changed from 7.65 nm in water to 7.85 nm in 144 g/l NaCl solution. A further increase of the NaCl concentration to 288 g/l resulted in a decrease in the repeat distance to 7.63 nm. Using this high NaCl concentration it is not clear whether there are still vesicles present in the solution. Although it is known that an increase in NaCl concentration results in a dehydration of the oxyethylene chains, it does not influence the swelling of the bilayers to a large extent. This might be due to a decrease in the hydrophilicity of the polyoxyethylene head group when adding NaCl resulting in a decrease in the interfacial area per molecule. This has also been shown by Kahlweit et al. (15), who showed that the phase behaviour of $C_{12}EO_5$ +H_2O is similar to the phase behaviour of $C_{12}EO_6$ + H_2O to which 10 w/w % NaCl has been added. It is remarkable that in the case of 7 molar NaCl solution still the swelling of the lamellae does not change to a large extent compared to the swelling of the lamellae of vesicles prepared in water, since it is known that the polyoxyethylene chains lie flat at the interface at concentrations above 5 molar (salting out effect).

3.3 Phase Diagram H_2O + $C_{12}EO_6$ (60%) Cholesterol (40%)

The phase diagram of the binary speudo system H_2O + $C_{12}EO_6$/cholesterol was determined. The molar ratio between $C_{12}EO_6$ and cholesterol was 3/2. This ratio was chosen because it is often used in preparing vesicles. The various phases were studied by small angle X-Ray scattering (Kratky camera, slit collimation) and freeze fracture electron microscopy.

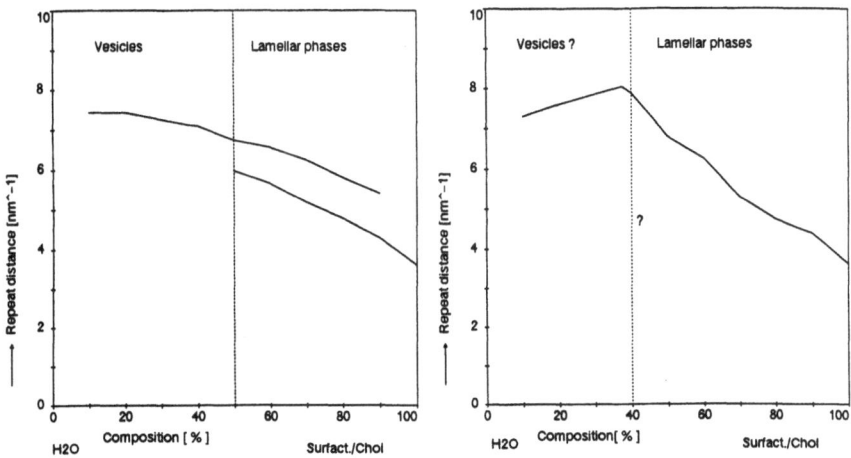

Fig. 7   Repeat distances of water + $C_{12}EO_6$/cholesterol mixtures at 20°C (a) and 40°C (b). The $C_{12}EO_6$/chol$_{molar}$ ratio was 3/2. This ratio has been kept constant.

Two coexisting lamellar phases were detected between 10 % w/w and 60 % w/w water.

The repeat distances of these lamellar phases are plotted in figure 7a. The relative intensities of the diffraction peaks of the two phases change as function of the water content At 10 % w/w and 20 % w/w H_2O the lamellar phase with a repeat distance of 4.8 nm is predominantly present (the other phase is only present for a few percent, while at 50 % w/w the lamellar phase with a repeat distance 6.8 nm is mainly present The two lamellar phases might be due to a difference in the cholesterol/surfactant ratios in the lamellae, which should also cause the difference in repeat distance. Upon heating to 40°C these phases transformed into one lamellar phase with a repeat distance which lies in between those found at 20°C (see figure 7b). This phase change appeared to be reversible. The lamellar phase is still present at 60°C and upon heating from 40°C to 60°C only a minor shift in the repeat distances was found. No crystalline cholesterol, which is characterized by a strong reflection at 3.35 nm, could be observed. This indicates that cholesterol is completely solubilized in the bilayers. A freeze fracture electron micrograph of the lamellar phase is shown in figure 8a. From the repeat distances obtained at 40°C the interfacial area per surfactant/choleterol molecule has been calculated. This temperature was chosen, since one lamellar phase is present. The densities of the polyoxyethylene moiety and the alkyl chain were estimated to be 1.11 g/cm^3 and 0.81 g/cm^3 (9). The densities of cholesterol and water were 1.067 g/cm and 1 g/cm^3 respectively (16). The interfacial area per surfactant molecule was calculated at 40°C, since at that temperature only one lamellar phase is present. The interfacial area increased at increasing water content. Above a water content of 40 % w/w probably phase separation takes place, therefore these values are not plotted. (see figure 8).

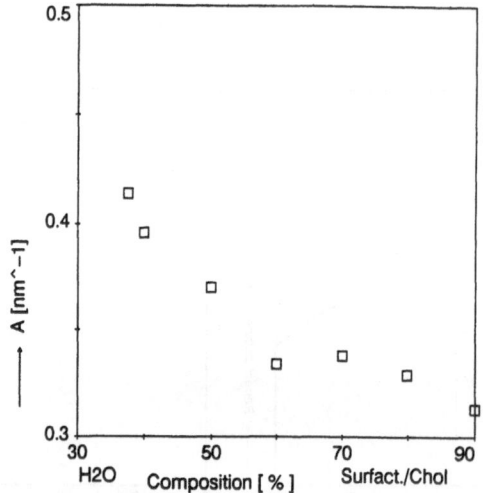

Fig. 8   The interfacial area per surfactant molecule of the system H_2O + $C_{12}EO_6$ (60% m/m)/cholesterol (40% m/m) at 40°C. The values are calculated from the densities and repeat distances (see figure 7b).

The interfacial area per surfactant molecule was lower compared to the values obtained in the lamellar phase of the H_2O + $C_{12}EO_6$ phase diagram at the same water content (16), which depending on the water content varied between 0.40 and 0.52 nm^2. This shows that at least at high surfactant contents cholesterol indeed decreases the interfacial area per molecule which favours the forming of vesicles.

At room temperature at a water content higher than 60 % w/w only one repeat distance was found. The peak showed a desmeared character in comparison with that obtained at 40 % w/w, which might indicate that the peaks are based on vesicles and not on a lamellar phase. During X-Rays the lamellae are often partly aligned along the axis of the glass capillary. In figure 9b the micrograph shows that indeed vesicles are present at 40 % $C_{12}EO_6$/cholesterol content. The micrograph also shows that the vesicles are located on a very rough background of which the structure is not yet understood. Comparing the phase diagram of $H_2O + C_{12}EO_6$ (16) with the phases observed in this study, it appears that when adding cholesterol to the surfactant water system only lamellar mesophases have been found, the hexagonal and viscous isotropic phases which are present in the $H_2O + C_{12}EO_6$ phase diagram completely disappeared.

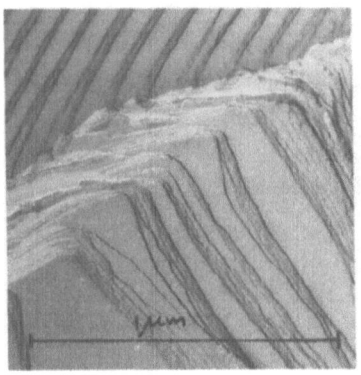 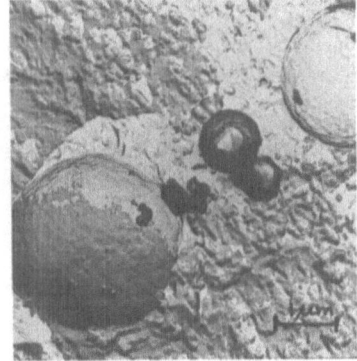

Fig. 9a

Electron micrograph of the lamellar phase of the phase diagram $H_2O + C_{12}EO_6$ (60% m/m)/cholesterol (40% m/m) at 20% w/w water.

Fig. 9b

Electron micrograph of the phase diagram $H_2O + C_{12}EO_6$ (60% m/m)/ cholesterol (40% m/m) at 60% w/w water.

References

1 Handjani-Vila, R.M., Ribier, A., Rondot, B., Vanlerberghe, G., J. Cosmet. Sci., 1, 303-314 (1979)
2. Baillie, A.J., Dolan, T.F., Alexander, J., Carter, K.C., Int. J. Pharm., 57, 23-28 (1989)

238

3. Kiwada, H., Nimura, H., Fujisali, Y., Yamada, S. Kato, Y., Chem. Parm.Bull., 33, (2), 753-759 (1985)
4. Echoyen, L.E., Hernandez. J.C., Kaifer, A.E., Gokel, G.W., Echoyen, L.J. Chem. Soc. Chem. Commmunications, 836-837 (1988)
5. Baillie, Florence, A.T., Hume, L.R., Muirhead, G.T., Rogerson, A., J. Phar. Pharmacol., 37, 863-868 (1985)
6. Bouwstra, J.A., Gooris, G.S., van der Spek, J.A., Bras, W., submitted to J. Invest. Derm.
7. Mitchell, D.J., Ninham, B.W., J. Chem. Soc., Faraday trans., 2, 77, 601 (1981)
8. Mitcheoll, J.D., Tiddy, G.J.T., Warning, L. Bistock, T., McDonald, M.P., J. Chem. Soc., Faraday Trans, 79, 975-1000 (1983)
9. Bouwstra, J.A., Jousma, J., van der Meulen, M.A., Vijverberg C.C., Spies, F., Junginger, H.E., Colloid & Polym. Sci., 1989, 267, 531-538
10. Hofland, H.E.J., Bouwstra, J.A., Ponec, M., Verhoef, J., Buckton, G., Junginger, H.E., submitted to J. Pharm. Pharmacol.
11. Ostrowsky, N., Sometti, D., in Physics of Emphiphiles: Micelles Vesicles and Micro Emulsions Ed. V. Degiogio, M. Corti, Elsevier, Amsterdam, 1983
12. Silver, B.L., The Physical Chemistry of Membranes, Alan & Unwin and The Salomon press, New York, 1985
13. Michels, B., Fazel, N., Cert. R. Eur. Biophys. J. 17, 187-189 (1989)
14. Alexander, X-Ray Diffraction Methods in Polymer Science, Wiley-Interscience, New York, 1969
15. Kahlweit, M., Strey, R., Haase, D., J. Phys. Chem. 89, 163-171 (1983)
16. Handbook of Chemistry and Physics, 64th edition, CRS Press
17. Jousma, H., Joosten, J.G.H., Junginger, H.E., Coll. & Polym. Sci. 266, 640-651 (1988)

NMR SELF-DIFFUSION STUDIES OF EMULSION SYSTEMS. DROPLET SIZES AND MICROSTRUCTURE OF THE CONTINUOUS PHASE

O. SÖDERMAN, I. LÖNNQVIST, B. BALINOV
Physical Chemistry 1
Chemical Center
Lund University
P.O. Box 124
221 00 Lund
Sweden

ABSTRACT. The application of the pulsed field gradient NMR-method for determining self-diffusion coefficients to emulsion systems is described. The method is first outlined and then two particular applications are described. The first one deals with the determination of the microstructure of the continuous phase and its relation to emulsion stability. The particular system studied in this regard is a surfactant/cosurfactant stabilized emulsion. Secondly, the determination of emulsion droplet sizes with the NMR method is outlined. Several examples of this approach will be given, including margarine and hydrocarbon gel emulsions.

1. Introduction

Various NMR-methods have played a central role in studies of thermodynamically stable surfactant systems, such as micellar solutions, microemulsions and liquid crystalline phases (Lindman, et al. (1987), Stilbs (1991)).

The situation is somewhat different for emulsion systems, where NMR-studies are sparse. The main application of the NMR-method appears to be in the determination of emulsion droplet sizes by means of the NMR pulsed field gradient method to determine self-diffusion coefficients (Packer and Rees (1971), Callaghan, et al (1983), Lönnqvist, et al. (1991)).

In the present communication, we will discuss this particular application of the NMR-method. In addition, we will show how the same NMR-technique can be used to obtain information about the microstructure of the continuous phase in emulsions. This information is important in trying to relate various aspects of emulsion stability to the micro structure of the continuous phase.

The disposition of this paper is as follows. First, the NMR-method for determining self-diffusion coefficients is introduced, and by choosing examples from thermodynamically stable systems we will show the usefulness of self-diffusion coefficients in determining microstructures of surfactant solutions. In particular, we will indicate how the outcome of the self-diffusion experiment can be related to droplet sizes in emulsions.

J. Sjöblom (ed.), Emulsions – A Fundamental and Practical Approach, 239–258.
© 1992 *Kluwer Academic Publishers.*

Secondly, we will discuss by invoking results from a particular surfactant/cosurfactant water/oil system the dependence of emulsion stability as well as droplet size on the microstructure of the continuous phase.

Thereafter, we will discuss the determination of droplet sizes with NMR. Several examples will be given, including one for which other methods to determine droplet sizes are few and those that exist are rather complicated to carry out. The particular example is margarine.

We will discuss a particular type of emulsion namely the so-called hydrocarbon gels (Hoffmann and Ebert (1988)). These are three-component systems containing a single chained surfactant (in the present study we have used hexadecyltrimethylammonium bromide (CTAB)), water and oil. Typically, these systems will contain up to 99 % oil.

We end the paper with a short note regarding the experimental details.

2. The NMR-method of determining self-diffusion coefficients

There are several accounts in the literature describing the NMR-experiments by which self-diffusion coefficients are determined. Therefore, we will only give a brief account of the experiment here. Readers interested in a more detailed account of the experiment as well as technical details may find these in (Callaghan (1984), Stilbs (1987)).

Traditionally, self-diffusion coefficients are determined by means of radioactive tracer techniques. The NMR method of determining self-diffusion coefficients is in fact related to the tracer techniques in that the spins are labelled at a certain time and their position after a certain time interval is monitored. This is achieved as follows. The protons (or any NMR-active isotope, for that matter) is labelled by its Larmor-frequency. This frequency in a "normal" NMR-experiment is constant over the NMR-sample. In the diffusion experiment, the static magnetic field is deliberately made spatially inhomogeneous by the addition of field gradients to the static magnetic field.

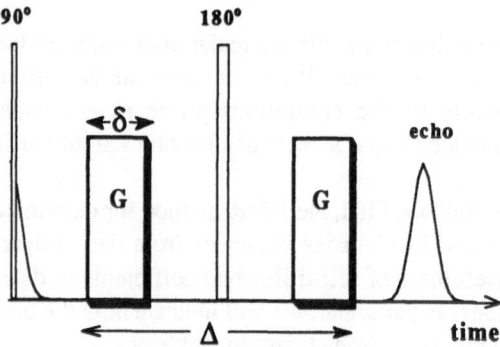

Figure 1. The pulse sequence used in the PGSE self-diffusion experiment.

The particular sequence of rf and field-gradient pulses (the experiment has been termed the Pulsed Field-Gradient Spin-Echo (PGSE) experiment) used in the experiment is shown in Fig. 1. Two gradient pulses are sandwiched on either side of the 180° pulse in a spin-echo experiment. In a spin-echo experiment the spins are refocussed by the 180° pulse only if the precession frequency is constant during the experiment. If the static magnetic field is made to vary over the sample volume, and the spins are performing random thermal motions, then the spins acquire a random phase shift and the refocussing of the echo is more or less incomplete. The time-interval during which the thermal motion is monitored in the PGSE experiment is equal to the time between the field-gradient pulses (Δ in Fig. 1).

Thus the goal of the experiment is really quite simple: It measures the mean displacement of the spin-bearing molecules during a time interval Δ the value of which is typically 50 ms or longer. Since the displacement is given by $< x^2 > = 2D\Delta$, one is monitoring displacements on a length scale of microns for molecules in low molecular-weight liquids. Thus any barrier for diffusion on this length-scale such as the confinements to a closed cavity (as would be the case for the molecules in an emulsion droplet) will affect the outcome of the experiment. We will return to this fact below.

For unrestricted (Gaussian) diffusion, i.e. when the quantity $(2D\Delta)^{0.5}$ is much smaller than the distance between any barriers, the amplitude of the signal following Fourier transformation of half of the echo in the PGSE-experiment is:

$$I = I_0 \exp \left(- \frac{2\Delta}{T_2} \right) \exp (-\gamma^2 G^2 D \delta^2 (\Delta - \delta/3)) \tag{1}$$

Here T_2 is the spin-lattice relaxation time, I_0 is the intensity without gradient, D the self-diffusion coefficient and γ is the magnetogyric ratio. The other quantities in Eq. 1 are defined in Fig. 1. Normally one chooses to perform the experiment by varying δ, keeping G and Δ constant. The diffusion coefficient is then readily obtained by fitting Eq. 1 to the raw data.

The usefulness of self-diffusion coefficients lies in the **direct** information they provide about the microstructure of microheterogeneous systems (Lindman, et al. (1987), Balinov, et al. (1991), Lindman and Stilbs (1987), Lindman, et al. (1989)). To illustrate this fact, consider the two idealized situations depicted in Fig. 2. In Fig. 2a, an ordinary oil-swollen micellar solution is shown while an idealized bicontinuous structure is shown in Fig. 2b. For the first case the diffusion coefficients of molecules confined to the micelles will be the same as those of the surfactant, which will be given by the diffusion of the micelle as such, while the diffusion of molecules comprising the continuous medium will be rapid. For the situation in Fig. 2b, both molecules in the hydrophobic environment and in the water will diffuse rapidly and will be only slightly reduced from the values of the diffusion in the corresponding bulk liquids (there will be some obstruction effects) (Anderson and Wennerström (1991)). Specific examples of these two cases will be given below.

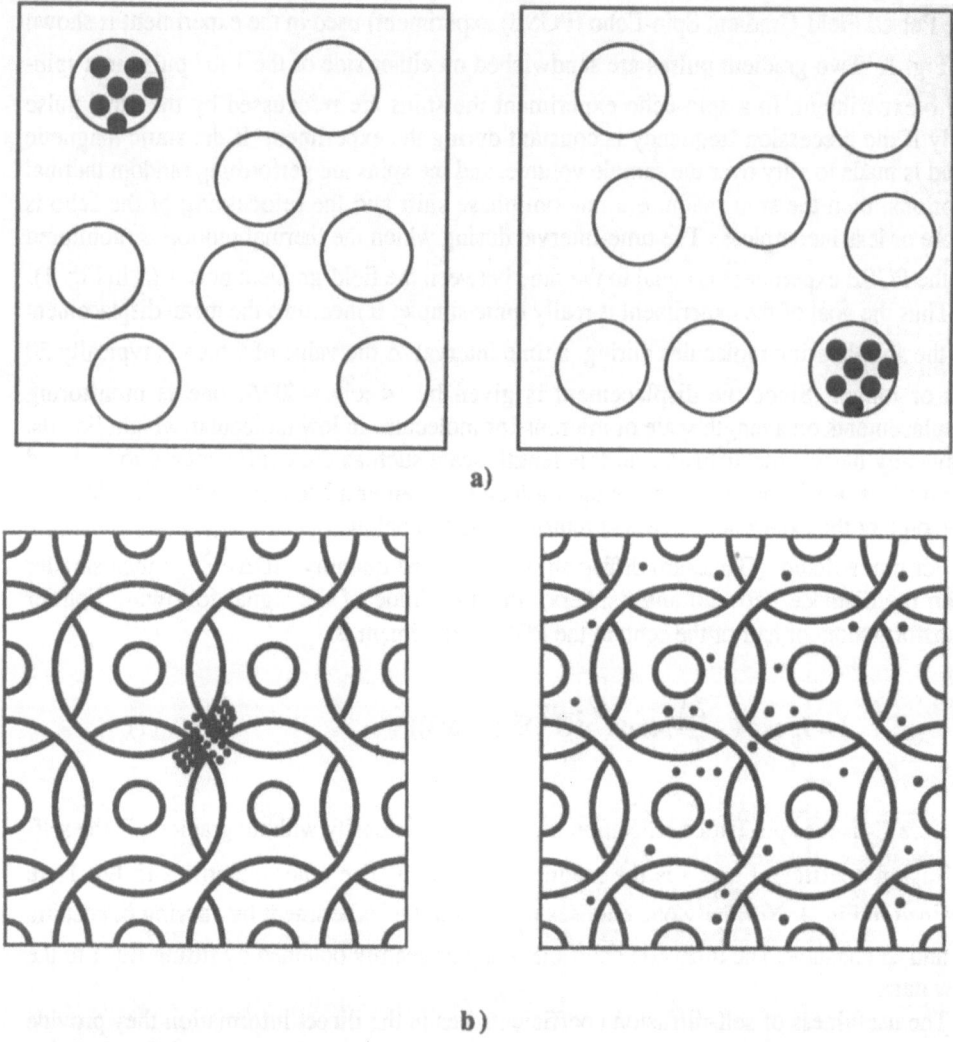

a)

b)

Figure 2. Figure 2a illustrates an oil-swollen micelle, while figure 2b illustrates a bicontinuous structure. The filled points represent spin-labelled molecules at an initial time (left figures) and the same molecules after a certain time-interval (right figure). The closed and bicontinuous structures will clearly give very different diffusion coefficients for the labelled molecules.

We now turn to a situation where $(2D\Delta)^{0.5}$ is of the same order or larger than the distance between any diffusional barriers in the system. For such a case the outcome of the experiment will depend on the geometry as well as the dimensions of the confinements. In the case of emulsion systems we will be dealing with molecules confined to spherical cavities. This situation has been treated by Murday and Cotts (1968) and the equation corresponding to Eq. 1 is, for a spherical cavity of radius R:

$$\ln\left(\frac{I}{I_0}\right) = -\frac{2\Delta}{T_2} - 2\gamma^2 G^2 \sum_{m=1}^{\infty} \frac{1}{\alpha_m^2(\alpha_m^2 R^2 - 2)} \text{ x} \tag{2}$$

$$\text{x} \left(\frac{2\delta}{\alpha_m^2 D} - \frac{2 + \exp(-\alpha_m^2 D(\Delta-\delta)) - 2\exp(-\alpha_m^2 D\delta)}{(\alpha_m^2 D)^2} - \right.$$

$$\left. - \frac{2\exp(-\alpha_m^2 D\Delta) - \exp(-\alpha_m^2 D(\Delta+\delta))}{(\alpha_m^2 D)^2} \right)$$

T_2 is assumed to be independent of R and α_m is the mth root of the equation: $2(\alpha R)J_{3/2}'(\alpha R) = J_{3/2}(\alpha R)$. It can be shown that Eq. 2 reduces to Eq. 1 if $(2D\Delta)^{0.5} \ll R$ and to $I/I_0 = \exp(-(1/5)\gamma^2\delta^2 G^2 R^2)$ if $(2D\Delta)^{0.5} \gg R$, where R is the droplet radius. In passing we note that Eq. 2 is derived for cavities that are not mobile during the experiment. However, a sphere of radius 1 µm will have a diffusion coefficient in water of about $2\cdot10^{-13}$ m^2/s at room temperature. Thus the contribution to the decay of the echo from the diffusion of the sphere as such is very small.

Depicted in Fig. 3 are simulated signal decay curves for molecules in spheres of different radii. Clearly, the decay curves do carry information about the droplet size, and as we shall see, they also carry information about the size polydispersity.

244

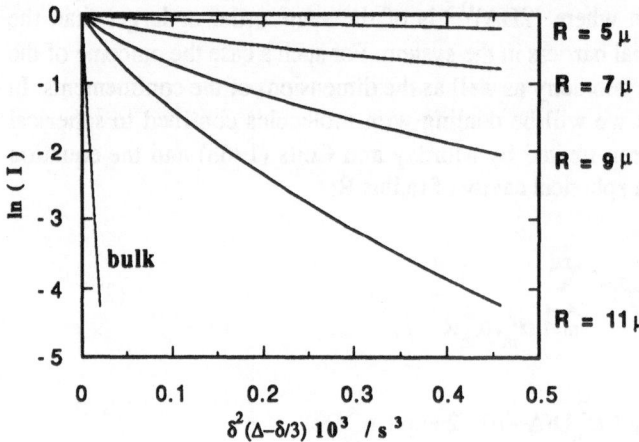

Figure 3. The natural logarithm of the echo attenuation I as function of $\delta^2(\Delta-\delta/3)$ for different radii of emulsion droplets according to Eq. 2, $\Delta=100$ ms, $\gamma^2 G^2=10^{14}$ m^{-2}s^{-2}, and $D=2\cdot10^{-9}$ m^{2}s^{-1}.

3. The microemulsion structure of the continuous medium in a SDS-emulsion and its relation to creaming and coalescence

3.1. PHASE DIAGRAM

In order to investigate the influence of the microstructure of the continuous phase on the stability of emulsions, we have chosen to investigate the system sodium dodecyl-sulphate (SDS)/glycerolmono(2-ethyl-hexyl)ether/decane/brine (3 wt% NaCl) (Shinoda, et al. (1984)). In this system emulsions can be made of the O/W or W/O type depending on the ratio between surfactant and cosurfactant. The total amount of surfactant and cosurfactant is kept constant at 5 wt%. The phase diagram is presented in Fig. 4. It is a cross-section of a five component phase diagram and it is difficult to predict the directions of the tielines in the two-phase areas where the emulsions are made. The samples are made with equal weight of brine and decane and thus along the vertical line in the middle of the diagram. The samples are prepared by weighing the components and then mixing them by gentle shaking by hand (keeping the time and amplitude of shaking as equal as possible between the samples). As emulsion stability depends critically on the droplets' size distribution, the emulsification process is critical. Nevertheless, we feel that some qualitative comparison can be made between the different samples prepared in this way.

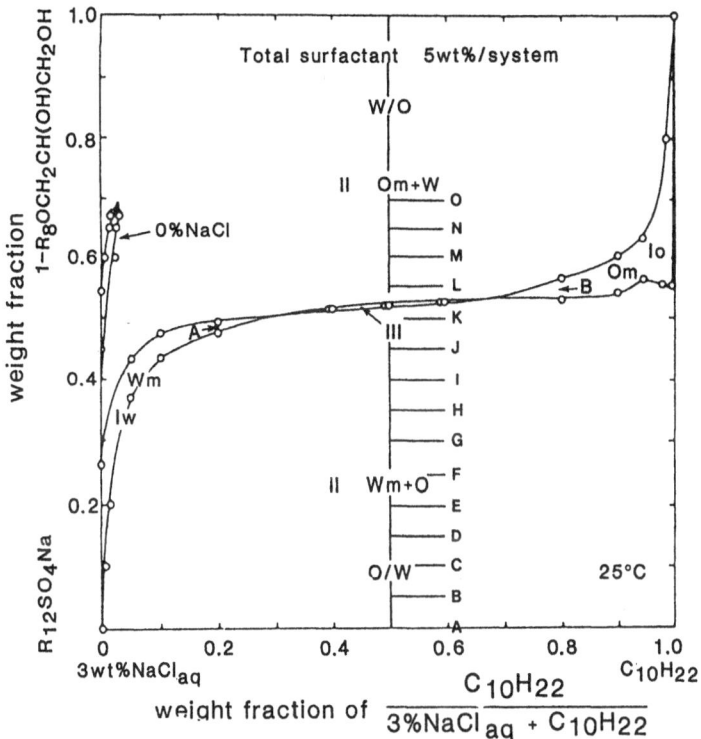

Figure 4. The phase-diagram for the SDS/glycerolmono(2-ethyl-hexyl)ether/decane/ brine (3 wt% NaCl) system at 25 °C. Total surfactant concentration is 5 wt%.

3.2. NMR DIFFUSION IN THE CONTINUOUS PHASE

The emulsions in this system are made in a two-phase area which for the O/W emulsions consists of an oil rich phase and a phase of normal micelles, while for the W/O emulsions it consists of a water rich phase and a micellar phase of reversed micelles. The micellar phase is the continuous medium for both types of emulsions. In order to determine the structure of this continuous medium we let the emulsion samples cream and separated the clear continuous medium from the creamed layer in each sample. The clear medium was then transferred to a 5 mm NMR tube and the diffusion coefficients for both the oil and the water were determined with the NMR self-diffusion method as described above. The result is shown in Fig. 5, where the reduced diffusion coefficients (D/D_0 where D is the actual diffusion coefficient and D_0 is the diffusion coefficient for the neat liquid) for the oil and the water are plotted versus the ratio between cosurfactant and surfactant. For the O/W emulsion where SDS is the only surfactant one finds that the continuous medium consists of small spherical micelles, that the water diffusion is fast (slightly lowered relative neat water due to obstruction effects (Jönsson, et al. (1986)) and that the oil diffusion is low and corresponds to a hydrodynamic radius of the oil-swollen micelle of about 50 Å (according to Stokes law). When the cosurfactant is introduced and its amount is increased the size of the micelle is increased as can be seen in the diagram by the lowered value of the reduced diffusion coefficient of the oil. Close to the three-phase area the continuous medium is bicontinuous as the value of the reduced diffusion coefficient is almost the same for both the oil and the water. When the amount of cosurfactant is increased further one passes over to the W/O emulsion area where the continuous medium is bicontinuous near the three-phase area and then changes to closed reversed micellar aggregates as can be seen from the now changed values of oil and water diffusion coefficients. Here, the hydrodynamic radius of the reversed micelles is about 70 Å.

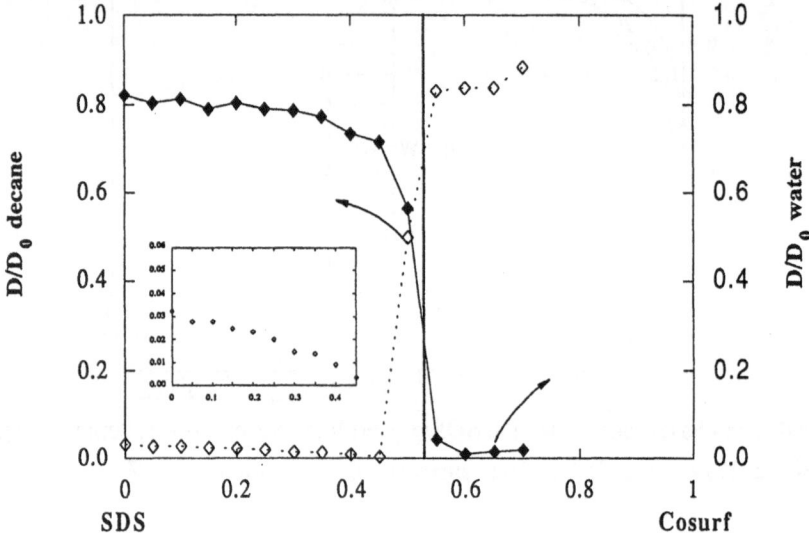

Figure 5. Microemulsion structure in the continuous phase studied by the diffusion coefficients (D) divided by the diffusion coefficients (D_0) for the neat liquid versus the ratio of cosurfactant/SDS. The vertical line in the middle of the diagram refers to the three-phase area (cf. Fig. 4). The insert gives the reduced diffusion coefficient for decane with the scale on the abscissa enlarged.

3.3. CREAMING AND COALESCENCE RATE

It is difficult to measure an absolute value for the creaming rate of an emulsion as there is often a broad size distribution of the emulsion droplets. For an isolated droplet in a continuous medium the creaming rate u follows the expression:

$$u = f(\Delta\rho, R^2, \frac{1}{\eta}) \tag{3}$$

where $\Delta\rho$ is the difference in density between the droplet and the continuous medium, R is the radius of the droplet and η is the viscosity of the continuous medium. It is obvious that different droplet sizes will give different creaming rates. However, this expression does not take into account any obstruction effects which will lower the creaming rate when the droplets become more closely packed. Thus we have measured the creaming mean "rate" as the time for three-fourths of the clear phase to appear ($t_{3/4}$), see Fig. 6. Inserted as comparison in the same diagram is the time of coalescence for W/O emulsions, measured as the time for one-half of the water rich phase to appear. The coalescence for O/W emulsions starts simultaneously throughout the creamed layer (after a much longer time, on the order of years) and can not be measured in this way.

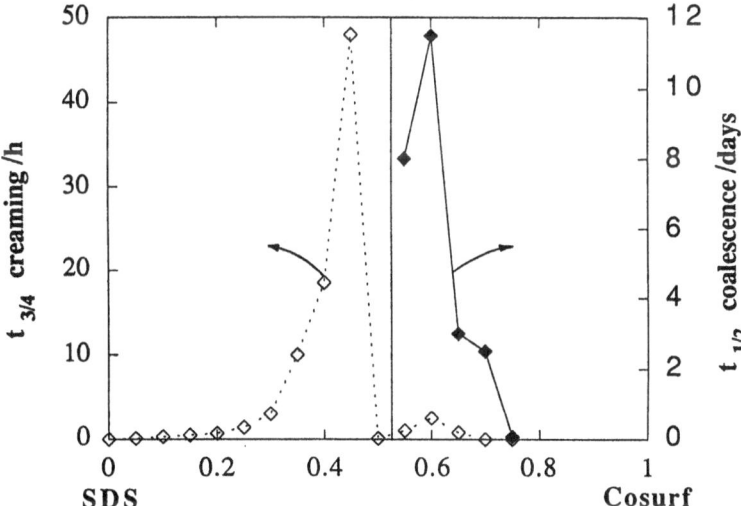

Figure 6. The time for one-half of the water rich phase to appear in W/O emulsions and the time for three-fourths of the clear phase to appear in a creamed emulsion versus the ratio of cosurfactant/SDS. The vertical line in the middle of the diagram refers to the three-phase area (cf. Fig. 4).

The creaming time for O/W emulsions increases up to the three-phase area where it suddenly decreases for the sample closest to this area. The increase of the creaming time can be ascribed to a decrease in the emulsion droplet size. NMR diffusion studies of the liquid inside the emulsion droplets show that the mean droplet size decreases towards the three-phase area. The experimental data for this diffusion are shown in Fig. 7, where a steeper initial slope indicates a larger droplet mean size (Lönnqvist, et al. (1991)) (see also Fig. 3 and the discussion below). The sudden drop in the creaming time near the three-phase area corresponds well with the change in the continuous medium from closed micellar aggregates to a bicontinuous microemulsion. In fact, this sample is difficult to emulsify and both creaming and coalescence occur fast here. On the W/O side of the three-phase area the same pattern is repeated but now in the other direction.

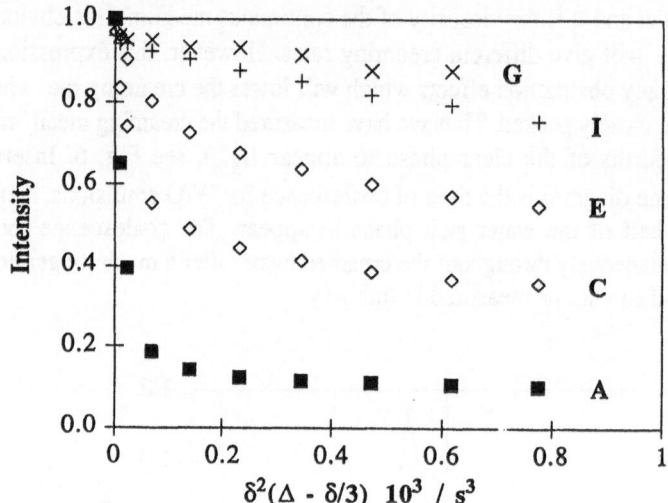

Figure 7a. The normalized NMR intensities (I/I_0) versus $\delta^2(\Delta-\delta/3)$ for the oil inside emulsion droplets in O/W samples. The letters refer to the sample compositions indicated in the phase diagram in Fig. 4.

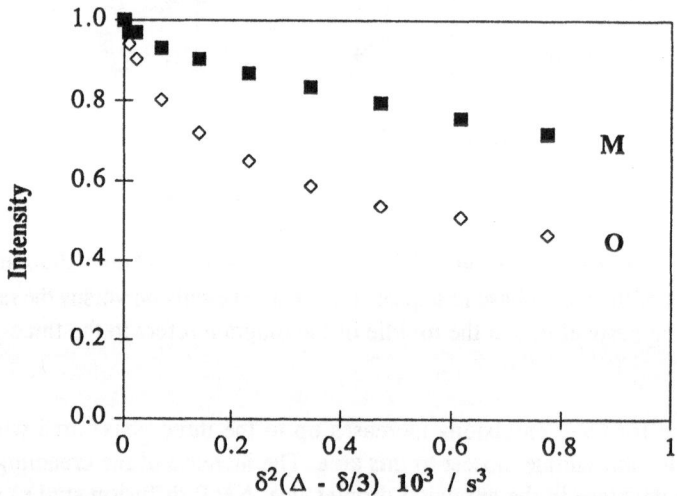

Fig 7b. The normalized NMR intensities (I/I_0) versus $\delta^2(\Delta-\delta/3)$ for the water inside emulsion droplets in W/O samples. The letters refer to the sample compositions indicated in the phase diagram in Fig. 4.

The surface tension will be the governing factor of the obtained emulsion droplet size in these emulsions gently shaken by hand. The general tendency for similar systems (Zhou, et al. (1987)) is that the surface tension between the two phases that are emulsified decreases towards the three-phase area from both sides. The free energy which is needed to create the surface of an emulsion droplet is a function of the surface tension, thus the droplet size decreases with decreasing surface tension.

The creaming time as well as the coalescence time have peaks in both the O/W and W/O area. The maximum in the creaming time is related to the emulsion droplet size and the presence of the bicontinuous microemulsion close to the three-phase area. The maximum in coalescence time is more difficult to interprete. The coalescence for O/W emulsions can not be measured directly but a photograph, see Fig. 8, shows that the coalescence first starts in both ends of the O/W area. This photograph is taken 2.5 years after the O/W emulsions were made and the great difference in coalescing time between the O/W and W/O emulsions is due to the charged droplet interface, which induces flocculation in the O/W emulsion. The fast coalescence near the three-phase area can be explained by the bicontinuous microemulsion between the emulsion droplets that offers a good possibility for the molecules inside the emulsion droplets to migrate to another emulsion droplet. The fact that the coalescence is more rapid at the high and low fraction of cosurfactant could maybe be explained by the large difference in curvature between the thermodynamically stable micelle and the unstable emulsion droplet. However, coalescence is of a complex nature and needs more research to be fully understood.

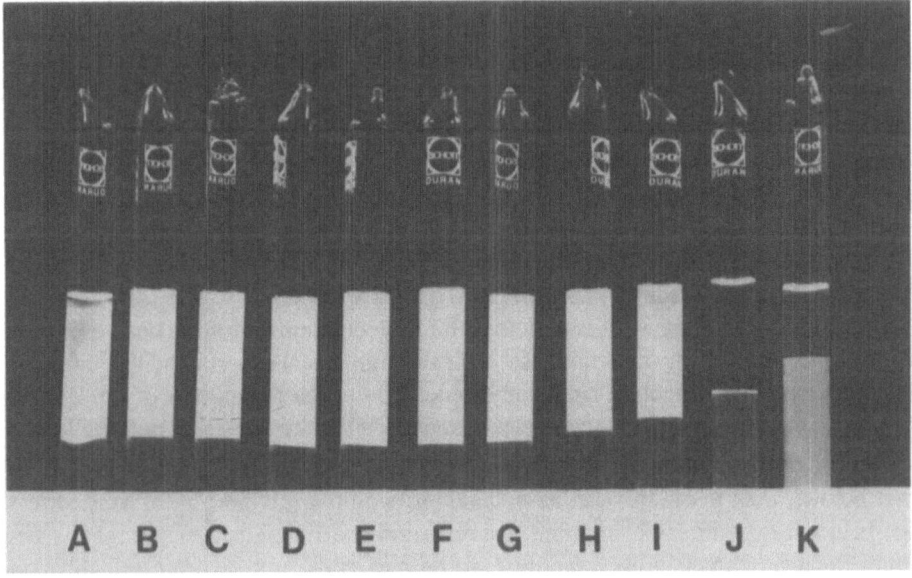

Figure 8. A photograph, showing the beginning of coalescence in O/W emulsions after 2.5 years. The letters refer to the sample compositions indicated in the phase diagram in Fig. 4.

4. Determination of droplet size

4.1. INTRODUCTION

One of the most important aspects in the characterization of emulsions is the determination of the size distribution of the emulsion droplets. Several different particle sizing techniques are available (Orr (1988)). Many of these methods do however require emulsion processing (dilution of the emulsion or the addition of stabilizers), while other systems are not suitable for commonly used methods. Thus, there seems to be a need for additional methods. A promising technique, that has been used surprisingly seldom, is based on measuring restricted self-diffusion of molecules within the droplets by means of the pulsed field gradient NMR-method (Packer and Rees (1971)). The method has been applied in the size determinations of emulsion droplets in different systems (O/W, W/O emulsions (Packer and Rees (1971), Lönnqvist, et al. (1991)), cheese (Callaghan, et al. (1983)) and margarine (van den Enden, et al. (1990))). The NMR method is rapid, non-perturbing and independent of the physical state of the sample. As an added bonus, the experiment will report on whether a particular emulsion is of the O/W or W/O type.

It was pointed out above, that the PGSE-technique monitors mean displacements of molecules in the direction of the field gradient. It follows that one may study restricted (non-Gaussian) diffusion. When barriers restrict the free diffusion one can obtain information on the geometry of the domains in which restricted diffusion occurs from the analysis of the NMR echo-amplitude. The NMR-echo amplitude has been calculated for different barrier geometries (Stejskal (1965)) and for some cases analytical expressions can be derived that predict the outcome of the diffusion experiment. The latter case has been experimentally tested and verified for the case of (spherical) emulsion droplets (Tanner (1966), Packer and Rees (1971), Callaghan, et al. (1983) and water in mica stacks (Tanner and Stejskal (1968)).

The echo attenuation for molecules confined to droplets of uniform radius is described by an equation due to Murday and Cotts (1968) (Eq. 2 above).

We have performed simulations by letting Eq. 2 generate data for the case of spheres uniform in size. By adding random noise, typical for an actual experimental situation, we then simulate an experiment on emulsion droplets. By a least-square fit of Eq. 2 to this data-set we can back out the correct droplet diameter. If the droplet size is not smaller than 1 μm (this value depends on the experimental setup and in particular on the value of G, see further below), then the radius can be obtained to within a few percent of the "correct" value. In a real experimental situation, the diffusion coefficient of spins in the disperse medium must be known. This quantity may be measured in a separate self-diffusion experiment on the separated disperse phase. Applying Eq. 2 to data from a polydisperse system will lead to deviations between the predictions of Eq. 2 and the actual observed data-set (see further below). Thus one can detect polydispersity, and in favorable conditions, one may indeed obtain information about the droplet size distribution (see further below). Packer and Rees (1971) and Callghan, et al. (1983) have extended Eq. 2 to include effects due to the emulsion polydispersity.

For a situation of polydisperse emulsion droplets, and provided that the exchange of the molecules between droplets is slow on the NMR time scale (which is essentially the value of Δ) then each size will follow Eq. 2, and the measured NMR echo intensity is a sum of the echo contributions from the various droplets weighted by the volume of each droplets fraction (number of spins in each fraction). Thus,

$$I_{poly} = \frac{\int_0^\infty R^3 \, P(R) \, I(R) \, dR}{\int_0^\infty R^3 \, P(R) \, dR} \tag{4}$$

where I is given by Eq. 2 and P(R) is the size distribution function. We are currently investigating different mathematical procedures for determining P(R) from the observed decay of I obtained at a given set of experimental conditions. A large set of data can be obtained by changing the gradient duration δ, the field gradient strength G, or the time between the gradient pulses Δ.

In the experiments described here the actual form of P(R) has not been determined. Rather we have assumed that a particular form of the distribution function is valid. A simple analytical solution of Eq. 4 was obtained by Callaghan, et al. (1983) assuming a normal distribution of the volume of the emulsion droplets and that $R << (2D\Delta)^{0.5}$.

There are several different types of distribution functions available (Orr (1983)) for describing the polydispersity. A wide variety of emulsions are represented by a log-normal distribution which is given by the form:

$$P(R) = \frac{1}{2R\sigma\sqrt{2\pi}} \exp\left(-\frac{(\ln 2R - \ln 2R_0)^2}{2\sigma^2}\right) \tag{5}$$

where R_0 is the diameter median and σ is the width of the distribution.

This form of P(R) has successfully been used to evaluate emulsion polydispersity from self diffusion data (Packer and Rees (1971)). On the basis of a log-normal distribution, a rapid determination of the emulsion polydispersity has been proposed (van den Enden, et al. (1990)) in which the measured echo attenuation is compared with tabulated data calculated beforehand.

By inserting Eq. 5 into Eq. 4, the resulting equation may be integrated numerically by changing the variables:

$$y = \frac{R - R_0}{R + R_0} \tag{6}$$

This reduces the integration limits in Eq. 4 from $(0,\infty)$ to $(-1,1)$ and a numerical procedure, such as the Gauss integration (Abramowitz and Stegun (1972)) , may be performed.

Thus, by applying Eq. 4 to the experimental data set, the parameters of the size distribution function can be evaluated and in the case of a log-normal distribution, R_0 and σ of Eq. 5 are evaluated.

Different types of size distributions may be compared in their ability to reproduce the observed values. The sensitivity of the method depends on the algebraic form of the chosen distribution function, the number of experimental data points, conditions at which the spin echo is measured and the technical characteristics of the NMR equipment. The main principle in designing an experiment is to find conditions where the spin echo is sensitive to the main fractions of the emulsion droplets. Smaller droplets (typically less than 0.2 μm) do not reduce the echo signal but the total amount of them may be determined by their contribution to the echo intensity. The recommended procedure for the best resolution of wide size distributions is to take several sets of data for the same sample and to analyse them simultaneously. Averaging the results from repeated calculations with different type distribution functions may also give reliable information.

We have also performed simulations for a polydisperse set of spherical droplets, using a log-normal distribution. Again, if the droplets' size distribution does not have substantial contributions from droplets smaller than 1 μm, the mean size can be obtained to within a few percent, while the parameter describing the width of the distribution is determined to within 20 %.

4.2. EXPERIMENTAL EXAMPLES OF SIZE DETERMINATIONS WITH THE NMR METHOD

4.2.1 *An O/W emulsion.* Given in Fig. 9 are the results of a NMR diffusion experiment on an emulsion consisting of equal amounts by weight of benzene and water, stabilized by 3 wt % of the non-ionic surfactant Triton X-100, which may serve as an illustration of the method. The broken line is a fit of a situation of monodisperse spheres (Eq. 2) to the data, while the solid line is a fit of a polydisperse situation with a log-normal distribution function (Eq. 4 and 5) to the data. As can be inferred from the data, the droplets are polydiperse. The values of the parameters of the distribution function are: $R_0 = 6.5$ μm and $\sigma = 0.3$. The obtained distribution function is displayed in the insert of Fig. 9. Microscopy investigations of the sample were carried out before the NMR-experiments were performed, indicating that the droplets are polydisperse and that the obtained droplet sizes are in accordance with the distribution function in Fig. 9.

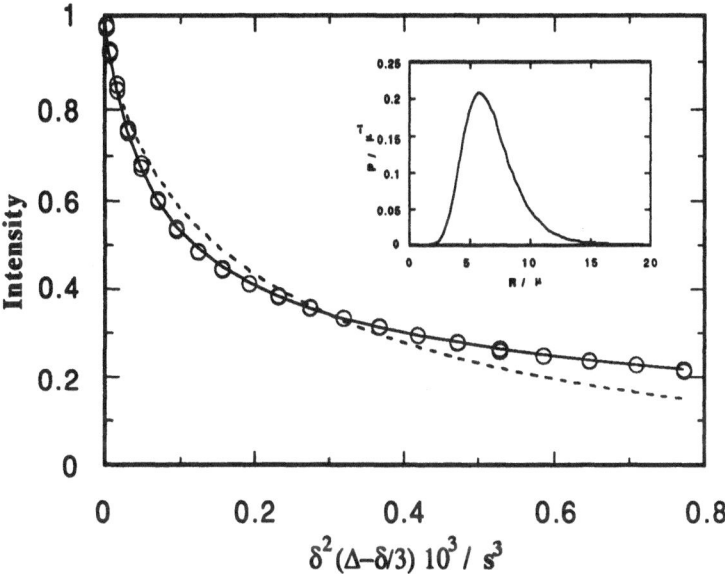

Figure 9. Echo intensity versus $\delta^2(\Delta-\delta/3)$ for an emulsion containing equal amounts by weight of benzene and water, stabilized by Triton X-100. The broken line is a fit of Eq. 2 to the data. The solid line is a fit of Eqs. 4 and 5 to the data, and the resulting distribution function is given in the insert.

4.2.2. Droplets sizes in hydrocarbon gels. The study of highly concentrated emulsions is of both scientific and practical importance. Such emulsions containing up to 99% hydrocarbon dispersed phase were described by Lissant (1966) and their properties are studied by Princen (1983) and Ebert, et al. (1988). The structure of these very stable and gel-like emulsions was studied by optical microscopy and the size distribution of the emulsion droplets was measured by a Coulter Counter Multisizer (Ebert, et al. (1988)). As the internal phase volume in the emulsion droplets is typically in excess of 90 %, the droplets are most likely deformed, a fact that poses additional difficulties in applying other conventional methods for particle sizing. We use the PGSE method for measuring the size distribution of the emulsion droplets for a 97.9 wt% heptane in water emulsion stabilized by 0.4 wt% CTAB. The result is shown in Fig. 10, where the observed echo attenuation is consistent with a monodisperse emulsion system with a droplet radius of about 7.3 μm (the solid line in Fig. 10 is a fit of Eq. 2 to the data). However, since the droplets are probably not spherical in shape, the determined size should be considered as an estimate of the dimensions of the closed oil domains.

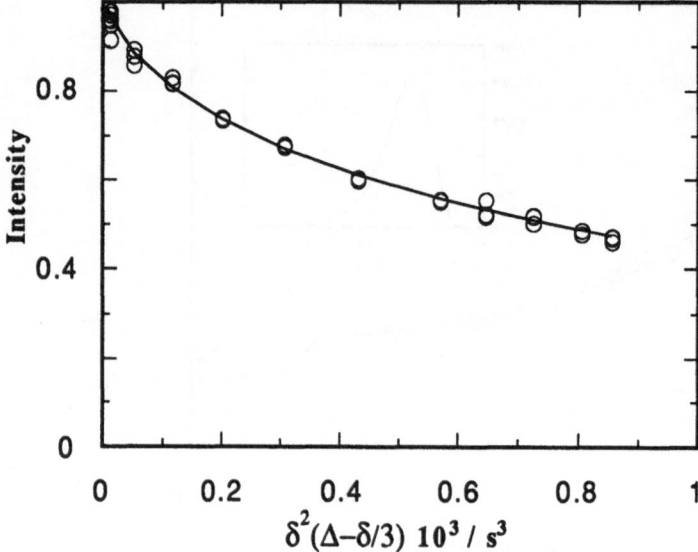

Figure 10. Echo intensity versus $\delta^2(\Delta-\delta/3)$ for an emulsion gel containing 97.9 wt% heptane, 0.4 wt% CTAB and 1.7 wt% water. The solid line is a fit of Eq. 2 to the data.

When an emulsion is prepared with cyclohexane (the emulsion contains 99% cyclohexane by weight and 0.2 wt% CTAB) as a disperse phase, the observed echo attenuation can not be explained with uniform droplets. When applying the log-normal size distribution function in interpreting the experimental data, the following values of the parameters were obtained: R_0= 4.4 μm and σ=0.35. Fig. 11 represents the agreement between the experimental data and the corresponding log-normal distribution. Most likely this discrepancy between the droplet sizes of the two different emulsion gels is due to the method of preparation (the emulsions were prepared according to procedures outlined in (Ebert, et al. (1988)). The emulsion gel based on cyclohexane, which is more concentrated in oil, has a significantly higher viscosity than the one based on heptane. It should be noted that the emulsion based on heptane is rather monodisperse with regard to droplet size. This would seem to offer a possibility to make very monodisperse emulsion droplets by diluting the concentrated emulsion with water. Such experiments are under way in our laboratory. Note also that the amount of surfactant needed for stabilizing a certain amount of disperse phase is very low.

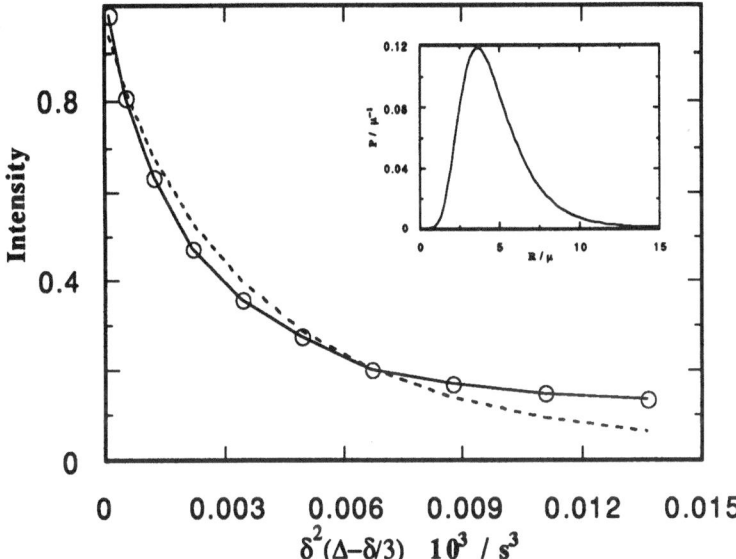

Figure 11. Echo intensity versus $\delta^2(\Delta-\delta/3)$ for an emulsion gel containing 99.0 wt% cyclohexane, 0.2 wt% CTAB and 0.8 wt% water. The broken line is a fit of Eq. 2 to the data. The solid line is a fit of Eqs. 4 and 5 to the data, and the resulting distribution function is given in the insert.

4.2.3. *Droplet Size in a "real-life" system - Margarine.*

A system where control of the droplet size is important is constituted by the low-calory spreads (margarine) that can contain rather large amounts of water. These systems are W/O emulsions, where the droplet size is difficult to determine with other methods, on account of the optical density of margarines. As pointed out above, NMR is not sensitive to the physical appearance of the sample, and can therefore be readily applied to margarines. Thus two margarines, containing 40 % and 60% fat respectively, of a common brand name were obtained. The margarines were transferred into 5 mm NMR tubes and self-diffusion experiments were performed at 17 °C. After prolonged storage at 30 °C, the systems phase separate into one water rich and one water poor phase. After centrifugation, the bulk diffusion coefficient of water was determined in the water rich phase, yielding a value of D=1.7 10^{-9} m²/s. This value was used when analyzing the data.

The two different margarines gave similar results. The results of the experiments for the margarine containing 40 % fat are given in Fig. 12. As can be seen the data are well described by a log-normal distribution function. The resulting values for the parameters R_0 and σ are: 0.36 µm and 0.82, respectively. A normal distribution of droplet sizes was also tried, however it gave a considerably poorer fit to the data. It might be worth noting that the total measuring time is of the order of 5 minutes for each sample.

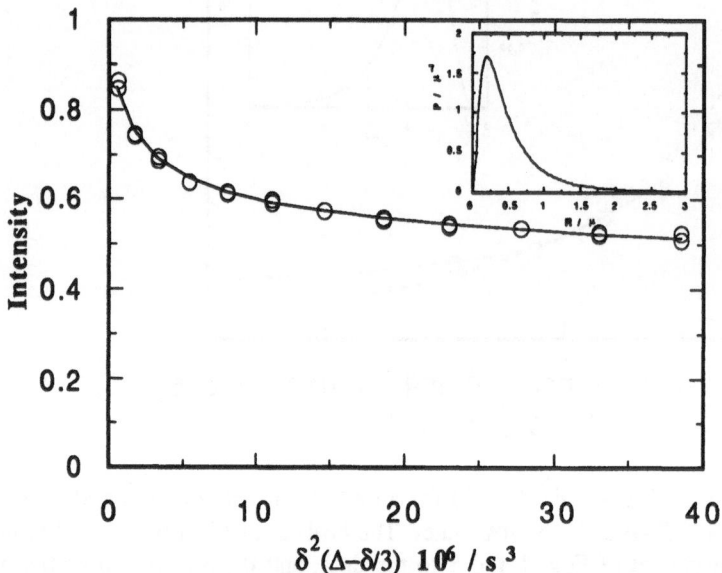

Figure 12. Echo intensity versus $\delta^2(\Delta-\delta/3)$ for a margarine, containing 40 % fat. The solid line is a fit of Eqs. 4 and 5 to the data, and the resulting distribution function is given in the insert.

There are two features of the data in Fig. 12 which deserve some comment. Firstly, as each droplet size in the distribution is in fact weighed by its volume, droplets with large radii give a large contribution to the decay of the echo. It follows that the fractions with comparatively large radii are rather well determined. In contrast, the distribution function towards smaller radii (say less than 0.5 μm or so) is less well determined. This means effectively, that the shape of the curve towards lower radii is uncertain, while the integral under the curve is rather well determined since the echo decay depends very little on the actual values of the radii for the small droplets (cf. Fig. 3). The actual radii which can be determined depend at least in part on the technical characteristics of the particular NMR-system used (see further below under Technical Details). In particular, by increasing the gradient strength is should be possible to measure smaller radii. In the literature, there are reports (van den Enden, et al. (1990)) where droplet sizes as small as 0.2 μm are determined by use of high gradient strengths (5 T/m).

We end this section by noting that although surface energy consideration would indicate that the droplets in margarine would tend to be spherical in shape, this is by no means obvious. As the mechanism of stabilization is complex in these systems, involving at least to some extent the presence of crystalline fat which may be enriched at the water/fat interface, non-spherical droplets may in fact be present. The self diffusion experiment clearly rules out the presence of continuous water channels, but it cannot discriminate between different geometries. For instance, if a situation of elongated water particles is at hand, then the experiment shows that these are closed.

5. Technical Details

The NMR self diffusion measurements discussed above were all performed on a JEOL FX-60 spectrometer, except for the emulsion gels and margarine experiments that were carried out on a homebuilt spectrometer, equipped with a Varian 2.3 T electromagnet. The gradient drivers were of "in-house" design and construction. The gradient strength for the XL-100 can be varied between 0 and 0.3 T/m. For the particular experiments reported on here the value used was 0.3 T/m. The time between 90°-180° pulses was typically 70 or 140 ms and the gradient pulse length was between 3 and 90 ms. The gradient strength on the JEOL spectrometer is variable between 0 and 40 mT/m.

The evaluation of the data and the different least-square analysis were performed on Macintosh computers, using the MATLAB package.

6. Acknowledgement

We are grateful to Hans Lilja for the design and construction of the NMR spectrometer used in this work, and to Lennart Nilsson for technical assistance. This project has been financially supported by the Swedish Board for Technical Development (STU).

7. References

Abramowitz, M. and Stegun, A. (1972) Handbook of Mathematical Functions, Dover Publications, New York, 887.

Anderson, D. M. and Wennerström, H. (1991) 'Self-diffusion in bicontinuous cubic phases, L3 phases, and microemulsions', J. Phys. Chem. 94, 8683-8694.

Balinov, B., Olsson, U. and Söderman, O. (1991) 'Structural similarities between the L3 and bicontinuous cubic phases in the AOT-brine system', J.Phys.Chem. 95, 5931-5936.

Callaghan, P. T. (1984) 'Pulsed field gradient nuclear magnetic resonance as a probe of liquid state molecular organization ', Aust.J.Phys. 37, 359-387.

Callaghan, P. T., Jolley, K. W. and Humphrey R. (1983) 'Diffusion of fat and water in cheese as studied by pulsed field gradient nuclear magnetic resonance', J.Colloid Interface Sci. 93, 521-529.

Ebert, G., Platz, G. and Rehage, H. (1988) 'Elastic and rheological properties of hydrocarbon gels', Ber.Bunsenges.Phys.Chem. 92, 1158-1164.

Hoffmann, H. and Ebert, G. (1988) 'Surfactants, micelles and fascinating phenomena', Angew.Chem.Int.Ed.Engl. 27, 902-912.

Jönsson, B., Wennerström, H., Nilsson, P.G., N. and Linse, P. (1986) 'Self-diffusion of small molecules in colloidal systems', Colloid Polym.Sci. 264, 77-88.

258

Lindman, B., Shinoda, K., Olsson, U., Anderson, D., Karlström, G. and Wennerström, H. (1989) 'On the demonstration of bicontinuous structures in microemulsions', Colloids Surf. 38, 205-224.

Lindman, B. and Stilbs, P. (1987) 'Molecular diffusion in microemulsions', in S. Friberg, and P..Bothorel (eds.), Microemulsions: Structure and dynamics, CRC Press: Boca Raton, pp. 119-152.

Lindman, B., Söderman, O. and Wennerström, H. (1987) 'NMR studies of surfactant systems', in R. Zana (ed.), Surfactant Solutions: New Methods of Investigation, Dekker, New York and Basel, pp. 295-358.

Lissant, K. J. (1966) 'The geometry of high-internal-phase-ratio emulsions', J.Colloid Interface Sci. 22, 462-468.

Lönnqvist, I., Khan, A. and Söderman, O. (1991) 'Characterization of emulsions by NMR methods', J.Colloid Interface Sci. 144, 401-411.

Murday, J. S. and Cotts, R. M. (1968) 'Self-diffusion coefficient of liquid lithium', J.Chem.Phys. 48, 4938-4945.

Orr, C. (1983) 'Emulsion droplet size data', in P. Becher (ed.), Encyclopedia of Emulsion Technology, Marcel Dekker,Inc., New York, Basel, pp. 369-404.

Orr, C. (1988) 'Determination of particle size', in P. Becher (ed.), Encyclopedia of Emulsion Technology, Marcel Dekker,Inc., New York, Basel, pp. 137-169.

Packer, K. J. and Rees, C. (1971) 'Pulsed NMR studies of restricted diffusion I. Droplet size distributions in emulsions', J.Colloid Interface Sci. 40, 206-218.

Princen, H. M. (1983) 'Rheology of foams and highly concentrated emulsions', J.Colloid Interface Sci. 91, 160-175.

Shinoda, K., Kunieda, H., Arai, T. and Saijo, H. (1984) 'Principles of attaining very large solubilization (microemulsions): Inclusive understanding of the solubilization of oil and water in aqueous and hydrocarbon media', J.Phys.Chem. 88, 5126-5129.

Stejskal, E. O. (1965) 'Use of spin-echoes in a pulsed magnetic-field gradient to study anisotropic, restriction and flow' J.Chem.Phys. 43, 3597-3603.

Stilbs, P. (1987) 'Fourier transform pulsed-gradient spin-echo studies of molecular diffusion', Progr.Nucl.Magn.Reson.Spectrosc., 19, 1-45.

Stilbs, P. (1991) Contribution presented at the 11th Scandinavian Symposium on Surface Chemistry, Bergen, Norway.

Tanner, J. E. (1966) Ph.D. thesis, University of Wisconsin.

Tanner, J. E. and Stejskal, E. O. (1968) 'Restricted self-diffusion of protons in colloidal systems by the pulsed-gradient spin-echo method' J.Chem.Phys. 49, 1768-1777.

van den Enden, J. C., Waddington, D., Van Aalst, H., Van Kralingen, C. G. and Packer, K. J. (1990) 'Rapid determination of water droplet size distributions by PFG-NMR', J. Colloid Interface Sci. 140, 105-113.

Zhou, J. S., Kamioner, M. and Dupeyrat, M. (1987) 'Is the low interfacial tension observed in two-phase microemulsion systems related to the composition of the adsorbed interfacial layer?' in H. L. Rosano, and M. Clausse (eds.), Microemulsion Systems, Dekker, New York, pp. 335-344.

ORGANIC IONS AS DEMULSIFIERS

MIKAEL JANSSON
Institute for Surface Chemistry
P.O.Box 5607
S-114 86 Stockholm
Sweden

ABSTRACT The effects of tetraalkylammonium ions on the stability of dilute o/w emulsions were investigated with turbidimetric measurements. Large tetraalkylammonium ions were found to induce flocculation of emulsions stabilized by ionic emulsifiers at concentrations 3 order of magnitudes lower compared to sodium ions. Emulsions stabilized by nonionic surfactant were also destabilized by low concentrations of organic ions. Inclusion complex formation of sodium ions with a crown ether, was found to drastically decrease the stability of the emulsion. All these observations demonstrate the importance of ion size and ion hydrophobicity on the stability of emulsions in presence of electrolyte. NMR self-diffusion measurements in micellar systems, monitoring ion-emulsifier interactions, and studies of the phase behaviour of the emulsifier/organic ion systems have been performed in an attempt to rationalize these observations.

1 . Introduction

Addition of salt is a very common step in industrial demulsification applications [1]. Most commonly, the chemical demulsification involves the use of inorganic salts. In the present work, the demulsifying efficiency of organic salts for o/w emulsions have been investigated. There are two main characteristics of organic ions which differ from those of inorganic ions. First, the hydrophobicity of organic counterions, which may increase the attractive interactions between ions and oil droplets. Secondly, the large sizes of many organic ions , which due to sterical effects may disturb the packing of emulsifiers at the oil/water interface. These two characteristics may drastically influence the emulsion stability. As demonstrated in this work, flocculation may be induced at concentrations several order of magnitude lower in presence of organic ions as compared to inorganic ions.

Flocculation rates of emulsions, after addition of organic and inorganic salts, have been followed by measuring the increase in sample turbidity, which is a consequence of aggregation of the emulsion droplets [2]. The organics salts which have been studied are tetramethylammonium (Me_4N) bromide, tetraethylammonium (Et_4N) bromide, tetrapropylammonium ($Prop_4N$) bromide and tetrabutylammonium (But_4N) bromide. The conditions which promote flocculation are directly related to interactions between the emulsion droplets. Therefore, in order to explain the influence of organic ions in these systems NMR self-diffusion measurements, providing quantitative information of the interaction between organic ions and aggregated surfactants, have been carried out using micellar model systems. The influence of organic ions on the phase behaviour of emulsifiers has also been investigated.

This study has been limited to the study of very dilute o/w emulsions (< 0.2 wt. % oil). Hence, a typical application in which these results may be relevant is the removal of oil residues from waste-water in industrial processes.

J. Sjöblom (ed.), Emulsions – A Fundamental and Practical Approach, 259–268.

2. Experimental section

Chemicals: The investigated surfactants were prepared from dodecanoic acid (Aldrich) which was mixed in equimolar amounts with tetramethylammonium hydroxide, tetraethylammonium hydroxide, tetrapropylammonium hydroxide, tetrabutylammonium hydroxide, (Sigma). The preparations were performed in water or ethanol solutions with solvent evaporation in vacuum. Tetraalkylammonium bromide salts (Merck), sodium bromide (Fisher) dodecane (Fluka), hexa-ethyleneglycol dodecyl ether, octa-ethyleneglycol dodecyl ether (Nikkol) and 1,4,7,10,13,16-hexaoxacylooctadecan (Sigma) were used as supplied.

Flocculation kinetics. Emulsion samples were prepared by mixing 97 wt % water and 3 wt % oil phase, which consisted of dodecane and 3 wt % of the emulsifier. Highly dispersed emulsions were obtained by using the high pressure homogenization technique termed "Microfluidization" [3] (Model 120 E, Microfluidics inc., Boston, Mass., USA). The homogenization pressure used was 800 bar, and the circulation time was 3 minutes. The stock emulsions were diluted 20-100 times to give a low particle concentration and a final turbidity of approximately 0.2 absorbance units. The flocculation rates were followed turbidimetrically at 540 nm using a Pye Unicam, SP8-200 spectrophotometer. The diluted emulsions were mixed with equivalent volumes of salt solutions directly in the cuvette before measurement. The initial increase in turbidity, quantified as the change in absorbance units with time, was used as a direct measure of the instability of the emulsion, in line with previous investigations. [4,5].

NMR: Self-diffusion coefficients were obtained with the Fourier transform modification of the NMR PGSE method, using a JEOL FX-100 NMR spectrometer, where the setup and procedures have been described previously [6,7]. The surfactants were dissolved in deuterated water directly in the NMR tubes. A two-site model was used to quantify the degree of micellar binding of organic ions and the monomer concentration of dodecanoate ions [8];

$$D_{obs} = p\ D_{mic} + (1-p)\ D_{free}$$

where D_{obs} is the observed diffusion coefficient, D_{mic} the micellar self-diffusion coefficient, D_{free} the diffusion coefficient in absence of surfactant aggregates, and p the fraction of ions associated to the micelles. D_{mic} was determined by measuring the self-diffusion coefficient of added hexamethyldisiloxane (HMDS), assuming complete micellar solubilization [9], and D_{free} was obtained from samples with concentrations below the cmc. The degree of counterion association, ß, was calculated from p_{ion}/p_{amph}, where p_{ion} and p_{amph} are the fractions of micellarly associated counterions and dodecanoate ions, respectively

Phase diagrams: Binary phase diagrams were determined by preparing samples by weight in sealed ampoules, allowing them to equilibrate at fixed temperatures in a thermostated bath and visually determine the type of phase(s) present.

3. Results and Discussion

3.1 FLOCCULATION KINETICS

The effect of adding different tetraalkylammonium bromide salts on the stability of emulsions stabilized by the ionic emulsifier sodium dodecanoate was investigated by turbidity measurements. The organics salts were compared to NaBr with regard to their demulsifying efficiency. The results are displayed in Figure 1.

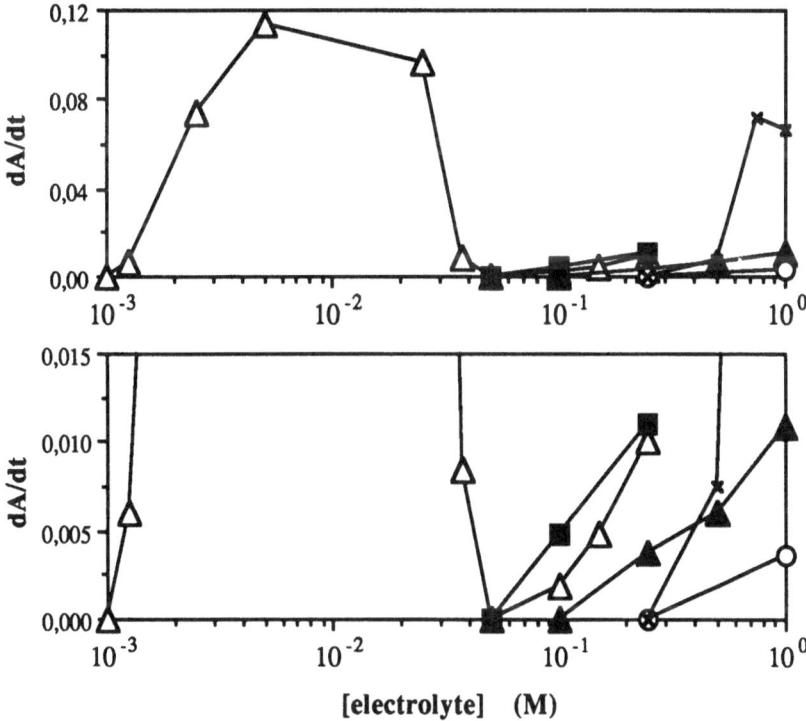

Figure 1. The initial flocculation rate, dA/dt (min^{-1}) of dilute o/w emulsions of water/dodecane/sodium dodecanoate versus the concentration of added electrolyte. The electrolytes are; (X) sodium bromide; (O) tetramethylammonium bromide; (▲) tetraethylammonium bromide; (■) tetrapropylammonium bromide; and tetrabutylammonium bromide; (Δ). The figures are identical, except for the scale on the dA/dt axis. The figure is taken from ref [10].

A general trend observed is that the concentration of electrolyte required to induce flocculation is directly correlated to the size of the tetraalkylammonium ion. Figure 1 shows that Me_4NBr is less efficient in destabilizing the emulsion than NaBr, that is higher concentrations of salt is required to induce flocculation of the emulsion. However, if the size of the organic ion is increased, the demulsifying efficiency increases significantly. An increase in size from $Prop_4N$ ions to But_4N ions, increases the demulsification efficiency dramatically. Two order of magnitude lower concentrations of But_4N ions is required to induce flocculation as compared to the $Prop_4N$ ions.

262

The flocculation rate reaches a maximum at 10 mM of But_4NBr. Above this concentration the flocculation rate is considerable lower. At 50 mM of salt the emulsion is stable over the time-scale of the experiment. This phenomena will be discussed further below.

The applicability of organic ions as demulsifiers is not limited to systems with ionic emulsifiers only. This is demonstrated in Figure 2, where the flocculation rates in presence of organic salt is shown for a system stabilized by $C_{12}E_6$ surfactants. As observed in the ionic emulsifier system, there is a dramatic difference between the demulsifying efficiency of But_4N ions and $Prop_4N$ ions. Although the flocculation rates are lower in Figure 2 than in Figure 1, the concentration dependence of the effect of But_4N ions on the emulsion stability is very similar in both systems.

But_4N ions are not efficient demulsifiers in all systems stabilized by oligo-ethylenglycol ether surfactants, however. Figure 2 shows that for emulsions stabilized by $C_{12}E_8$ surfactant, the emulsion stability is not affected by additions of But_4NBr below concentrations of 0.1 M of electrolyte.

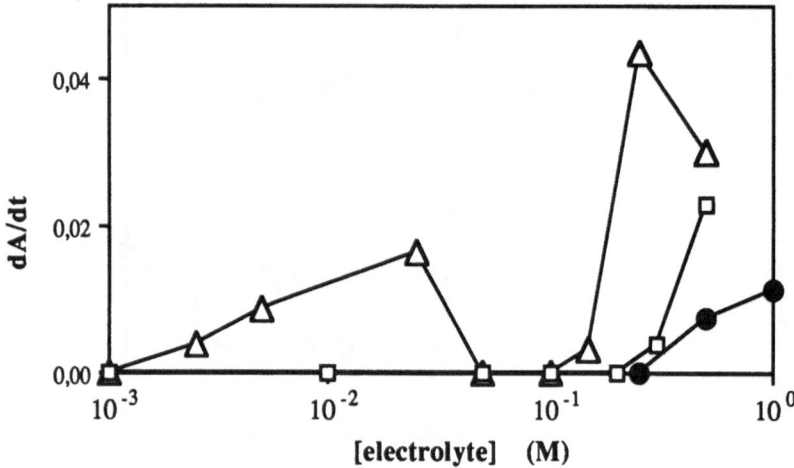

Figure 2. The initial flocculation rate, dA/dt (min^{-1}) of dilute o/w emulsions of water/dodecane/$C_{12}E_6$ or $C_{12}E_8$ emulsifiers versus the concentration of added electrolyte. For the $C_{12}E_6$ system the electrolytes are: (●) tetrapropylammonium bromide; and (Δ) tetrabutylammonium bromide. For the $C_{12}E_8$ system the electrolyte is tetrabutylammonium bromide (□). Addition of NaBr had no effect on the turbidity.

The importance of ion size /hydrophobicity with regard to the demulsifying efficiency of salts was further demonstrated by studying the effect of large cryptate molecules, which may form complexes will inorganic ions, on the emulsion stability. The effective size of the inorganic ion was increased by complexation with the macrocyclic cryptate [11]. Flocculation rates were measured in presence of mixtures of sodium bromide and the crown ether 1,4,7,10,13,16-Hexaoxacyclooctadecane (18C6). Figure 3 shows that the NaBr/18C6 mixture is about 10 times more effective than NaBr with regard to the destabilization of the emulsion. The macrocyclic compound, in absence of sodium ions, had no effect on the emulsion stability.

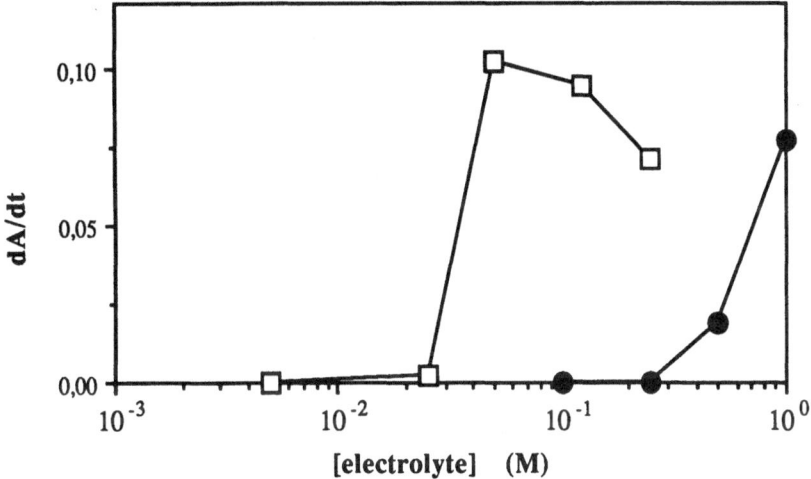

Figure 3. The initial flocculation rate, dA/dt (min^{-1}) of dilute o/w emulsions of water/dodecane/sodium dodecanoate versus the concentration of added electrolyte. The electrolytes are; (●) sodium bromide; and (□) 1:1 mixture of sodium bromide and the crown ether 18C6. The figure is taken from ref. [10].

3.2 ORGANIC ION-EMULSIFIER INTERACTIONS

In order to rationalize the flocculation data, interactions between organic ions and emulsifiers have been studied in micellar systems, using NMR self-diffusion measurements and phase diagram determinations.

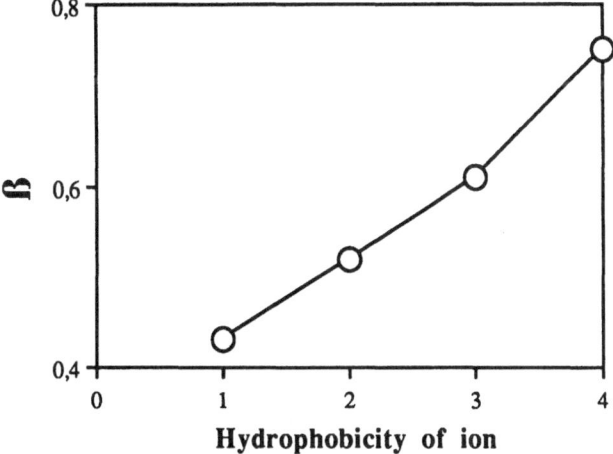

Figure 4. The degree of counterion association, ß, in different tetraalkylammonium dodecanoate micellar system, determined by NMR self-diffusion measurements. The hydrophobicity of the ion is here defined as the number of methyl or methylene segments in one of the alkyl chain of the ion, i.e. 1 correspond to tetramethylammonium ions. The ß-values correspond to average values observed at surfactant concentrations of 0.1 M- 0.2 M.

For emulsions stabilized by the ionic emulsifier, sodium dodecanoate, electrostatic interactions have to be considered when analysing the flocculation data. As seen from Figure 4, the degree of counterion association increases with the size of the ion. This is due to that the attractive interaction between organic ions and micellized emulsifiers increases with the hydrophobicity of the ion [12]. Consequently, the electrostatic potential at the aggregate surface is lowered in presence of more hydrophobic ions, decreasing the repulsive double-layer interactions which acts as a barrier for flocculation. This may partly explain the higher demulsifying efficiency of the larger organic ions. The lower demulsifying efficiency of Me_4NBr compared to NaBr may be a consequence of weak ion-droplet interactions in the former system. The fraction of micellarly bound counterions in the Me_4N system, 0.42, is lower than theoretically predicted from Poisson-Boltzmann calculations [13], and experimentally observed [14], for inorganic counterions.

The enormous difference in demulsifying efficiency between $Prop_4N$ ions and But_4N ions can not be explained with differences in electrostatical interactions only, however. This is evident from the fact that But_4N ions are very efficient demulsifiers in systems with both ionic and nonionic emulsifiers. The hydrophobicity of the ion may be an important factor in this context. The adsorption of hydrophobic ions at the oil/water interface may give rise to attractive hydrophobic interactions between emulsion droplets. Since the tetraalkylammonium ion has a tetrahedrical configuration, it is not possible that all four alkyl chains of the adsorbed ion are confined within the hydrocarbon domain. Hence, the adsorption of these ions may change the hydrophilic/hydrophobic nature of the droplet surface, which may decrease the emulsion stability. An observation which supports this line of argument, is that a cloud point, which is normally associated with nonionic surfactants, has been observed in mixtures of sodium tetradecyl sulfate surfactants and But_4NBr salt [15].

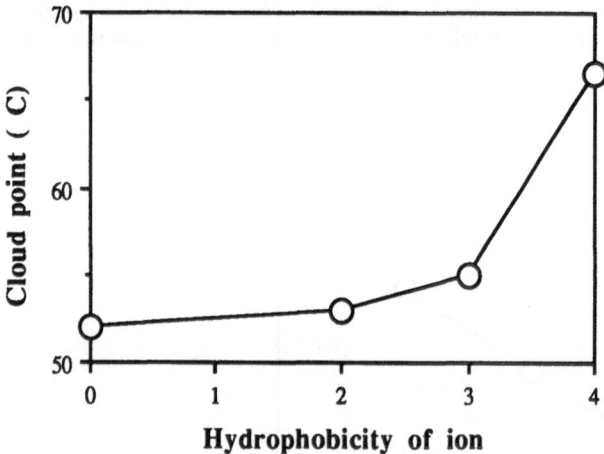

Figure 5. Cloud point observed in $C_{12}E_6$ solutions in presence of different tetraalkylammonium bromide salts. The concentrations of surfactant and salt are 0.1 M. The hydrophobicity of the organic ion is here defined as the number of methyl or methylene segments in one of the alkyl chain of the ion. The cloud point given at zero hydrophobicity corresponds to a 0.1 M $C_{12}E_6$ + 0.1 M NaBr system.

The adsorption of But_4N ions to the oil/water interface of emulsions stabilized by $C_{12}E_6$ surfactants is indicated from the cloud-point data displayed in Figure 5. The cloud-point increases with the amount of adsorbed ions due to that repulsive double-layer forces are introduced between the micelles [16]. The effects of adding salt to the surfactant solutions are rather small for ion sizes

265

below that of But$_4$N ions, which demonstrates that the interaction between emulsifiers and the smaller organic ions is rather weak. This may explain the low demulsifying efficiency of the smaller and less hydrophobic organic ions in the C$_{12}$E$_6$ emulsifier system.

But$_4$N ions do not induce flocculation at low concentrations in the C$_{12}$E$_8$ system (Figure 2). This may be due to that the interaction between ions and emulsifiers is lower in the C$_{12}$E$_8$ system compared to the C$_{12}$E$_6$ system. NMR self-diffusion measurement demonstrate that the fraction of But$_4$N ions bound to micelles are lower in the former system [10].

A more empirical approach to explain the influence of organic ions on the stability of emulsions is to study the phase behaviour of the emulsifier in presence of organic ions. Friberg et. al. have shown that the tendency of surfactants to form lamellar liquid-crystalline phases is directly related to the ability of stabilizing emulsions [17,18]. Due to the inability of large ions to effectively screen surfactant head-group repulsions when the aggregate curvature is low, large counterions inhibits the formation of liquid-crystalline phases. This is demonstrated by the fact that both the lamellar and the hexagonal liquid-crystalline regions observed in the phase diagram of sodium dodecanoate/water system [19], disappears in presence of the crown-ether 18C6 (Figure 7). A similar phase behaviour is observed if sodium ions are exchanged with But$_4$N ions [20]. One line of argument would then be that , due to the promotion of higher curvatures of organic ions, the emulsifier is less suitable for stabilizing the emulsion in presence of these large ions.

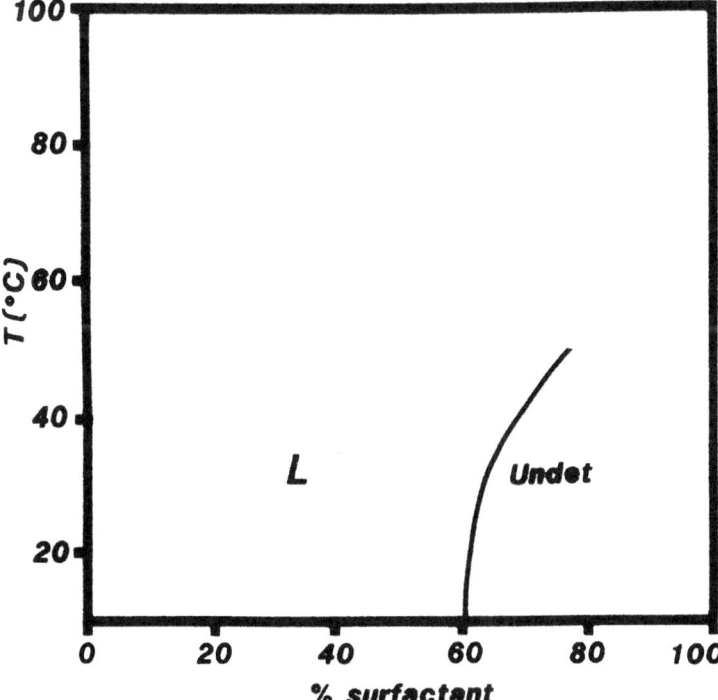

Figure 6. Phase diagram of sodium dodecanoate/1,4,7,10,13,16-hexaoxacyclooctadecane (18C6)/water. The ratio between sodium dodecanoate and 18C6 was held constant at 1. The existence region for the solution phase L is shown.

Promotion of higher surface curvatures by large organic ions is also observed in nonionic micellar systems. $C_{12}E_6$ micelles are large and rod-shaped at 25 °C [21], which is reflected in low self-diffusion coefficients of the surfactants. Figure 7 shows that addition of But_4N ions to a $C_{12}E_6$ solution increases the micellar self-diffusion coefficient, corresponding to a decrease in size of the $C_{12}E_6$ micelles. The presence of $Prop_4N$ ions affect the micellar size to a much smaller degree, which is compatible with the lower demulsifying efficiency of tthese ions compared to the But_4N ions.

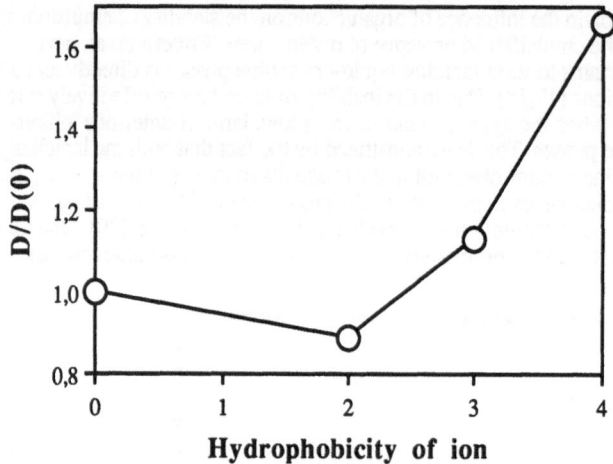

Figure 7. The diffusion coefficient of $C_{12}E_6$ micelles in presence of different tetraalkylammonium bromide salts relative the micellar diffusion coefficient of a pure $C_{12}E_6$ system. The concentrations of surfactant and salt were 0.1 M. The hydrophobicity of the organic ion is here defined as the number of methyl or methylene segments in one of the alkyl chain of the ion. The D/D(0) value given at zero hydrophobicity corresponds to a 0.1 M $C_{12}E_6$ + 0.1 M NaBr system.

3.3 EMULSION STABILIZATION BY ORGANIC IONS

Both emulsions stabilized by ionic and nonionic emulsifiers were stable at intermediate concentrations of But_4NBr (Figure 1 and 2). This is probably due to that the adsorption of But_4N ions at the oil/water interface, at significantly high concentrations, gives rise to an excess of ions at the surface, producing repulsive double-layer interactions. This is indicated by the data in Figure 8 which shows the effect of substituting a faction of the But_4N ions with sodium ions. The emulsion becomes instable in the concentration interval 0.05 M-0.1 M, in contrast to what is observed for the pure But_4NBr and NaBr electrolytes. Since sodium ions mainly interact electrostatically with the aggregate surface, no excess of ions is obtained at the oil/water interface in the mixed electrolyte system. Consequently, the emulsion is not stable in this system.

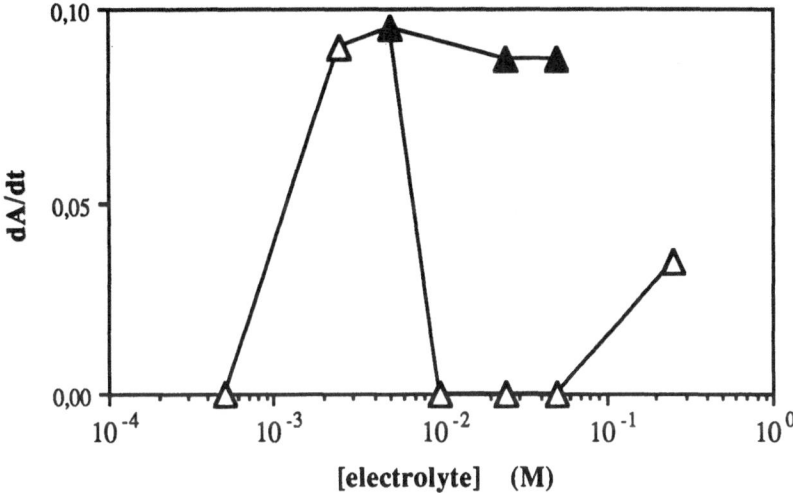

Figure 8. The initial flocculation rate, dA/dt (min^{-1}) of dilute o/w emulsions of water/dodecane/sodium dodecanoate versus the concentration of total amount of added electrolyte. The electrolytes are; (Δ) tetrabutylammonium bromide; and ($\blacktriangle$) a mixture of sodium bromide and 0.05 M tetrabutylammonium bromide.

4. Conclusions

Large tetraalkylammonium ions may destabilize dilute o/w emulsions stabilized with both ionic and nonionic emulsifiers.

The size and hydrophobicity of the ions are important parameters in this context. If a critical ion size/hydrophobicity is exceeded, the organic ions may show a very high demulsifying efficiency.

5. Acknowledgements

Thanks are due to Angela Jönsson (phase diagrams), Kristina Ryden (flocculation kinetics) and Peter Skagerlind (NMR) for their experimental work performed within this project.

6. References

1.   Lissant, K.J. (1983) 'Demulsification: industrial applications', Marcel Dekker, Inc., New York.
2.   Egusa, S. (1982) J.Colloid Interface Sci., 86, 135.
3.   Cook, E.J. and Lagrace, A.P. (1985) U.S. Pat. 4533254 .
4.   Bergenståhl, B., Fogler, S. and Stenius, P. (1986), in G.O. Phillips, D.J. Wedlock and P.A. Williams (eds), Gums and stabilizers for the food industry, Vol 3, Elsevier Applied Science, London, Vol 3.

5.   Bergenståhl, B. (1989) in G.O. Phillips, P.A. Williams and D.J. Wedlock (eds), in Gums and stabilizers for the food industry, Vol 4,
      IRL Press, Oxford.
6.   Stilbs, P. (1987) Prog.Nucl.Magn.Reson.Spectrosc., 19, 1.
7.   Jansson, M. and Stilbs, P. (1985) J.Phys.Chem., 89, 4868.
8.   Hertz, H.G. (1967) Ber.Bunsenges.Phys.Chem., 75, 195.
9.   Stilbs, P. (1982) J.Colloid Interface Sci., 87, 385.
10. Jansson, M., Eriksson, L. and Skagerlind, P. (1991) Colloids and Surfaces, 53, 157.
11. Evans, D.F., Sen, R. and Warr, G.G. (1986) J.Phys.Chem., 90, 5500.
12. Jansson, M. and Jönsson, B. (1989) 93, 1451.
13. Gunnarsson, G., Jönsson, B. and Wennerström, H. (1980) J.Phys.Chem., 84, 149.
14. Lindman, B., Puyal, M., Kamenka, N., Rymden, R. and Stilbs, P. (1984)
      J.Phys.Chem., 88, 5048.
15. Yu, Z. and Xu, G. (1989) J.Phys.Chem., 93,7441.
16. Nilsson, P.G. and Lindman, B. (1984) J.Phys.Chem., 88, 5391.
17. Friberg, S. and Mandell, L. (1970) J.Am.Oil.Chem.Soc. 47, 149.
18. Friberg, S. (1971) Kolloid Z.Z.Polym., 244, 333.
19. McBain, J.W. and Lee, W.W. (1943) OilSoap (Chicago), 20, 17.
20. Jansson, M., Jönsson, A., Li, P. and Stilbs, P (1991) Colloids and Surfaces, In press.
21. Jansson, M. and Rymden, R. (1987) J.Colloid Interface Sci., 119, 185.

SURFACE FORCES AND EMULSIFIERS

Per Stenius[1,3], Björn Bergenståhl[1] and Per Claesson[1,2]

1) Institute for Surface Chemistry
Box 5607
S-114 86 Stockholm
Sweden

2) Department of Physical Chemistry
The Royal Institute of Technology
S-10044 Stockholm
Sweden

3) Present address:
Department of Forest Products Technology,
Helsinki University of Technology,
Vuorimiehentie 1 A,
SF- 02150 Espoo,
Finland

ABSTRACT. The development of precise methods to measure the force/distance dependence of the surface forces between solid or liquid surfaces immersed in liquids and covered with adsorbed emulsifiers (polymers and low molecular weight surfactants) has revealed that the conventional description of these forces given by van der Waals interactions, double layer interactions and steric interactions between intermixing polymer layers is very far from complete. In particular, this is the case at relatively short distances between the adsorbed layers.
Between hydrophobed surfaces in aqueous solution long-range strongly attractive forces occur that are reduced by surfactant adsorption but can be reversibly restored if conditions are created so that the surfactants desorb. Strong repulsive interactions occur at short distances between layers of monoglycerides, polyoxyethene alkyl ethers ($C_{12}E_5$) and dodecyldimethylphosphine oxide when adsorbed on hydrophobed surfaces. A weak minimum in the interaction curve occurs that at elevated temperatures may become attractive.
Attempts have been made to clarify to which extent these forces influence emulsion stability. The temperature dependence of the forces correlates with the phase behaviour of the surfactants: the increasing attraction shows up in binary phase diagrams as a decreased stability of lamellar phases or as a lower consolute solution temperature. For $C_{12}E_5$ it is well known that O/W emulsions tend to coalesce as temperature is increased, but coalescence generally occurs at much higher temperatures than the one at which the interactions between adsorbed layers on a solid hydrophobic surface becomes attractive. This probably reflects the effect of the oil on the lower consolute temperature.

1. Introduction

The surface forces between solid or liquid surfaces immersed in liquids and covered with adsorbed emulsifiers (polymers and low molecular weight surfactants), although essentially interdependent, are conventionally separated into a number of components. Of these, the long-range electrostatic interactions, the van der Waals interactions and the repulsive and attractive interactions caused by the restrictions in the freedom of movement of adsorbed or non-adsorbed polymers in the vicinity of surfaces are relatively well understood. However, the development of precise methods to measure directly the force/distance dependence of the molecular interactions between surfaces has revealed that the description given by these interactions is in many respects insufficient. In particular, our understanding of the interactions between hydrophobic surfaces and between adsorbed layers of nonionic surfactants of the type extensively used as emulsifiers and dispersants is incomplete.

J. Sjöblom (ed.), Emulsions – A Fundamental and Practical Approach, 269–281.
© 1992 *Kluwer Academic Publishers.*

In this paper we review our studies of the interactions between layers of typical emulsifiers adsorbed on a solid surface consisting of bare or hydrophobed mica. For the evaluation of the importance of the observed interactions in emulsions it would be of great interest to compare the measurements to measurements of forces in thin liquid films. Although such forces can now be measured to a high precision, the data available for direct comparison with results from the surface forces apparatus are so far limited. We will therefore restrict ourselves to a very limited comparison with measurements of forces in lamellar liquid crystalline phases, as evaluated by the osmotic stress method. Of partcular interest is the temperature dependence of the forces, and its relationship to the tendency of solutions of non-ionic surfactants to phase-separate on heating.

2. Experimental

For details on experimental procedures, the reader is referred to the original papers cited in the description of results.

2.1. PREPARATION OF SURFACES

The substrate used in all surface forces measurements was green muscovite mica from Brown Mica Co, Sydney. Hydrophobed mica surfaces were prepared by depositing a double-chained cationic surfactant, dimethyldioctadecylammonium (DDOA) ions on the mica by the Langmuir-Blodgett technique, using a computerized Langmuir trough system developed by KSV Instruments, Helsinki, Finland. It has been shown that this surfactant forms a close-packed layer with a contact angle above 90^0 on the mica [1-3]. In the studies of the polyoxyethene/polypropene block copolymers the surfaces were hydrophobed by depostition of a double-chained surfactant with fluorocarbon tails (N-(α-trimethyl-ammonioacetyl)-O,O'-bis(1H,1H,2H,2H, perfluorodecyl)-L-glutamate chloride). The packing density and surface properties of this surface have been characterized in detail [4].

Surfactants were deposited on the bare or on the hydrophobed mica by direct adsorption or by the Langmuir.Blodgett technique, as described below for each surfactant.

2.2. SURFACE-FORCE MEASUREMENTS

The forces acting between the mica surfaces immersed in aqueous solution were measured as a function of surface separation using a surface force apparatus of the type developed by Israelachvili [5]. In this apparatus the two micas surfaces are glued onto optically polished silica discs and mounted in a crossed cylinder configuration. The surface separation is determined interferometrically to within 0.2 nm using fringes of equal chromatic order. The force (F) is determined within a detection limit of 10^{-7} N from the deflection of a spring supporting the lower surface. From the force, the free energy of interaction per unit area between flat surfaces (G) can be calculated according to the Derjaguin approximation

$$\frac{F(D)}{R} = 2\pi G(D)$$

where R is the local geometric mean radius of the cylinders, which is assumed to be much larger than the surface separation D.

The swelling pressure in the lamellar and gel pases of monoglyceride/water systems were measured by determination of the relative water vapour pressure in equilibrium with the phases, using head-space gas chromatography. The phase structures (the thickness of the water layers) were determined by X-ray diffraction. For details, see ref. [6,7].

2.3. CHEMICALS

Dimethyldioctadecylammonium bromide (DDOA), dodecylammonium chloride (DAHCl) and octylammonium chloride (OAHCl) were obtained from Eastman Kodak Co. The OAHCl was further purified by recrystallization. (N-(α-trimethylammonioacetyl)-O,O'-bis(1H, 1H, 1H, 2H, perfluorodecyl)-L-glutamate chloride) was obtained from Sogo, Ltd Japan. Pentaoxyethenedodecylether ($C_{12}E_5$) was obtained from Nikkol Co, Japan and dimethyldodecylamineoxide (DDAO) from Merck Ag. Dimethyldodecylphosphineoxide (DDPO) was a sample supplied by dr R.G. Laughlin, Procter & Gamble, Cincinnati, Ohio. Monopalmitin was obtained from NuChek-Prep Inc., Elysian, MN, USA. Monoolein was supplied by dr Niels Krog, Grinsted A/S, Braband, Denmark. The polyoxyethene/-polypropene oxide block copolymer was a sample (Proxanol 268) synthesized by the Research Institute of Semiproducts and Dyestuffs, Moscow, USSR. All other chemicals were of analytical reagent grade.

The water was purified by decalcination, prefiltration and reverse osmosis followed by filtration trough charcoal, ion exchange beds and a final 0.2 μm filter.

3. Results and Discussion

3.1 HYDROPHOBED MICA

Freshly cleaved muscovite mica carries $2.1 \cdot 10^{14}$ exchangeable cations/cm^2. It has been shown for the deposition of the DDOA [8] as well as for the fluorocarbon surfactant [4] that deposition from an insoluble monolayer on the Langmuir balance at sufficiently high surface pressure results in the formation of monolayers on the mica with a maximum molecular density corresponding closely to the number of exchangeable cations,. Thus, the surfactants are attached to the surfaces by ionic bonding forming close-packed layers that are stable at low electrolyte concentrations. The contact angles of water at these layers are 90° (DDOA) and 115° (fluorocarbon surfactant).

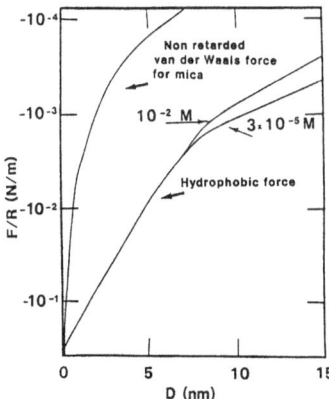

Figure 1. Comparison between the hydrophobic interaction between DDOA-covered mica surfaces in dilute electrolyte solution and the non-retarded van der Waals interaction (calculated using a Hamaker constant of $2.2 \cdot 10^{-20}$ J) From ref. [1].

In water and in dilute electrolyte solutions, very long range and strong attractive forces are observed between both types of surfaces [1, 4]. As shown by figure 1, the range of the force is orders of magnitude larger than expected from the van der Waals forces in the mica/DDOA/water system. Similar forces were earlier observed for adsorbed layers of cetyl trimethyl ammonium bromide [9] and also (later) in very dilute dodecylammonium chloride solutions [2]. The investigations of forces between surfactant layers give some further insights into how this force depends on surface charge and the presence of hydrophilic groups.

3.2. FORCES BETWEEN LAYERS OF ALKYLAMMONIUM IONS

3.2.1. Dodecyl- and octylammonium ions on bare mica. Figure 2 shows the distance dependence of the forces acting between mica surfaces immersed in solutions of DAHCl. In pure water (not shown in the figure) the total interaciton is due to a long-range repulsive double-layer force and an attractive van der Waals force that becomes predominant at a distance of about 3.5 nm. The long-range interaction is well described by the DLVO theory, assuming an area per surface charge of 27 nm^2. This is much larger than the area per exchangeable cation on the mica surface (0.48 nm^2), indicating considerable adsorption of H$^+$ ions. At the DAHCL concentrations shown in figure 2 the double layer force is weak or undetectable for several hours after injection of the surfactant. However, in 40μM DAHCl a small surface charge is built up after two days of immersion [3]. The uncharged surfaces jump into contact at a distance of 10 nm, which is larger than the distance predicted for non-retarded van der Waals forces (7 nm).

Thus, the adsorption of DAH$^+$ ions at these very low concentrations gives rise to a attractive hydrophobic force, although much weaker than between the close-packed DDOA layers, although in both cases the surfaces are completely or almost completely neutralized by adsorbed ammonium ions and H$^+$. The hydrophobic force appears to be strongly dependent on the density of hydrocarbon on the surface, which is much less for DAH$^+$ than for DDOA. The force between the DAH$^+$ layers is similar to the one earlier observed for CTAB [9].

At concentrations > 100 μM repulsive forces between the DAH$^+$ layers are again observed. The net surface charge (which is now positive) increases and finally, at concentrations > 6 mM the surfaces no longer jump into adhesive contact. Instead, a very strong repulsive barrier occurs at 5.4 nm. This indicates that a bilayer has formed on each surface, with a formal thickness of ≈ 2.7 nm/bilayer. This thickness very likely includes a thin film of water. The formation of the bilayer is reversible, i.e. dilution results in desorption of the surfactant.

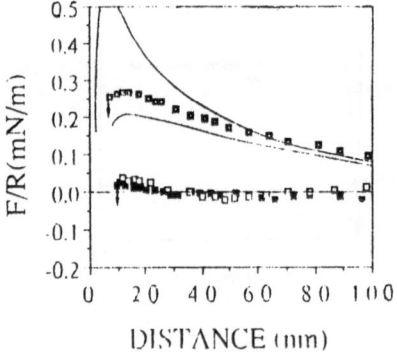

Figure 2. The interaction forces between mica surfaces in 9 μM (■) and 40 μM (□) DAHCl solutions. After 1-2 days of immersion a small charge builds up on the surfaces in 40 μmolar solution (uppermost curve). Upper boundary of shaded area: interaction at constant surface charge density according to DLVO theory. Lower boundary: interaction at constant potential. From ref. [3].

For OAH+ ions, complete charge neutralization was never observed. The surface charge density even at the highest concentrations investigated (1 mM) was fairly low and the surfaces jumped into contact at distances larger than predicted by the DLVO theory. The distances between the surfaces in contact correspond to an adsorbed layer thickness on each surface in the region of 0.4 - 0.5 nm. Thus, the OAH+ ions form layers with a much looser packing than the DAH+. This is also evident from the adhesion between the layers (figure 3). For both surfactants, adhesion increases with increasing concentration, but the adhesion is considerably weaker between the OAH+ layers.

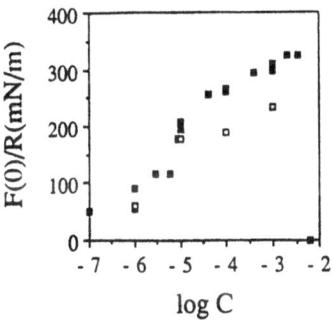

Figure 3. The adhesion between mica surfaces in different DAHCl (■) and OAHCl (□) solutions. The ordinate represents the force required to pull apart the surfaces in contact. From ref [3].

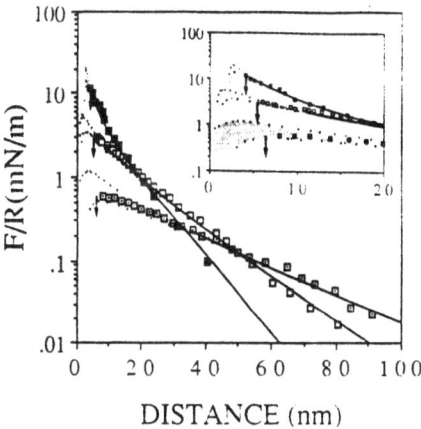

DISTANCE (nm)

Figure 4. Interactions between hydrophobed mica surfaces in DAHCl solutions. (▣) 10 μM, (□) 60 μM, (■) 100 μM solution. Arrows indicate jump into contact. Theoretical boundaries according to DLVO theory for interaction at constant charge and constant potential are also given. Inserted is a more detailed picture of the forces just before jump into contact. From ref [2]

3.2.2. Dodecylammonium ions on hydrophobed mica. Figure 4 shows the forces between hydrophobed mica immersed in DAHCl solutions. At all concentrations shown in the figure, the surfaces jump into contact at distances larger than predicted by the DLVO theory. However, as the DAH+ concentration increases, the long-range repulsive interactions increase. They are in good agreement with the DLVO theory, assuming an increasing charge density on the surfaces. At concentrations above 0.1 mM the surfaces no longer jump into adhesive contact. Instead a strong repulsive barrier is present which is located further out from the contact between the hydrophobed mica surfaces with increasing DAH⁺ concentration. In 6 mM solution the barrier occurs at a distance of 3.8 ± 0.3 nm. The extended length of a DAH⁺ ion is about 1.8 nm. Thus, when condensed surfactant monolayers perhaps also containing some hydration water are formed, a very strong repulsive force occurs at short distances [2].

Figure 5 shows the force needed to separate the surfaces after bringing them into adhesive conntact in different DAHCl solutions. Up to 5 μM solutions the adhesion is the same as for hydrophobed mica in pure water. However, as the DAH⁺ concantration increases, the adhesion begins to decrease. This indicates that some DAH⁺ ions are trapped between the surfaces, in accordance with the observed increase in the distance between the surfaces on contact.

In summary, DAH⁺ ions are adsorbed on both bare mica and on hydrophobed mica. At sufficiently high concentrations, on both surfaces, double layers of surfactants are formed, in the case of bare mica consisting entirely of DAH⁺, on hydrophobed mica of DAH⁺ on top of a DDOA layer. Between these double layers, short-range strongly repulsive interactions occur that prevent adhesive contact. The behaviour of the surfaces at lower DAH⁺ concentration is, not surprisingly, quite different. On bare mica, the DAH⁺ (and OAH⁺) ions are oriented with the hydrocarbon chain towards the watertso that adhesion between the surfaces increases with increasing surfactant concentration in soloution. For hydrophobed mica, on the other hand, adhesion is very strong in pure water and is decreased (finally to zero) as appreciable amounts of DAH⁺ are adsorbed.

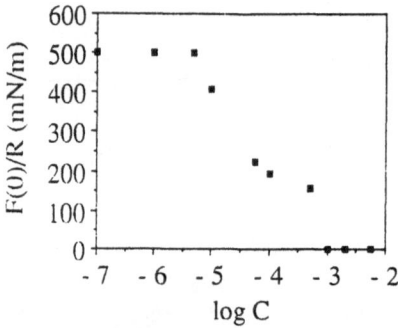

Figure 5. The adhesion between hydrophobed mica surfaces at different DAHCl concentrations. The ordinate represents the force required to pull apart the surfaces in contact. From ref [2].

3.3. DIMETHYLPHOSPHINE OXIDE

3.3.1. Forces between DDPO layers on mica. The short-range force-distance dependence for hydrophobed mica surfaces immersed in solutions of DDPO is shown in figure 6 [15]. At the lowest temperature investigated (18 °C) a very weak double layer force is observed at large distances. A weak attractive minimum is shifted to increasingly smaller separations and becomes deeper with increasing temperature. However, when subtracting the double layer repulsion observed at 18 °C the temperature dependence of the remaining force becomes rather small.

At even shorter distances a very steeply rising repulsive force occurs. The repulsion is so strong that on further compression the surfaces essentially behave as "hard walls". In the figure, zero distance is taken as the location of this wall, which is formed by a layer of adsorbed DDPO molecules on the hydrophobed mica. The thickness of the layer increases as temperature increases.

3.3.2. Comparison with phase diagram of DDPO. The change in the attractive minimum and in the adsorbed layer thickness may both be taken as indiciative of a decreased head group repulsion as temperature increases. The phase diagram for DDPO-water system includes a hexagonal phase at low temperatures (up to *ca.* 25 °C) and a lamellar phase (stable up to *ca.* 95 °C) [10, 11]. There is also a miscibility gap with a lower consolute boundary at 39 °C.

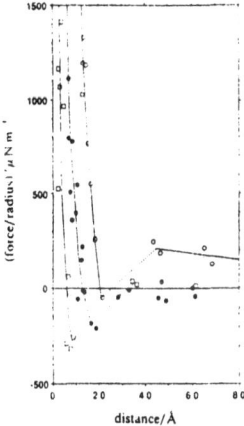

Figure 6. The forces between layers of dimethyldodecylphosphine oxide (DDPO) adsorbed on hydrophobed mica immersed in an 0.02 M aqueous DDPO solution. Zero separation is defined as the position of the surfaces under a load of $F/R = 10^5$ μN m^{-1}. From ref. [15].

The chemical composition of the interface between the aqueous solution and the adsorbed DDPO layer is similar to that of the interface between surfactant aggregates and water. At the same chemical potential, however, the two systems differ in curvature and very likely also in area per molecule and surfactant mobility. This also applies to the more mobile surfactant layers in surfactant-stabilized emulsions. Nevertheless, the forces in all systems are expected to be similar enough so that the surface-force measurements would provide information about the forces acting between surfactant aggregates as well as emulsion droplets. Indeed, the observation that the range of the repulsive force between DDPO layers, measured from the hard wall, decreases with temperature, agrees with the fact that the DDPO water solution phase separates on heating. The closer packing of the adsorbed layer as temperature increases is also in agreement with the higher temperature of existence of the lamellar phase than that of hexagonal phase.

3.4. DIMETHYLDODECYLAMINE OXIDE

Figure 7 shows the forces between mica sheets with adsorbed layers of DDAO [12]. This surfactant is partially protonated in water ($pK_a = 4.65$). This enables DDAO to adsorb onto a negatively charged mica surface in rather well-defined mono- and bilayers. At the concentration shown in figure 7, 4 mM, bilayers are adsorbed on each surface. A well-defined hard wall such as observed for DDPO on hydrophobed mica does not occur. However, if zero separation is defined as the position when the DDAO layers are compressed to the same load as the DDPO layers ($F/R = 10^5$ μN m^{-1}) the adsorbed layer thickness (2 nm) is found to be independent of temperature. A weak attractive minimum occurs outside the steeply rising repulsive part of the force/distance curve. This minimum is also temperature independent.

The phase diagram of the DDAO-water system [11] also indicates that the forces in this system are considerably less temperature-dependent than those operating in the other two systems. The lamellar phase is stable up to 170 °C, the hexagonal phase is stable up to 150 °C and no lower consolute temperature is observed.

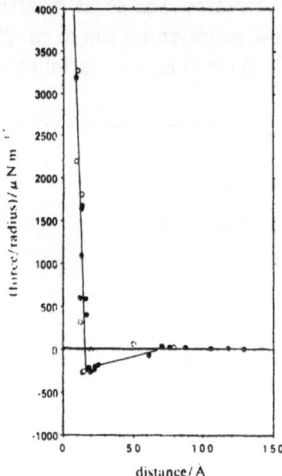

Figure 7. Forces between layers of dimethyldodecylamine oxide (DDAO) adsorbed on mica immersed in a 4 mM aqueous DDAO solution. Zero separation is defined as the position of the surfaces under a load of $F/R = 10^5$ μN m^{-1}. Open symbols: T = 14 °C, filled symbols: T = 27 °C. From ref [12].

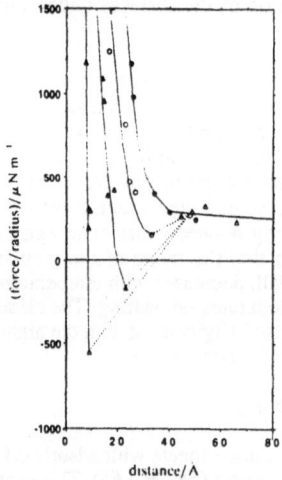

Figure 8. Forces between pentaoxyethene dodecyl ether (C$_{12}$E$_5$) layers adsorbed on hydrophobed mica. The concentration of C$_{12}$E$_5$ in solution is 60 μM dm^{-3}. Zero separation is defined as the position of the surfaces under a load of $F/R = 10^5$ μN m^{-1}. From bottom upwards the curves represent forces measured at 37 °C,. 30 °C, 20 °C and 15 °C. From [14].

3.5. PENTAOXYETHENE DODECYLETHER

3.5.1. Forces between C$_{12}$E$_5$ layers on hydrophobed mica. The adsorption of C$_{12}$E$_5$ onto bare mica does not lead to any detectable layered structure [13]. However, on hydrophobed mica a well-defined layer is formed. The interaction between two such layers is strongly temperature dependent, as shown in figure 8.

At a compressional load of $F/R = 10^5 \mu N\ m^{-1}$ the thickness of the adsorbed surfactant layer increases from 1.4 ± 0.4 nm at 20 °C to 2.6 ± 0.4 nm at 37 °C. These thicknesses do not change under a tenfold increase of the load, i.e. the adsorbed layer behaves like a hard wall .

Outside the hard wall, as shown in figure 8, a monotonically rising repulsive force occurs at 15 °C. At 20 °C a weak minimum is observed at a separation of $\approx$ 3 nm. On further increase of the temperature the minimum becomes deeper and it is shifted towards a smaller separation.

The temperature dependence of the thickness of the hard wall and the short-range interaction forces are both similar to the corresponding behaviour of DDPO. The increase in layer thickness implies that the adsorption increases and that the area per molecule decreases with increasing temperature, again indicating a reduced intralayer repulsion.

3.5.2. Comparison with phase equilibria and effect on emulsion stability. For $C_{12}E_5$, there is also a clear correspondence between forces observed in the surface forces apparatus and the phase equilibria of the $C_{12}E_5$ system. Figure 9 shows the phase diagram of this system, including determinations of the area per polar group in the lamellar phase. At low temperatures, this area increases strongly with increasing water content. When a water-rich lamellar phase is heated, the area per molecule is reduced, which is clear evidence for a reduced head-group repulsion at higher temperatures, just as was observed also in the surface force measurements. Additional evidence for this is also given by the presence of a miscibility gap with a lower consolute temperature at 26 °C.

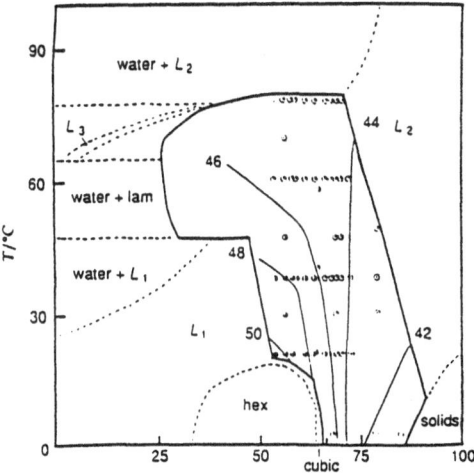

Figure 9. Phase diagram for the system pentaoxyetheneoxide dodecyl ether ($C_{12}E_5$)/water. The solid lines in the lamellar phase region connect points with the same area per molecule. The numbers at each line give the area in Å^2. From ref [16, 17].

The coincidence between the l.c.t. and the temperature at which the forces between the adsorbed $C_{12}E_5$ layers become attractive is probably fortuitious. The mobility and packing of the surfactant on the hydrophobed mica will certainly not be the same as in surfactant aggregates (micelles or lamellar phase). The actual transition temperature from stability to instability will depend on these factors and, in emulsions, also the degree of solvent penetration in the emulsifying layer. As an illustration, some studies of emulsion stability are presented in figure 10 [18]. This figure shows the flocculation rate as a function of temperature for a dilute emulsion stabilized by $C_{12}E_5$. The phase inversion temperature of this system is 48 °C. No dramatic changes in emulsion stability corresponding to this

278

temperature are observed and the system is stable at much higher temperatures than the l.c.t. of the binary system. However, the stability of the emulsion decreases with increasing temperature, which conforms with a decreasing repulsion between the adsorbed surfactant layers.

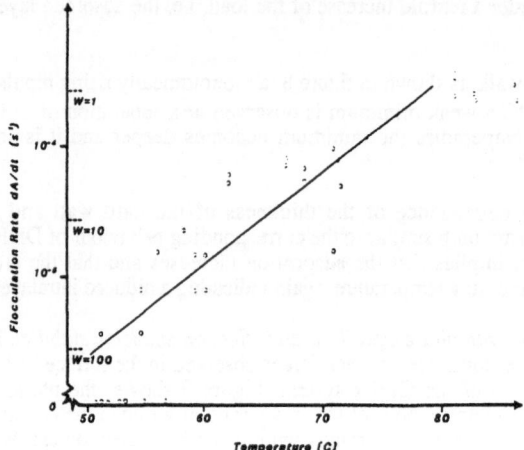

Figure 10. The flocculation rate of dilute tetradecane emulsions stabilized by $C_{12}E_5$. The flocculation rate was measured as the rate of change of emulsion turbidity. From ref. [18]

3.6. MONOGLYCERIDES

3.6.1. Forces between bilayers of monoglycerides. The swelling pressure between the surfactant layers in liquid crystalline phase formed by monoglycerides with water have been investigated by the "osmotic stress" method. In this method, the repulsive pressure between the layers is calculated from the vapour pressures of the water, determined either by head space chromatography or isopiestically. Repulsive pressures have been measured in the lamellar phase of monoolein and in the lamellar and gel phases for monopalmitin. It is generally found that the pressure decreases exponentially according to the equation

$$\Pi = \Pi_0 \exp(-d_w/\lambda)$$

where d_w is the water layer thickness between the surfactant layers. Some decay lengths (λ) and pressures at zero distances (Π_0) are given in table 1.

TABLE 1. Repulsive forces in liquid crystalline phases formed
by monoglycerides: Decay lengths and pressures at zero distances [7]

Compound	λ nm	Π_0 N/m^2
Monopalmitin, lamellar (60 °C) [7]	0.24	$1.1 \cdot 10^8$
Monopalmitin, gel (23 °C) [7]	?	$0.26 \cdot 10^8$
Monoolein, lamellar (30 °C) [7]	0.18	$1.4 \cdot 10^8$
Monoolein, lamellar (23 °C) [7]	0.20	$1.3 \cdot 10^8$
Monocaprylin, lamellar (20 °C) [19]	0.13	$1.8 \cdot 10^8$
Monoelaidin, gel (40 °C) [19]	0.13	$5.1 \cdot 10^8$
1-O-decylglycerol, lamellar (20 °C) [20]	0.22	$1.2 \cdot 10^8$

These parameters can be compared as characteristic parameters of the force between the layers. It is striking that the type of monoglyceride or the temperature has relatively little influence on the repulsive pressure. Moreover, the ether analogue of a monoglyceride, 1-

O-decylglycerol, behaves in the same way [20]. The decay length is also about the same as for phospholipids [21] and is of the same order of magnitude as the size of a water molecule.

For both phospholipids (dipalmitoyl phosphatidylcholine) [21] and monopalmitin the lamellar phase swells more than the gel phase. For monopalmitin, the pressure at zero distance is much less in the gel phase than in the lamellar phase.

3.6.2. Forces between adsorbed layers of monoglycerides. The swelling pressures have been directly compared to the interaction force between monoolein and monopalmitin layers measured by the surface force method. For details of the data treatment required to facilitate this comparison, see ref. [7]. The results are shown in figure 11.

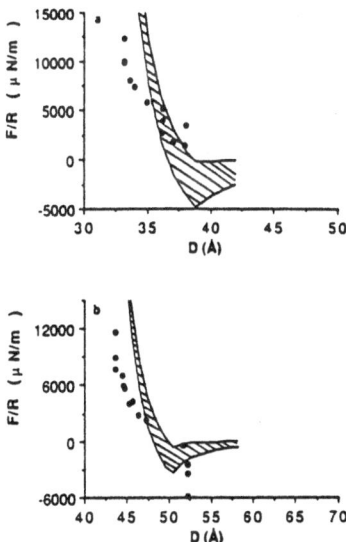

Figure 11. Surface forces between (a) monoolein and (b) monopalmitin (•)[22] layers on hydrophobed mica and the corresponding interaction energies per unit area measured by the osmotic pressure technique [21]. The distance D is the water layer thickness $d_w + 44$ Å for monopalmitin and $d_w + 33$ Å for monoolein.

The two methods give qualitatively similar results both for monopalmitin and monoolein. However, the repulsion observed in the surface forces measurements is weaker than obtained in measurements of osmotic pressures. Two explanations have been advanced [7]: (i) a depletion of molecules in the contact zone in the surface forces apparatus and (b) at these small separations repulsive fluctuation forces may contribute to the pressure between bilayers in the lamellar or gel phases but not between adsorbed layers. The importance of such fluctuation forces for free bilayers has recently been discussed by Israelachvili and Wennerström [23].

3.7. CONCLUSIONS

For all the surfactant bilayers investigated (ionic, nonionic, partially ionizable), a short-range repulsive interaction ("hydration force") is observed between adsorbed surfactant layers on mica, whether the mica has been previously hydrophobed or not. The force occurs provided the layers are sufficiently well anchored to the surface so that surfactants are not desorbed or able to diffuse away from the contact zone as the surfaces come close to each other. Outside this repulsion there is in many cases a weak attractive minimum which is consistent with the expected van der Waals forces in the system. The temperature dependence of the repulsive force is strongly dependent on the nature of the polar end group. It can be can be qualitatively correlated with the phase behaviour of the corresponding surfactant/water systems. A quantitative correspondence is not expected in view of the higher mobility of surfactant layers in free aggregates.

Between hydrophobed mica surfaces in pure water and also between hydrophobed surfaces immersed in surfactant solutions at concentrations which are too low for the formation of closely packed surfactant monolayers a long-range attractive force of considerably larger strength than the expected van der Waals force ("hydrophobic force") occurs. The strength of the hydrophobic force decreases rapidly as the area per surfactant molecule decreases. It is noteworthy that it can be observed (in very dilute solutions) even when a long-chain surfactant is adsorbed from solution directly on bare (not hydrophobed) mica.

The origin of both the hydration and hydrophobic force is obviously a fascinating topic for further theoretical and experimental research.

Very few correlations of measured surface forces with the effects of the surfactants as emulsifiers have been reported. For C12E5 there is a qualitative correspondence between the temperature dependence of the stability of a o/w emulsion stabilized by this surfactant and the temperature dependence of the forces observed in the surface forces apparatus, but the influence of the oil phase cannot be directly simulated in the surface force measurements.

The hydration forces are probably usually very short-range and hence they are expected to play an important role as factors governing coalescence. This is another topic for further research.

4. Acknowledgements

The results presented in this paper were produced by the skilful assistance of students and co-workers at the YKI-KTH Surface Forces Group in Stockholm and at the Department of Applied Mathematics, ANU, Canberra: Johan Berg, Hugo Christenson, Christina and Peter Herder, Paul Persson, Isabelle and Erwoan Pezron.

5. References

1. Claesson. P.M., Blom, C.E., Herder, P.C. and Ninham, B.W. (1986) 'Interactions between Waterstable Hydrophobic Langmuir-Blodgett Monolayers on Mica", J. Colloid Interface Sci. 114, 234-242.
2. Herder, P.C.(1990) 'Forces between Hydrophobed Mica Surfaces Immersed in Dodecyl-ammonium Chloride Solution', J. Colloid Interface Sci. 134, 336-345.
3. Herder, P.C.(1990) 'Forces between Mica Surfaces Immersed in Dodecyl- and Octylammonium Chloride Solutions', J. Colloid Interface Sci. 134, 346-356.
4. Claesson. P.M., Herder, P.C., Berg, J.M. and Christenson, H.K. (1990) 'The State of Fluorocarbon Surfactant Monolayers at the air-Water Interface and on Mica Surfaces', J. Colloid Interface Sci. 136, 541-551.
5. Israelachvili, J.N. and Adams, G.E, (1978) 'Measurements of Forces between Two Mica Surfaces in Aqueous Electrolyte Solutions in the Range 0 - 100 nm', J. Chem. Soc. Farad. Trans. I, 74, 975-1001.
6. Persson, P.K.T. and Bergenståhl, B. A. (1985) 'Repulsive Forces in Lecithin Glycol Lamellar Phases', Biophys. J. 47, 743-746
7. Pezron, I., Pezron, E., Bergenståhl, B.A. and Claesson, P.M.(1990) 'Repulsive Pressure between Monoglyceride Bilayers in the Lamellar and Gel States', J. Phys. Chem., 94, 8255-8261.
8. Herder, P.C., Claesson, P.M. and Herder, C.E. (1987) 'Adsorption of Cationic Surfactants on Muscovite Mica as Quantified by Means of ESCA', J. Colloid Interface Sci. 119, 155-167.

9. Israelachvili, J.N. and Pashley, R.M. (1984) 'Measurement of the Hydrophobic Interaction between Two Hydrophobic Surfaces in Aqueous Electrolyte Solutions', J. Colloid Interface Sci. 98, 500-574.
10. Lutton, E.S. (1966)J. Am. Oil. Chemists Soc., 43, 28
11. Laughlin, R.G. (1978) 'Solvation and Structural Requirements osf Surfactant Hydrophilic Groups', Adv. Liq. Cryst. 3, 41-98.
12. Herder, C. E., Claesson, P.M. and Herder, P. C. (1989) 'Interaction between Amine Oxide Surfactant Layers Adsorbed on Mica', J. Chem. Soc. Farad. Trans. 85, 1933-1943.
13. Rutland, M.W. and Christenson, H. (1990) 'Effect of Nonionic Surfactant on Ion Adsorption and Hydration Forces', Langmuir 6, 1083-1087.
14. Claesson, P.M., Kjellander, R., Stenius, P. and Christenson, H.K., (1986) 'Direct measurements of Temperature-Dependent Interactions between Nonionic Surfactant Layers ', J Chem. Soc. Farad. Trans. I, 82, 2735-2746.
15. Herder, C.E. (1991) 'Adsorption of Dimethyldodecylphosphineoxide on Mica and the Interaction between the Adsorbed Layers', J. Colloid Interface Sci. 143, 573-580.
16. Mitchell, D.J., Tiddy, G.J.T., Waring, T., Bostock, T. and McDonald, M.P. (1983) 'Phase Behaviour of Polyoxyethylene Surfactants with Water', J. Chem. Soc. Farad. Trans. I, 79, 975-1000.
17. Claesson. P.M., Eriksson, J.C., Herder, C. E., Bergenståhl, B.A., Pezron, E., Pezron, I. and Stenius, P. (1990) 'Forces between Nonionic Surfactant Layers', Farad. Disc. Chem. Soc. 90, 129-142
18. Bergenståhl, B.A. and Claesson, P. M.(1990) 'Surface Forces in Emulsions', in K. Larsson and S.E. Friberg (eds.) Food Emulsions, 2. ed., Marcel Dekker, New York, 41-95.
19. McIntosh, T.J., Magid A.D. and Simons, S.D. (1989) 'Repulsive Interactions between Uncharged Builayers. Hydration and Fluctuation pressures for Monoglycerides', Biophys. J. 55, 897-904
20. Bergenståhl, B. and Persson, P. Institute for Surface Chemistry, Stockholm, unpublished results.
21. Lis, L.J., McAlister, M., Fuller, N. Rand, P.R.P. and Parsegian, V.A. (1982) 'Interactions between Neutral Phospholipid Bilayer Membranes', Biophys. J. 37, 657.
22. Evans, E. and Needham, D. (1987) Physical Properties of Syrfactant Bilayer Membranes: Thermal Transitions, Elasticity, Rigidity, Cohesion and Colloidal Interactions', J. Phys. Chem. 91, 4219-4228.
23. Pezron, I., Pezron, E., Claesson, P.M. and Bergenståhl, B.A. (1991) 'Monoglyceride Surface Films: Stability and Interlayer Interactions', J. Colloid Interface Sci. 144, 449-457.
24. Israelachvili, J.N. and Wennerström, H. (1990) 'Hydration or Steric Forces between Amphiphilic Surfaces?', Langmuir 6, 873-676

DESTABILIZATION OF WATER-IN-OIL EMULSIONS

D.T. WASAN
Department of Chemical Engineering
Illinois Institute of Technology
Chicago, Illinois 60616
U.S.A.

ABSTRACT. In the emulsion coalescence process in demulsification, the approach of two droplets under the capillary pressure acting normal to the interface, causes liquid to be squeezed out of the film into the bulk. The rate of drainage of the film is strongly dependent upon adsorption kinetics of the demulsifier, its interfacial activity and solubility, and interfacial rheological properties such as interfacial tension gradient or elasticity, and interfacial viscosity. The predictions of our recent theoretical model for the drainage of film in the drop-drop coalescence process are reviewed which indicates that the interfacial activity of the demulsifier, in addition to its partitioning ability, must be high enough to suppress the interfacial tension gradient. This accelerates the rate of film drainage, thus promoting coalescence. Recent experimental data on demulsification of water-in-oil emulsions using oil soluble demulsifiers are presented which verify the predictions of the theoretical model.

1. Introduction

Methods currently available for demulsification can be broadly classified as chemical, electrical and mechanical. Chemical demulsification is the most widely applied method of treating crude oil emulsions. It involves the use of chemical additives to accelerate the emulsion breaking process.

The breaking of a petroleum emulsion is of considerable importance and a necessity for several reasons. Obviously, the quality of the crude oil is highly dependent on residual contents of water and water-soluble contaminants present. Even small amounts of these components can cause unwanted effects in pipelines and in refineries. The formulation of the optimum demulsifier for a specific petroleum emulsion is a complicated undertaking. In petroleum systems, asphaltenes and resinous substances comprise a major portion of the interfacially active components of the oil (1,2). They are large polyaromatic and polycyclic condensed ring compounds containing heteroatoms such as N, S and O. Chemically, they represent the pentane or hexane insoluble portion of the oil.

The addition of polyfunctional demulsifying agents to petroleum emulsions serve the same purpose as polymeric electrolytes in aqueous systems. These compounds are adsorbed at the interfaces of droplets and promote coalescence. It is well to remember,

283

J. Sjöblom (ed.), Emulsions – A Fundamental and Practical Approach, 283–295.
© 1992 *Kluwer Academic Publishers.*

however, that the actual demulsifier polymers responsible for emulsion destabilization are usually not very soluble in either the oil or water phase. We are thus faced with the problem of delivering them to the interface. To achieve this, the demulsifier is usually dissolved in a solvent which is miscible with oil. It is then necessary to add the demulsifier solution and mix it thoroughly with the emulsion before the oil phase has a chance to dilute the solvent and cause the demulsifier to sequestrate. Thus, in the use of any demulsifier, it is necessary to be sure that the treatment chemical is added in such a way that local excesses are avoided so that the demulsifier can contact the individual droplets. It is also necessary to allow time for the demulsifying agent to enter the surfactant film and affect its interfacial properties. Additional time is required for the droplets to contact, coalesce and settle.

Demulsifying agents are most always preferentially oil-soluble blends consisting of high molecular weight polymers (3). These blends are expected to contain four classes of components (a) flocculants - which are typically large slow-acting polyethers (b) coalescers - which are typically lower molecular weight polyethers, (c) wetting agents, and (d) solvent/cosolvent mixtures which aid dispersion of the additives into process streams. Such demulsifiers are mixtures of blends of polymerized alkoxylated polyglycols, ethylene oxide/propylene oxide block copolymers, polyglycolesters, polymerized oils or alkanolamine condensates (e.g., oxyalkylated phenols and polyamines.)

The process of demulsification, itself, is complex and cannot be thought as the reverse of emulsification (4). It involves settling, flocculation and coalescence of drops and it requires disrupting the stabilized layers of the interface.

According to Bancroft (5), the stability of any emulsion is largely due to the nature of the interfacial film that is formed. The stability of this film is strongly dependent upon the surfactant adsorption/desorption kinetics, solubility and interfacial rheological properties such as interfacial tension gradient or elasticity and interfacial viscosity. In what follows, we will first review the predictions of our comprehensive theoretical model for the drainage of emulsion films associated with drop-drop coalescence. The model accounts for flows in plane-parallel film and dispersed phase, the partitioning of surfactants, interfacial viscosities, interfacial elasticity and mass transfer involving both the bulk and interfacial diffusion. The predictions of this model are then compared with our recent experiments on demulsification of water-in-oil emulsions using oil-soluble demulsifiers.

2. A Model for Film Drainage

The overall coalescence process in demulsification can be conveniently divided into 1) movement of two single (non-interacting) droplets, 2) deformation of mutual approaching droplets and formation of a plane-parallel film (Figure 1), and 3) thinning of that film to a critical thickness at which the film becomes unstable, ruptures, and the two droplets unify to form a single larger droplet.

The approach of two droplets under the capillary pressure acting normal to the interface causes liquid to be squeezed out of the film into the bulk. This liquid flow results in the convective flux of surfactant in the sublayer (Figure 2). Therefore, the surfactant concentration at the interface is increased in the direction of that flow. The other fluxes associated with the drainage process shown in Figure 2 include: 1) bulk flux in the droplet, 2) bulk flux in the film phase, and 3) interfacial diffusion flux caused by the concentration gradient at the interface. The bulk fluxes can be conveniently divided into two subsequent step: 1) diffusional flux up to a layer adjacent to the film interface, and 2) adsorption flux from this layer onto the interface.

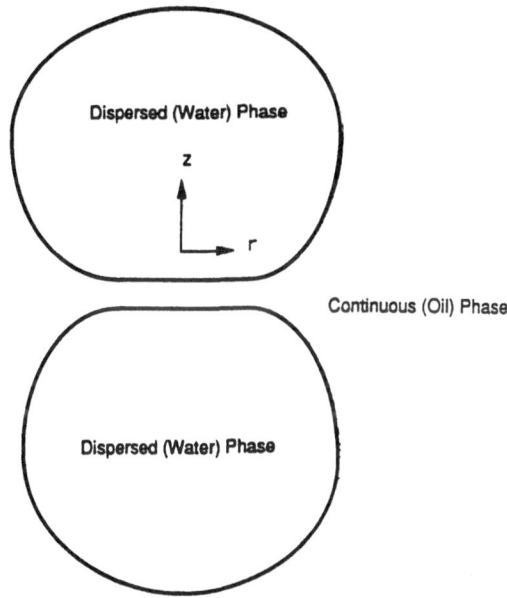

Figure 1. Mutual approach of two droplets and subsequent formation of plane-parallel film.

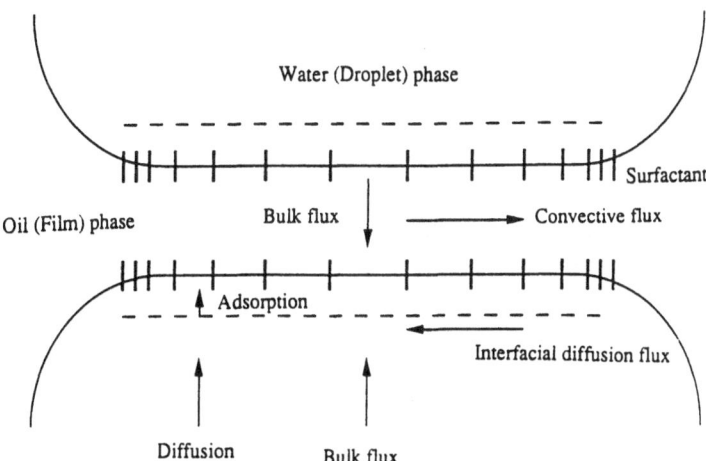

Figure 2. Surfactant mass balance at the film interface.

The tangential stress balance at the film interface is shown in Figure 3. The non-uniform surfactant distribution leads to interfacial flow which in turn gives rise to interfacial stresses. The difference in concentration along the interface results in difference of the local values of interfacial tension which produce a force (equal per unit length to the gradient of interfacial tension) opposite to liquid flow (Marangoni-Gibbs effect). In addition, the surfactant monolayer may undergo dilating and shearing deformation which also produces interfacial stresses. The sum of the above stresses and the tangential bulk stress from the liquid in the droplet must counterbalance the tangential bulk stress from the film liquid which causes interfacial flow (6).

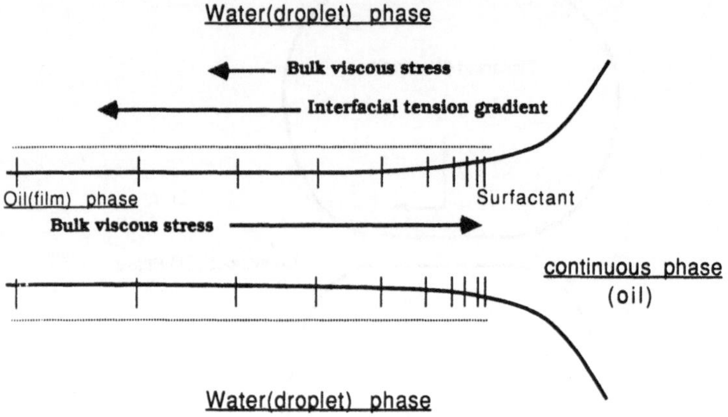

Figure 3. Tangential stress balance at the film interface.

Reynolds (7) was the first to study the rate of approach between surfaces separated by a draining thin film. His analysis assumed that the two surfaces were both flat and rigid. His result of the rate of drainage is given by:

$$V_{RE} = -\frac{dh}{dt} = \frac{8F\lambda^3}{3\Pi\mu R^4} \tag{1}$$

As pointed out by many researchers (8), Reynolds' equation (1) represents a most conservative prediction. Due to the mobility of the interfaces, the rate of film thinning and thus the coalescence rate, can be several times greater than predicted by this equation.

More recently, we have developed a generalized model which accounts for the effect of the mobility of the interfaces on film thinning phenomena by considering the kinetics of adsorption-desorption of surfactants, partitioning of surfactants, interfacial and bulk diffusion, interfacial rheological properties, and flow in both film and bulk phases. The interfacial mobility is given by us (6) as follows:

$$\frac{V}{V_{RE}} = 1 - \frac{3\Pi}{h^2 F} \sum_{n=1}^{N} A_n \frac{e^{-\lambda_n h}}{1 + \lambda_n h} J_0(\lambda_n) \tag{2}$$

In Figure 4, the effect of the interfacial tension gradient upon interfacial mobility is shown in terms of the dimensionless elasticity number E_s. The effect of interfacial tension gradient on the film drainage time is depicted in Figure 5. At high values of tension gradient (i.e., high E_s), bulk and interfacial diffusion cannot counterbalance the interfacial tension gradient (the Marangoni-Gibbs' effect), and, hence, the velocity of thinning (or the drainage time) is essentially given by the Reynolds' equation. However,

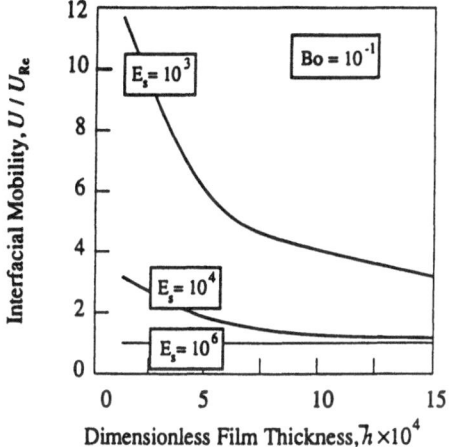

Figure 4. Interfacial Mobility, or dimensionless drainage velocity, versus dimensionless film thickness, at three values of the dimensionless interfacial elasticity (Ref. 8).

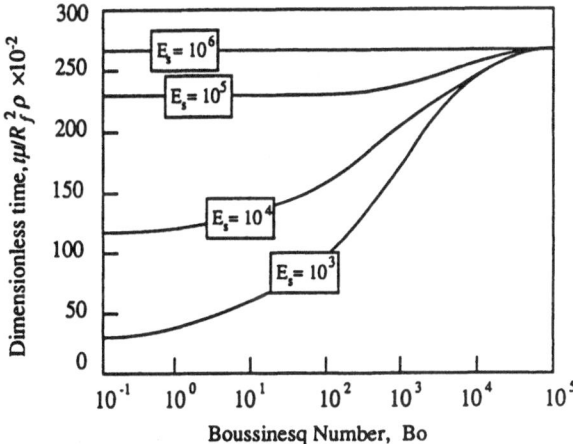

Figure 5. Dimensionless drainage time for the film to drain from a dimensionless thickness $\bar{h}_i$, to the thickness $\bar{h}_f$, versus Boussinesq Number, at various values of the dimensionless interfacial elasticity (Ref. 8).

for smaller values of E_s, even at a moderate interfacial viscosity (i.e., moderate Boussinesq Number), the thinning or approach velocity is several times greater than Reynolds' velocity. An increase in interfacial viscosity results in decreased interfacial mobility and, hence, higher drainage time. Thus, the thin film drainage model predicts that at low values of interfacial viscosity (i.e., Boussinesq Number < 10), the Marangoni-Gibbs' effect will impart the more significant influence upon film drainage and, thereby, on the droplet coalescence rate (6,9). Therefore, these theoretical findings clearly suggest that in the emulsion coalescence process in demulsification, not only the interfacial viscosity must be sufficiently low but, also, the interfacial tension gradient must be minimized.

288

Two factors that significantly influence the magnitude of interfacial tension gradients in the thin film are surface diffusion and surfactant adsorption. In Figure 6, drainage time is plotted versus the dimensionless interfacial viscosity (or the Boussinesq Number), varying the surface diffusivity number. As the number increases, corresponding to increased surface diffusion, the drainage time diminishes, which is evidence of the fact that the film surfaces become more mobile. The cause for this effect may be understood by reference to Figure 2. If a large surface diffusion occurs in the film surface, the surfactant may diffuse back into the film, diminishing the surface concentration gradient and, thereby, canceling the Marangoni effect.

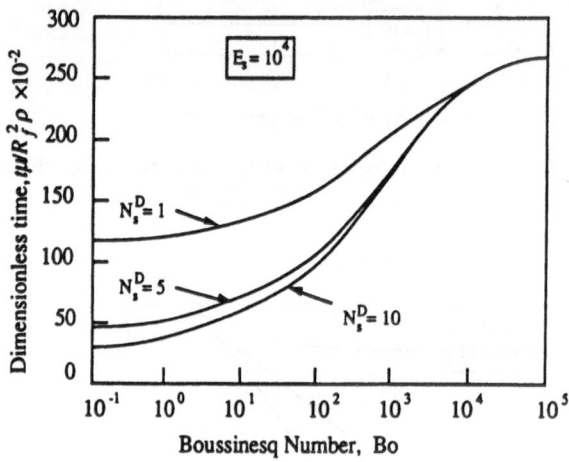

Figure 6. Dimensionless drainage time for the film to drain from a dimensionless thickness h_i, to the thickness h_f, versus Boussinesq Number at various values of the surface diffusivity number (Ref. 8).

A similar effect may be achieved by increasing the rate of surfactant adsorption to the film surfaces from the dispersed phase (the influence of adsorption from the film phase is negligible, as the film-phase surfactant transfer process is diffusion controlled). The adsorption effect is illustrated in Figure 7, which reveals that an increased adsorption rate results in a diminished time of drainage, again owing to a cancellation of Marangoni flow.

Figures 8 and 9 compare the effect of selective surfactant solubility on interfacial mobility and drainage time, respectively, for the cases where 1) surfactant is soluble only in the film phase, and 2) surfactant is soluble only in the drop phase. It is observed that solubilized surfactant in the dispersed phase is most effective in destabilizing droplets with the consequence that emulsions with surfactant soluble in the droplet phase are generally less stable than emulsions with surfactant soluble in the continuous phase, all other things being equal. This conclusion, which is clearly illustrated in Figure 9, is known as "Bancroft's Rule". The higher interfacial mobility (Figure 8) and lower drainage times (Figure 9) are obtained for the case where surfactant is soluble in the droplet phase, while lower interfacial mobility and higher drainage times are obtained when surfactant is soluble only in the film (continuous) phase.

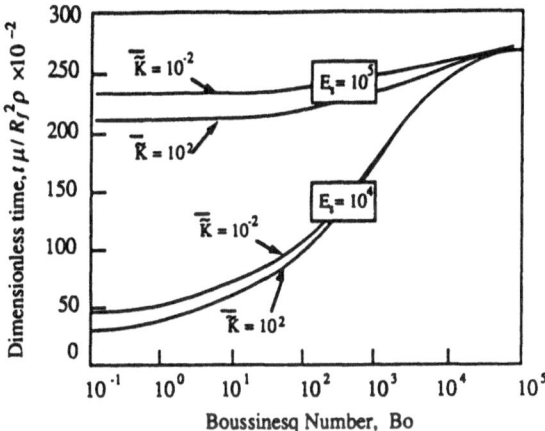

Figure 7. Dimensionless drainage time for the film to drain from a dimensionless thickness h_i, to the thickness h_f, versus dimensionless interfacial viscosity, at various values of the rate of surfactant adsorption (Ref. 8).

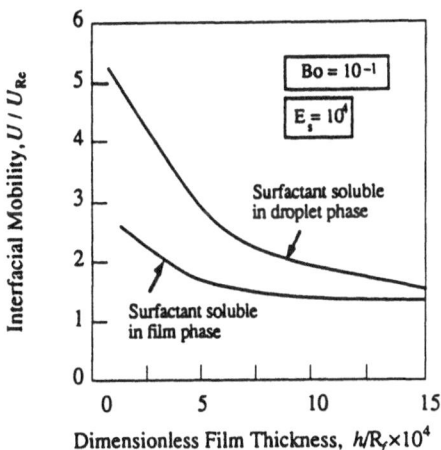

Figure 8. Interfacial mobility versus dimensionless film thickness, for surfactant soluble in film and droplet phases only (Ref. 8).

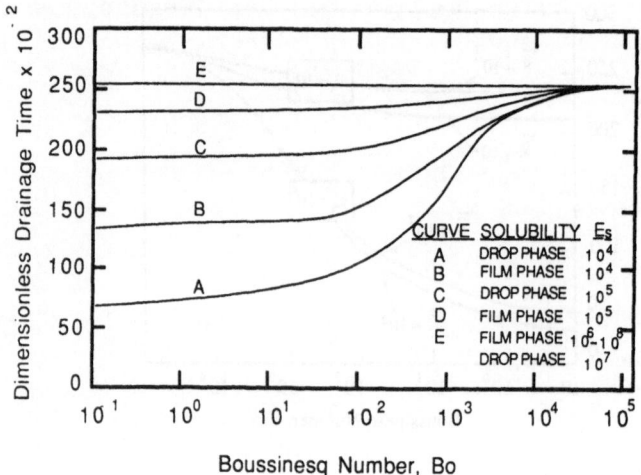

Figure 9. Dimensionless drainage time versus Boussinesq Number for surfactant soluble in film and droplet phases.

In summary, our model (6,9) for a draining thin film in the emulsion/coalescence process predicts that good demulsifiers should display high interfacial activity, i.e., increased surface diffusion and increased adsorption rate and should be able to partition into the dispersed phase, in order to minimize the magnitude of Marangoni stresses.

3. Experimental

Bottle stability tests for a 30% water-in-oil (w/o) emulsion were conducted in the crude oil and model systems. The categories of demulsifier studied were alkoxylated polyol, alkoxylated resin, ethoxylated polyol, blend of phenolic resins, dispersed mixtures of alkylaryl sulfonate, polyamines and EO/PO co-polymers (10,11). These demulsifiers were chosen because of the differences in their effectiveness and chemical structure. A heavy crude oil from California was used in the present studies. It had a viscosity of 2,000 cp at $25^{o}C$, density of 0.78 g/cc and asphaltene content of 3.2 wt% (11). The crude oil was centrifuged to remove all the finely suspended particles naturally present in the oil. The model system consisted of the addition of asphaltenes derived from the crude oil to an alkane-aromatic solution of 70% heptane and 30% toluene. This solution containing asphaltenes was called heptol.

In addition to the bottle tests for determining the effectiveness of emulsion breakers, coalescence of drops in demulsification experiments was also characterized by measuring the drop size distribution as a function of time using the photomicrographic method. Experiments were also conducted to measure both the static and dynamic interfacial tensions, interfacial activity and adsorption of emulsion breakers, their partition coefficients and interfacial shear viscosities at oil-water interfaces. The details of the measurements for some of the systems studied by us are again given in our recent paper (10). These properties of demulsifiers were measured to verify the results of the theoretical findings as discussed above under Section 2.

Figure 10 shows the results of our most recent photomicrograph observations (11) for both the flocculation and coalescence kinetics for water-in-crude oil emulsions using a series of different demulsifiers which are EO/PO copolymers (e.g., RE-1747, RE-1748, RE-1250 and RE-1751) and ethoxylated resins (e.g., RE-1756). Figure 10 shows a plot of the normalized number of water droplets (N/N_0) as a function of time (t) for these various emulsion breakers (EB). The concentration of each of the demulsifiers used was 100 ppm.

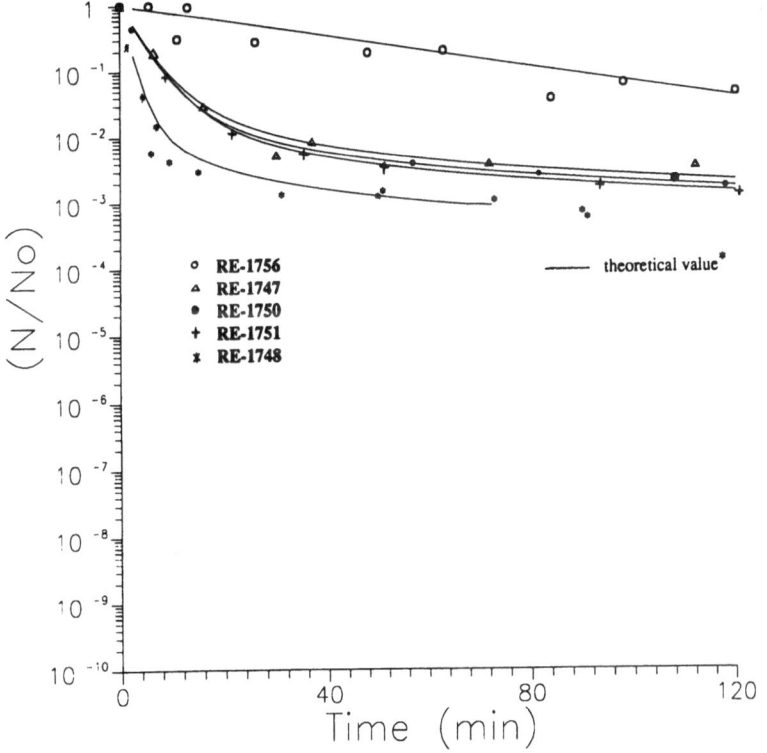

Figure 10. Dimensionless droplet number versus time for various demulsifiers in the water-in-crude oil emulsion.

The coalescence rate coefficient (K) and the flocculation coefficient (a) were calculated using our recent theoretical model for emulsion stability (12). This plot shows that, initially, the demulsification process is flocculation rate controlled, however, with the passage of time, the process becomes coalescence rate controlled (See Table 1).

TABLE 1. Coalescence rate constant and flocculation coefficient for various demulsifiers in water-in-crude oil emulsions.

EB	1748	1750	1747	1751	1756
K	0.746	0.306	0.298	0.320	0.003
$a \times 10^8$	0.765	0.925	1.423	1.974	8.580

Furthermore, the bottle tests for these systems confirmed the one-to-one correlation between these coalescence constants (K) and the demulsifier effectiveness. That is to say, that the best emulsion breaker (RE-1748) had the highest coalescence rate constant.

Table 2 lists the coalescence rate constants and the interfacial activities for the four emulsion breakers. The interfacial activity was determined from the slope of the interfacial tension isotherms. In this case, the crude oil containing emulsion breaker was pre-equilibrated with water for 3-4 days. The equilibrated water was first separated from the crude oil and it was further contacted with "pure" crude oil. The static interfacial tension of this pre-equilibrated water against pure oil was measured as a function of the water dilution ratio. This procedure of measuring interfacial activity is different from that used by us previously (10). It is noted that there is a perfect one-to-one correlation between the interfacial activity (δi) of the emulsion breaker, the coalescence rate constant and its performance. The dynamic interfacial activity was also measured using the maximum droplet pressure method (13) for the best and worst emulsion breaker and it was found that the dynamic interfacial activity was also highest for the best emulsion breaker, RE-1748, and lowest for the worst emulsion breaker, RE-1756.

TABLE 2. Coalescence rate constants and interfacial activities for various demulsifiers in water-in-crude oil emulsions.

demulsifier	RE-1748	RE-1753	RE-1868	RE-1756
δi	1.97	1.05	0.63	0.15
coalescence rate constant (K)	0.74566	0.56465	0.31643	0.00277
performance	1	2	3	4

In addition to measuring the coalescence rate constants and interfacial activity, partition coefficients were also determined for various demulsification systems. In these experiments, the model oil containing asphaltenes ($1g/L$) in heptol was used. Table 3 lists the partition coefficients (Kp) for a series of emulsion breakers which are dispersed mixtures of alkalaryl sulfonate, phenolic resins and polyamines. The partition coefficient is defined as a ratio of the concentration of demulsifier in the water phase to that in the oil phase. In this set of experiments, a model oil containing emulsion breaker was contacted with fresh water and the pre-equilibrated oil was then separated and contacted with fresh water. It is seen that the best emulsion breaker (RP-4011) also has the highest partition coefficient. Indeed, this emulsion breaker also has the highest interfacial activity and there is one-to-one correlation between the emulsion breaker performance and the partition coefficient and the interfacial activity.

TABLE 3. Partition coefficients and interfacial activities for various emulsions in water-in-model oil emulsions.

demulsifier	RE-4011	R-77	RP-2327	RP-0484	no EB
K_ρ	0.40	0.19	0.19	0.16	0.0
δ_i	2.11	1.03	0.65	0.55	0.05
order of performance	1 (0.5 min)	2 (2 min)	3 (10 min)	4 (days)	5

We also measured the interfacial shear viscosity for the best (RP-4011) and the worst (RP-0484) emulsion breaker in the model oil system using our viscous traction instrument (14) and found that the interfacial viscosity was much higher for the worst demulsifier than for the best one.

Results for other emulsion breakers mentioned above and other oil systems have been previously presented by us in our recent publications (10). These data also support the observations discussed above.

4. Conclusions

Experimental results reported previously by us (10) and others (15), as well as our most recent observations on emulsion coalescence processes in demulsification using different types of oil soluble emulsion breakers for both water-in-crude oil and model oil emulsions clearly lead us to conclude that there is a one-to-one correlation between the performance of an emulsion breaker and its interfacial activity, and partition coefficient in the drop phase. Therefore, good demulsifiers should possess high interfacial activity, high rates of diffusion/adsorption and should be able to partition effectively into the water phase in water-in-oil emulsions. These emulsion breaker properties will assure that the interfacial tension gradients are suppressed, thus minimizing the Marangoni-Gibbs' effect. Our experimental findings to date are consistent with the predictions of the theoretical model for a draining thin film in the emulsion/coalescence process as presented in this paper.

5. Acknowledgement

Financial support was provided by the National Science Foundation and U.S. Department of Energy. Emulsion breakers and oils were supplied by Baker Performance Co. and Petrolite Corp. Special thanks are due to Young Ho Kim, my graduate student, for carrying out the experimental work summarized in this paper.

6. Notation

A_d = Adsorption Number = $\dfrac{\rho^s}{R\rho_0}$

A_n = unknown constants in equation [2]

BO = Boussinesq number = $\dfrac{\mu^s}{\mu a}$

D = bulk diffusion coefficient

E_s = Elasticity number = $E_o\dfrac{R_f}{\mu D}$

E_0 = Gibbs elasticity = $\left(\dfrac{\partial \sigma}{\partial \ln \rho^s}\right)_0$

F = force

h = half film thickness

J_o = Bessol function of order zero

λ_n = nth root of $J(\lambda) = 0$

μ = film viscosity

μ^s = surface shear viscosity

N_s^D = Surface diffusivity number = $\dfrac{D^s}{D}$

$R_f = R$ = film radius

ρ^s = surface concentration

ρ_0 = bulk concentration

σ = Surface tension

t = time

V = velocity

V_{RE} = film drainage velocity given by the Reynold's equation

7. References

1. Nordli, K.G., Sjoblom, J., Kizing, J. and Stenius, P., (1991) "Water-In-Oil Emulsions from the Norwegian Continental Shelf," *Colloids and Surfaces*, 57, 83.

2. Sjoblom, J., Söderland, H., Johansen, E.J., Skyarvo, I.M., (1991) "Water-In-Crude Oil Emulsions from the Norwegian Continental Shelf." *Colloids and Surfaces*.

3. Jones, T.J., Neustadter, E.L. and Whittington, K.P., (1978) "Water-In-Crude Oil Emulsion Stability and Emulsion Destabilization by Chemical Demulsifiers," *J. Canadian Pet. Tech.*, 101.

4. Menon, V.B. and Wasan, D.T., (1985) "Demulsification", in P. Becher (ed.) <u>Encyclopedia of Emulsion Tech.</u>, 2, 1.

5. Bancroft, W.D., (1913) "Theory of Emulsification," V, *J. Phys. Chem.*, 17, 501.

6. Malhotra, A.K. and Wasan, D.T., (1987) "Effects of Surfactant Adsorption Description Kinetics and Interfacial Rheological Properties on the Rate of Drainage of Foam and Emulsion Films," *Chem. Eng. Comm.*, 55, 95.

7. Reynolds, O., (1886) "On the Theory of Lubrication and Its Application to Mr. Beauchamp Tower's Experiments, Including an Experimental Determination of the Viscosity of Olive Oil, *Phil. Trans, Roy. Soc.*, (London), A177, 157.

8. Edwards, D., Brenner, H. and Wasan, D.T., (1991) <u>Interfacial Transport Processes and Rheology</u>, Butterworth-Heinemann Publishing, Stoneham, Massachusetts, pp. 295-299.

9. Zapryanov, Z., Malhotra, A.K., Aderangi, N. and Wasan, D.T., (1983) "Emulsion Stability: An Analysis of the Effects of Bulk and Interfacial Properties on Film Mobility and Drainage Rate," *Int. J. Multiphase Flow*, 9, 105.

10. Krawczyk, M.A., Wasan, D.T. and Shetty, C.S., (1991) "Chemical Demulsification of Petroleum Emulsions Using Oil-Soluble Demulsifiers," *Ind. Eng. Chem.*, 30, 367.

11. Kim, Young-Ho, Mechanisms of Demulsification, (1991) Ph.D. thesis in progress, Illinois Institute of Technology, Chicago, Illinois.

12. Borwankar, R.P., Lobo, L.A. and D.T. Wasan, (1991) "Emulsion Stability - Kinetics of Flocculation and Coalescence," *Colloids and Surfaces*, (submitted for publication).

13. Kao, R.L., Edwards, D.A., Wasan, D.T. and Chen, E., (1991) "Measurement of Dynamic Interfacial Tension and Interfacial Dilatational Viscosity at High Rates of Interfacial Expansion Using the Maximum Bubble Pressure Method. Part II. Liquid-Liquid Interface," *J. Colloid Interface Sci.*, (in press).

14. Wasan, D.T., Gupta, L., and Vora, M., (1971) "Interfacial Shear Viscosity at Fluid-Fluid Interfaces," *Am. Inst. Chem. Eng. J.*, 17, 1287.

15. Berger, P.D., Hsu, C. Arendell, J.P., (1987) "Designing and Selecting Demulsifiers for Optimum Field Performance Based on Production Fluid Characteristics," *Soc. Pet. Eng.*, 16285, 457.

Index